RURAL TRANSPORT SERVICES

RURAL TRANSPORT SERVICES

A Guide to their Planning and Implementation

HENRI L. BEENHAKKER
with S. CARAPETIS, L. CROWTHER
and S. HERTEL

LONDON AND NEW YORK

First published 1987 by Westview Press

Published 2019 by Routledge
52 Vanderbilt Avenue, New York, NY 10017
2 Park Square, Milton Park, Abingdon, Oxon OX14 4RN

Routledge is an imprint of the Taylor & Francis Group, an informa business

Rural transportation services
Includes index.
1. Rural roads—Developing countries. I. Beenhakker, Henri L.
HE 336.R87 1987 338.1'09172'4 87-8226

Typset by lnforum Ltd, Portsmouth

ISBN 13: 978-0-367-28642-2 (hbk)

ISBN 13: 978-0-367-30188-0 (pbk)

Contents

Tables

Preface

This book considers the problem of providing maximum access to transport services, and to roads for the rural population of the world's developing countries when limited funds are available. Access is a key factor in both social and economic development. It promotes social intercourse and opens up markets for both the rural and urban populations. Access connotes the ability to travel and to transport goods. The components of access include both the infrastructure and the transport modes or aids that use the infrastructure.

Such access is highly prized by all rural communities. It provides them with the means to market their products, and to purchase goods with the proceeds. In extreme cases it may indeed be their first introduction to a cash economy and it is always an essential component in the economic and social development which the people of developing countries are seeking. It includes the feeder roads that connect the rural areas to the country's main road networks and the tracks, trails and paths that provide local access to those feeder roads. In this book this system of feeder roads, tracks, trails and paths is designated as the Rural Access Infrastructure (RAI).

The rural poor have long utilized transport aids such as head baskets, carts, and pack and draught animals to assist in their transport activities. This text refers to these transport means as traditional transport aids. It refers to trucks and buses as conventional transport aids.

Access problems are solved only when both infrastructure and transport aids are available. Infrastructure and transport aids, and ways of evaluating their combination to meet specific transport requirements, must therefore be considered simultaneously.

This book is intended for decision-makers, planners, engineers and technicians in developing countries, aid agencies, non-government organizations, and consultants involved in transport and rural development activities. It offers a rationale for modifications to planning and implementing RAI project components. Guidelines and procedures for identifying transport problems and their solutions rather than arbitrary preconceived standardized solutions are suggested.

Although the problems and needs of no individual country are addressed directly, data from actual projects are included to illustrate specific points. The flexibility built into the guidelines is intended to be used innovatively for solving particular problems. No guidelines can be a substitute for imagination and resourcefulness on the part of the people involved in project preparation.

The text draws from a wide range of sources and fields such as engineering, economics, finance and statistics, and develops new methodology for, for instance, cost evaluation of spot improvements, simplified soil sampling, stage construction of water crossings, and analysis of the organizational framework for decentralized road maintenance. It is intended to describe comprehensively the multifaceted approach to the provision of rural infrastructure and transport services and to suggest a variety of solutions to specific problems.

In principle, the first part of the book is intended for policy makers, the second part for planners, and the third part with its annexes for personnel involved in project preparation. However, everyone's activities are interrelated so specific information related to complementary activities is referenced between parts. Thus the book can be of use to intended readers without the need to read it from beginning to end.

For example, those readers interested in RAI maintenance will find a policy statement about maintenance in Chapter 1, and a suggested strategy to implement that policy in Chapter 2. Further detailed technical information concerning guidelines and recommendations related to maintenance activities are presented in Chapters 7 and 15, and their supporting Annexe 7. In other words, the reader may pick and choose among sections of the book to investigate maintenance related information or other areas of interest without digesting the whole publication.

The following table outlines the major topics covered and their general location in the text:

SUBJECT	CHAPTER
Accessibility	4, 9
Construction	14
Cost Evaluations	17
Drainage	12, 13
Economic Analysis	19
Infrastructure Improvements	6, 11–13
Institutional Issues	16
Local Labour	14–16
Maintenance	7, 15
Objectives, Policies and Strategies	1, 2
Planning	3–8
Spot Improvements	13
Storage	8, 18
Transport Aids	5, 10

The preparation of this book has been financed by the Danish bilateral aid agency DANIDA and the World Bank. We have benefited from discussing the subject matter of this book with a large number of professionals active in the field of rural infrastructure and transport services, both within and outside DANIDA and the World Bank. It is impossible to acknowledge all the individuals who have helped us, but special appreciation should be expressed to R. Millard and S. Thriscutt, consultants, A. S. Rasmussen of DANIDA, G. A. Edmonds and R. P. Sikka of the International Labour Organization,

J. Hine and S. Yerrell of the Transport and Road Research Laboratory, C. G. Vandervoort of the U.S. Department of Transportation, and C. Cook, P. Long, P. Morris and L. Y. Pouliquen of the World Bank. As the early drafts were developed, Heather Mattheisz displayed a cheerful patience as she produced efficient transcripts of illegible copy and incomprehensible symbols. Kathy Rosen redrew all the illustrations found throughout the text. None of the people credited is, however, responsible for any errors or omissions.

Washington, D.C.
October 1986

H. L. BEENHAKKER
S. CARAPETIS
L. CROWTHER
S. HERTEL

PART I

CHAPTER 1

The Policy Maker's Role

1.01 According to 1983 estimates [103, 104], over 70% of the population in the world's developing countries is rural. In low income countries, agriculture accounts for about 37% of Gross Domestic Product (GDP) and involves over 70% of the labour force. Agriculture's priority is obvious as a significant area of attention for any programme attempting to improve the well being of the rural poor. In view of the relatively low productivity and efficiency levels in many of these countries, agriculture also represents a major potential avenue for increasing GDP growth.

1.02 Lack of a reasonably reliable and economic means of personal and goods transport in rural areas is thought to be a major constraint to rural development. Given the current budget constraints in developing countries, the prospects for additional funding to improve transport for the poorer segment of the rural population in the near future are quite bleak. Therefore ways and means must be considered to maximize the number of rural poor that will benefit from the limited transport investment resources that will become available.

1.03 Most developing countries have already made considerable progress in providing a basic network of primary, secondary and tertiary traffic arteries. In the last decade increasing attention has been focused on extending highway systems into agricultural areas, thus allowing the rural population better access to the existing through traffic network. This attention has resulted in only limited lengths of transport infrastructure improvements, in part because feeder roads have been designed to the same standards as the rest of the highway system. Consequently, feeder roads have frequently been overdesigned for the low traffic volumes that they have generated, and related costs of construction are high.

1.04 There are still large areas of potentially productive agricultural land beyond these improved feeder roads, representing a significant untapped source for economic development. These remote areas are mostly populated by some of the poorest and underemployed rural groups. They must rely on networks of paths, trails, tracks and unimproved roads that connect their homes, farmlands and villages to the improved feeder roads. Such networks often provide only intermittent, unreliable access for slow, low-capacity, and predominantly traditional forms of travel and transport. Consequently, the population that these networks serve often get no, or few, benefits from agricultural, industrial and social programmes carried out in more accessible areas.

1.05 In both literature and practice, there is inconsistency in the definitions

Table 1. Definitions
Highway, Road and Rural Access Infrastructure Classification

Primary Highways:	Those roads whose main function is to form the principle avenue of communications between major regions of a country, including connection between regional capitals and neighbouring countries.
Secondary Highways:	Those roads, not being Primary Highways, whose main function is to form the principal avenue of communication for movements: – between regional capitals and adjoining regions and their capital centres; – between regional capital centres and key towns; – between key towns.
Tertiary Roads:	Those roads, not being Primary or Secondary Highways, whose main function is to form an avenue of communication for movements: – between important centres and the Primary and Secondary Highway network and/or key towns; – between important centres.
Rural Access Infrastructure (RAI):	Any road, track, trail or path not being part of the Primary, Secondary or Tertiary highway network, whose main function is to provide access to abutting property. For purposes of this study they are further sub-classified as: – *feeder road:* a road that links a zone of access to the highway network and is normally accessible to motor vehicle traffic, but does not provide for through traffic, i.e., not a tertiary road; – *track:* normally a single lane cleared seasonal road, generally unimproved, traversable at times by light four-wheel drive vehicles, pickup trucks, animal-drawn carts, pack animals and connected to a road of higher classification; – *trail:* a narrow track suitable only for two-wheeled or pedestrian and pack animal traffic; – *path:* a narrow cleared way for walking traffic and, in some cases, bicycles and motorcycles.
Zone of Access (ZOA):	The area which uses the rural access infrastructure as its main connection to the primary, secondary or tertiary road system. The boundaries of the zone of access are determined by the criterion that, at the margin, more than 50% of travel and transport gravitates to that infrastructure.
Zone of Influence (ZOI):	The area subject to change as a result of alterations in the area's infrastructure conditions. Its boundary may be defined by physical obstructions, ZOI of adjacent infrastructure, or an assumed distance.
Level of Accessibility:	The accessibility of various types of transport over an infrastructure link, measured in terms of access reliability during the year for each type of transport.
Travel and Transport:	Where used in this text, travel denotes the movement of people by their own independent means, while transport implies the conveyance of either goods or people.

of terms used in rural road planning and implementation. Table 1 presents the definitions adopted in this book. Although primary, secondary and tertiary roads are not dealt with in this presentation, their definitions have been included to contrast them with rural access infrastructure.

I. NATIONAL, SECTORAL AND PROJECT OBJECTIVES

1.06 All development projects should relate to appropriate countrywide, national objectives. These objectives are broad statements of the goals a country hopes to achieve through investing the limited resources available for economic development purposes. They ideally embrace the two major areas of planning activity that must be considered on a national scale: maintaining and improving existing facilities and services; and constructing new facilities and introducing new production and service techniques [7]. Examples of national objectives include: diversification of sources of national income; mobilization of foreign exchange earnings; maximization of output; development of human resources; and less dependence on foreign assistance and on foreign imports, particularly of food.

1.07 Maximization of output in the transport sector means that the allocations for spot improvements and maintenance should, in general, receive higher priority than those for new construction, while the social services sector should concentrate on projects with field level reinforcing effects. This objective may be based on observations of underutilization of existing infrastructure, and/or a large portion of the populace living at the subsistence level. It is precisely these observations (Paras 1.03 and 1.04) which have led to the preparation of this text, which emphasises spot improvements and maintenance as well as field level reinforcing effects. Naturally, maximization of output should not be carried out to such extremes that the requirements of new physical infrastructure and human resources are completely ignored. For instance, decentralization of project planning and preparation is preferred to centralization in view of field level reinforcing techniques; however, decentralization requires more trained manpower, which is often a scarce resource in developing countries.

1.08 Development of human resources in the transport sector pertains to a more effective contribution to RAI improvements and maintenance and increased participation in the development process by several elements of society. It may be based on considerations of rural underemployment and unequal distribution of the benefits of development between social groups and economic classes of the population. Translation of the fruits of economic gains into social improvements leads not only to a more equal distribution of the benefits but also to a population better equipped to contribute to the development process themselves. Consequently, considerations of local participation in the planning, construction and maintenance of RAI (Chapter 7) and the development of local contractors (Chapter 14) are important subjects to be addressed when analysing the desirable improvements in rural infrastructure and transport services.

1.09 Less dependence on foreign assistance is commendable in view of a country's concern for its future generations living with the problem of debt service. In addition, foreign aid from diverse sources and supplier export credits may result in a proliferation of different makes and models of construction equipment, thus compounding spares support and servicing problems.

Public works organizations with road maintenance fleets of many different makes and models may be confronted with the problem that the number of each item is well below the minimum fleet size at which it is economic to maintain separate stocks of spares and tools and to train mechanics. In view of this problem, this text discusses the pros and cons of labour-based techniques and equipment selection (Chapter 14). In short, a country should adopt sensible plans and procedures which minimize the need for foreign exchange, allocate its expenditure wisely, and maximize the use of local resources. The development of these plans and procedures are the principal subjects of Parts II–IV of this text.

1.10 Since national objectives are broad, they may sometimes conflict. For instance, an implication of emphasizing maximization of output may be the simultaneous de-emphasis of other matters such as the development of human resources or a more equitable distribution of benefits by regions. In addition, the pursuit of diverse objectives is often difficult in light of the constraints imposed by the scarcity of resources. Policy makers must tenatively rank mutually consistent national objectives in descending order of their relative importance when developing national policies and strategies. In order to allow for flexibility of planning at the sectoral level, this ranking may be used to reduce the total number of initially selected objectives by eliminating the lowest ranked ones.

1.11 National objectives form the basis for sectoral objectives. Sectors, in the context of this book, include areas of various functional concerns, such as transportation, agriculture and education. Sectoral objectives are statements of the goals of the particular sectors. They become, in effect, a more detailed translation of national objectives because limitations on sectoral activities call for a more detailed analysis of specific actions and their impacts. Ideally, developing national and sectoral objectives should be an iterative process, with the more detailed evaluation of sectoral objectives acting as further input to their national counterparts. This translation serves to make the national objectives more practicable to deal with, to ensure consistency between national and sectoral objectives, and to indicate the mutual dependence of the functions of national and sectoral planning and consequent coordination between and among various governmental agencies (Table 2).

1.12 The following is an example of consistency among national and sectoral objectives and existing limitations. If the national objective pertains to mobilization of foreign exchange earnings, and if local farmers could replace grain imports given better access and affordable methods of transporting their inputs and outputs, sectoral objectives should relate to the construction of suitable infrastructure, the development of transport aid technology, and the establishment of credit to purchase suitable transport aids.

1.13 'Freelance' planning in the various sectors should be avoided since it may lead to a number of inconsistencies and duplication of efforts. An example of such duplication relates to agricultural planners and transport planners who do basically the same thing. That is, the former make forecasts of domestic demand for different products, imports, possible developments of agricultural

Table 2. National and Sectorial Objectives and Project Activities

National objectives	*Sectorial objectives*	*Project activity orientation*
Foreign exchange savings/earnings	– Promote agricultural and industrial productivity	– Increase access to farms – Coordinate transportation and production investments
	– Increase fuel conservation	– Encourage use of labour-based construction and maintenance – Encourage use of traditional transport means
	– Reduce imports of new vehicles and spare parts	– Increase resource allocation to infrastructure maintenance/ rehabilitation – Increase utilization of locally manufactured transport aids
Domestic resource mobilization/ budgetary savings	– Reduce government subsidies for transportation	– Establish cost accounting and improve financial management – Establish better management and administrative practices – Increase role of private sector in transport services – Emphasize decentralization of transport infrastructure responsibilities and administration – Mobilize self-help through community participation
	– Ensure utilization of least cost transport mode	– Improve intersectoral coordination and planning – Improve economic analysis of project investments – Encourage increased usage of transport aids at the local level
Income distribution/ employment	– Increase access of poor to transport	– Encourage investment and maintenance for rural access – Ensure access to affordable low-cost transport – Develop appropriate design, maintenance, and investment for different types of rural accessibility
	– Generate employment	– Produce development of disadvantaged regions – Encourage employment-creation activities, short-term (e.g. labour-based construction) and long-term (e.g. labour-based maintenance) construction skill training – Encourage general development
Supply/deliver social services	– Encourage access to services	– Reduce cost of delivering all social services – Increase access within and out of development area – Foster local institutions – Increase or facilitate the opportunities for social, cultural and recreational activities

Table 2. *(contd.)*

National objectives	*Sectoral objectives*	*Project activity orientation*
Output maximization	– Emphasize rehabilitation, spot improvements and maintenance rather than new infrastrucure	– Carry out surveys to locate necessary spot improvements – Increase resource allocation for rehabilitation, spot improvements and maintenance
	– Promote free entry into transport industry	– Deregulate transport activities – Allow for free choice of transport mode
	– Promote use of simple transport aids	– Introduce demonstrations of such use
Human resource development	– Increase local participation	– Encourage use of labour-based construction and maintenance – Emphasize decentralization of transport infrastructure responsibilities and administration
	– Training	– Foster local institutions
Foreign assistance reduction	– Promote use of local resources	– Encourage use of labour-based construction and maintenance – Encourage appropriate design, maintenance, and investment for different types of rural accessibility
	– Standardize equipment fleets	– Establish pool of spare parts – Limit need for specialized and sophisticated equipment – Encourage better equipment management and accountability

processing industries, needs for feeder-road systems, etc., on the basis of their own assumptions about long-term developments in population, incomes, etc. Transport planners, however, often commence their studies with their own long-term projections, sectoral developments, regional developments, etc., in order to be able to design their networks [7]. Institutional arrangements for the preparation of rural development projects should not allow such freelance planning to take place (Chapter 16). In addition, an examination of the economic viability of RAI investments calls for a close cooperation between agricultural and transport planners (Chapter 3).

1.14 Development programmes are action orientated packages designed to achieve well-defined objectives or sets of allied objectives. A programme can consist of one or more interrelated development projects to be carried out by one or several organization units such as different ministries, different agencies within a ministry, and/or different agencies within different ministries. Project components may be assigned to the various units in line with their specific organizational objectives. This text is orientated towards the rural transport project components of development programmes, regardless of the organizational unit responsible for their accomplishment.

1.15 Project component activities are justified only to the extent they contribute to the achievement of sectoral objectives, which must in turn be linked to national objectives. Table 2 gives examples of linkages between national and sectoral objectives and project activities. Equally important, however, is synergism, the cooperative action of discrete agencies to produce a total effect greater than the sum of their effects taken independently, or the '2 + 2 = 5' effect. Policy makers should insist that project planners consider these intersectoral linkages as outlined in Para 3.02.

II. POLICIES

1.16 National policies are statements of the ways and means of achieving national objectives through allocating resources between different sectors and different functions, such as transport and agriculture. Policies are in fact agreements that outline the courses of action a government will follow to achieve its national objectives. Each policy implies the policy makers have evaluated the national goals for the future; have defined the issues involved; have determined the government has the capability to act; and have decided that the actions are both feasible and timely. Sectoral policies carry similar implications for sectoral objectives.

1.17 This section contains policy statements compatible with achieving the national and sectoral objectives related to the improvement of rural access infrastructure and transport services. These policies, when consistent with a country's national and sectoral objectives, may increase the number of beneficiaries from the employment of limited resources available for rural transport and development programmes. The policy statements emphasize the common roots of the transport and agriculture sector objectives. For purposes of this book, no distinction is made between national and sectoral policies. For ease in later reference, policy statements are designated by a P followed by a number.

1.18 **P-1: To obtain maximum use of limited resources, access should be tailored to specific area needs.**

The high cost of providing all-weather motor vehicular access can rarely be economically justified in rural areas except on feeder roads that carry large traffic volumes or high value perishable produce. However all-weather pedestrian access can be provided for a fraction of the cost of all-weather motor vehicle access. Therefore this text considers access to be transport mode specific.

1.19 To facilitate implementation of this policy, three types of access are suggested:

(i) *All-weather access* implies that no infrastructure closures are acceptable. Travel and transport are possible 24 hours per day throughout the year. Those feeder roads that require all-weather access need not be evaluated for any lower level of access; however, all feeder roads in a specific project may not have the same accessibility requirements.

(ii) *Reliable access* implies that intermittent closures are acceptable if they do not constrain the area's economic and social development. Reliable access over a specific infrastructure link will vary by transport mode and is a function of climatic conditions, local soil characteristics, topography and the incidence of transport demand to seasonal travel restrictions.

(iii) *Minimum access* implies longer closures are acceptable for transport activities if sufficient access is provided to permit the local population to travel and transport their basic essentials, albeit with some difficulty. Without minimum access, the local population cannot exist, so obviously some form of minimum access is already available in populated areas. It is of specific concern in areas where development and resettlement are common objectives. However, in many settled areas an improved standard of minimum access can be achieved at relatively low cost. Better minimum access standards free the local population to pursue more productive activities and should be a concern in all development planning.

1.20 **P-2: To increase a feeder road's impact on the rural population's socio-economic welfare, improved access to the feeder road should be considered when evaluating feeder road improvements.**

Specific access needs (P-1) vary throughout a project area in accordance with the possible activities in individual sub-areas. For example, when new cash crops such as palm nuts, cotton, maize or rubber trees are proposed in a development project, the inhabitants outside the project area may find a market among the primary cash crop farmers for vegetables and other food crops. Thus, there must be suitable access between sub-areas for marketing activities. This secondary development expands the project's impact over a larger population segment.

1.21 The zone of influence of a feeder road is often considered as a relatively narrow band adjacent to each side of the feeder road. The feeder road's zone of access is often considerably larger than its zone of influence since it considers the tracks, trails or paths leading to the feeder road. Examination of bottlenecks on these forms of rural infrastructure is important, since related improvements costs are frequently lower than costs of increasing feeder road density. In addition, an approach based on rural access needs in a zone of access is important for the success of both the primary and secondary development mentioned in Para 1.20.

1.22 **P-3: To better satisfy local transport needs, the use of transport aids other than conventional trucks and buses should be encouraged.**

Much of the transport demand in developing rural areas is associated with day-to-day living, social activities, and occupational requirements such as farming. These activities often do not require motorized transport since the distances are short, the loads small, and transport speed of secondary importance.

1.23 In many countries, the rural poor's only alternative to motor vehicle transport is walking. Past tendencies to emphasize the construction and improvement of motorable roads assume the private sector will provide transport services with motor vehicles. Although a variety of proven simple traditional aids, such as carrying poles, backpacks, bicycles, pack and draft animals, carts and trailers exist, their use is frequently limited to a few countries or local areas within a country. Many countries, particularly in Africa, are unaware of the versatility and applicability of these aids. In other cases, although such aids are known, they are unavailable because of import restrictions or government policies and attitudes against them.

1.24 Each of these transport aids has its particular features, capacity and limitations which affect costs of ownership and operation, operating methods, the ability of different people to use them, and their infrastructure requirements. However, their introduction and use can improve economic and social development by increasing accessibility and transport capacity while, at the same time, reducing infrastructure investment requirements. In addition, their introduction may lead to the development of small, local industries providing new employment opportunities.

1.25 **P-4: To encourage foreign exchange savings, appropriate construction technology should be used whenever possible.**

Feeder roads, tracks, trails and paths need to be built at minimal cost if the maximum rural population is to be served with limited investment resources. Since foreign exchange is often in short supply, it is also important to consider the foreign exchange costs normally associated with road building, i.e., the costs of imported equipment, spare parts, tools and machines required for maintaining that equipment, and the petroleum products necessary for equipment operation. Appropriate technology is defined as one that is both efficient and least costly. To apply it, one should examine the potential use of local labour, equipment, materials, and animals as compared to the use of equipment normally associated with highway construction works.

1.26 A policy of tailoring access to specific needs makes the use of labour-intensive techniques attractive, since improvement works are often small and scattered over large geographical areas. To encourage appropriate technology, design standards should be kept simple and formal engineering design kept at a minimum level. Most developing countries have yet to realize the significant reductions in the use of hard gained foreign exchange possible by emphasizing appropriate technology.

1.27 **P-5: To obtain long term benefits from adequate rural access infrastructure, no construction should begin until an effective road maintenance organization is in place.**

Maintenance is a major problem for all transport infrastructure because of inherent institutional, financial and human resource constraints. Whether RAI maintenance is under the jurisdiction of local, district, regional or national authorities, feeder road maintenance has been mostly inadequate, and in many cases, completely non-existent. What little maintenance is undertaken is frequently limited to emergency repairs and removing life-threatening obstructions.

1.28 It is important to recognize and accept the role of maintenance activities in preserving assets and access reliability, and in reducing travel and transport costs. Although maintaining RAI requires small-scale operations, maintenance needs continue throughout the life of the facility. Maintenance activities require capable and effective management and supervision so they will be undertaken at the most appropriate time to reduce infrastructure damage. In many countries, the crucial times occur immediately following the rainy season, and just prior to harvest when that occurs at a different time. Without sufficient identifiable manpower and budgetary resources, maintenance activities are least likely to be able to compete with other development activities during those critical times. The resulting lack of timely maintenance adversely affects the access needed to foster other development activities and dissipates the previous resource investment in improving access.

1.29 In the context of feeder roads, tracks and trails the clear distinction between construction and maintenance tends to disappear since many of the operations required in their upkeep improve the services they provide. Typical examples lie in placing gravel over short sections of a road where earth surfaces begin to fail, in providing culverts where cross-drainage becomes obviously necessary and in repairing unsafe side slopes. This is in fact no different from the stage construction whereby the minor road networks of more developed countries have evolved. Any road maintenance for RAI should be planned and financed with this objective in mind.

1.30 **P-6: To encourage the continued use and maintenance of improved RAI, its planning, construction and maintenance should be decentralized to the greatest extent possible consistent with the availability of qualified staff and the degree of local and regional control over resources.**

Currently, most funds allocated for RAI construction and maintenance flow from or through central government agencies. Since central government agencies are remote from the local problems and often unaware of the real needs of farmers, decisions about allocations of necessary funds are seldom based on an analysis of specific area needs. Moreover, maintenance allocations from central governments are rarely more than token, insignificant amounts. To better respond to local needs, it is therefore desirable to decentralize RAI planning, construction and maintenance as much as possible. Indeed, once local communities are convinced that RAI feeder roads, tracks, trails and paths are local facilities with inherent local responsibility for their upkeep, that task may prove a powerful incentive in the emergence of responsible local government organizations. In view of limited capabilities prevailing in most developing countries, decentralization should, however, be approached with extreme caution and be based on an analysis of training requirements.

1.31 The long-term prospects for a viable RAI development programme are closely tied to the creation of an institutional system that is sensitive to local issues and needs concerning transport; has the capacity and ability to adjust to both the pace and nature of development; and is capable of mobilizing local resources to the extent that they are available. Above all, the institutional arrangements should be self-sustaining because they provide, and are seen by the local population to provide, in an efficient and timely way, facilities and services needed by the local community.

1.32 **P-7: To increase the resources available to develop and maintain an expanded RAI system, the government should encourage the use of local resources to the greatest extent possible.**

Since RAI links are low cost, low technology infrastructure, any serious attempt to upgrade local RAI networks can add considerable lengths of useable infrastructure in a relatively short period of time, even within the constraints of a restricted budget. Therefore lack of physical and human resources may hamper RAI improvement and subsequent maintenance as much as budgetary constraints.

1.33 The unique features of an RAI network make the use of local sector resources attractive. The local sector is here defined as both local contractors and community groups such as local community development associations, cooperatives, religious groups, merchants' associations and non-profit organizations. Maximum use of the local sector is recommended since it alleviates problems related to limited trained manpower employed by the government, and allows for quicker involvement by the local population. This involvement will often lead to better solutions to area-specific access problems. It is, however, not suggested that local groups will build or maintain RAI networks at no cost to the government. Except for simple maintenance activities related to track, trails and paths, the local population is generally not interested in working at RAI improvements and maintenance without compensation.

1.34 **P-8: To encourage agricultural development in project sub-areas where providing suitable access is not feasible, improved on-farm or local storage for non-perishable produce should be evaluated.**

The purpose of tailoring access to specific area needs (P-1) and introducing transport aids (P-3) is to provide appropriate accessibility to a greater portion of the population of a development area at the least possible cost. However, it is not always possible to justify or pay the costs of appropriate access to all the project sub-areas where development is feasible. The provision of storage facilities can in this case alleviate problems related to lack of reliable access. Only the storage of non-perishable produce is considered since perishable products normally require refrigerated storage facilities. These facilities are beyond the scope of this handbook since its approach is limited to simple technology.

1.35 Improved storage may provide motivation for that segment of the population left in semi-isolation to join in development activities. Storage by itself does not offer the same socio-economic benefits as accessibility, but it is often a feasible, less costly alternative to RAI improvements. Agricultural products may be stored during periods of no access due to heavy rains, and moved out of the area when access is possible. Storage facilities have the additional advantage of spreading the demand for transport services over a longer period. In addition, they allow farmers to keep their products until better prices can be obtained.

1.36 **P-9: To reduce the cost of preparing rural access project components, simplified methods and procedures should be used wherever appropriate.**

Project preparation represents a resource investment. It requires the resource most often lacking in many developing countries, i.e., professional or trained manpower. Reducing infrastructure construction costs by increasing costs of such manpower can be counter-productive unless it results in a definite reduction in overall resource investment or a net increase in benefits.

1.37 Much of the evaluation methodology used for determining the benefits derived from future development is based on estimates of future growth. These 'best estimates' become increasing unreliable as the forecast time period is expanded. Simple rules of thumb are sometimes as accurate as detailed analyses. Until development activities begin to impact the activities within a specific project area, the data base from which the ultimate success of the development may be predicted is usually weak. Limiting access requirements to those required for development to begin in projects where benefits will mature slowly, preserves resources until a better data base can be developed. In such situations, it is also more important to preserve development options than to proceed on unfounded optimistic assumptions. This emphasizes the importance of the stage construction approach referred to in Para 1.29. Simplified methods and procedures can determine the magnitude of probable development in the future with sufficient accuracy to identify and respond to immediate access needs. Ultimate access needs become more evident as development proceeds.

1.38 Chapter 2 suggests strategies for achieving the policy statements outlined in Paras 1.18 to 1.37. Specific strategies relate to specific policies. To relieve the reader of the necessity of reviewing all of Chapter 2 to find the strategies related to a specific policy statement, Table 3 gives the relevant strategies. A similar table at the end of Chapter 2 locates information relevant to the various strategies throughout the remainder of the book.

Table 3. Policies and Their Strategies

Policy number	*Strategy numbers*
P-1	S-1 to S-5
P-2	S-6
P-3	S-7 to S-11
P-4	S-12 to S-15
P-5	S-16
P-6	S-17 and S-18
P-7	S-19 and S-20
P-8	S-21
P-9	S-22 and S-23

III. CONCLUSIONS

1.39 Policy makers set the tone of a country's development efforts. They collaborate in defining and prioritizing the country's national and sectoral objectives and translating these into policies. Policies determine the country's approach to spending that portion of its resources allocated to development activities. The policy maker's major effort should be directed to ensuring the compatibility of or linkage between national and sectoral objectives.

1.40 Proper linkage occurs when the actual resource expenditures are limited to those necessary to accomplish the agreed objectives, both national and sectoral. Resource expenditures occur as project activities; policies create the setting for those activities. Therefore any modification to the way in which resources are to be expanded in providing or improving access to rural areas, either in constructing or upgrading rural infrastructure or improving transport services, must be preceded by modified policies. The policy makers must be convinced policy modifications are in the country's best interests, are within the country's capabilities, and are conducive to linking project activities to national and sectoral objectives. Once policies are modified, the policy makers must see that the modifications are implemented. Without leadership, implementation oftens falls victim to inertia. Policy makers are the engines of improvement only when they provide active leadership.

CHAPTER 2

Strategies

2.01 Strategies are methods the policy maker selects to implement the type of policies described in Chapter 1. To be effective, they must function within the current capabilities of the individual country. As for the formulation of policies, the principal guideline to be followed when formulating strategies is whether, with the given limitations of the country's absorptive capacity, they can result in the realization of national and sectoral objectives.

2.02 Strategies further refine policies by defining specific actions that should be instituted to accomplish agreed upon policy decisions. The discussion of maximization of output as a national objective (Para 1.07) illustrated the initial refinement of a general concept or objective into policy statements providing direction for courses of action that are within a country's capabilities. That is, such an objective leads to the formulation of policies stressing economizing measures for construction, policy P-4, and maintenance, policy P-5. Based on policy P-4, Section I of this chapter further refines this policy decision into possible activities by formulating strategies related to the use of local labour and materials, method rather than performance construction criteria, performance budgeting, systems to monitor costs, and reduced formal design activities. Another example of developing a concept into a defined activity is the evaluation of the national objective of mobilization of foreign exchange earnings, leading to the policy regarding the use of transport aids other than conventional trucks and buses, policy P-3, which in turn leads to a strategy of promoting local manufacturing facilities for appropriate transport aids.

2.03 The discussion of national, sectoral and project objectives in Chapter 1, already mentioned the importance of considering possible synergy effects. This should also be done when strategies are formulated. For instance, the national objective of output maximization results in the formulation of a strategy for the composition of investments and required actions which will reinforce the success of project components. Complementary investments and actions are determined by the demand for correlative and intermediate products and services, the supply of identified excess capacity, and the by-products of existing activities. Consequently, a strategy dealing with investments in transport aids, and another one regarding the elimination of regulations inhibiting free entry into the transport industry are formulated in this chapter.

I. STRATEGIES FOR IMPROVING RURAL TRANSPORT

2.04 The strategies described in the following paragraphs can be used to implement the policies suggested in Paras 1.18 to 1.37. Their principal relationship to a suggested policy is outlined in Table 3. It should be recognized, however, that some strategies are appropriate for more than one policy. For instance, it could be argued that the strategy relating to the elimination of speed as a design parameter applies to both the policy regarding the maximum use of limited resources (P-1) and the policy related to foreign exchange earnings (P-4). In addition, the strategies described here should not be regarded as an exhaustive set of all strategies pertaining to improving rural transport. Neither is it suggested that all of these strategies are recommended for any individual country. The principal role of policy makers and planners is to ensure that policies and strategies are compatible and consistent with national and sectoral objectives. The following paragraphs state individual strategies and offer reasons for their adoption and use. The remaining part of this text, Chapters 3 to 19 and Annexes 1 to 11, discuss technical details and methodology for executing these strategies together with additional reasons for their introduction. The strategies are designated by an S followed by a number for reference purposes.

2.05 **S-1: Eliminate speed as a design parameter.**

Some components of geometric design speed criteria are inappropriate for low volume feeder roads because they increase cost without improving access. Design speed is a concept used to standardize the relationship among the geometric elements of a road. Geometric design includes all the visible dimensions of a road's horizontal and vertical alignments and cross-sectional elements. An increase in design speed dictates longer horizontal and vertical curves, flatter gradients and wider road surfaces. The use of a specific design speed on the arterial highway network is essential to provide a constant level of service over a specific road segment so a motor vehicle operator unfamiliar with the road is never surprised by a sharp curve or slowed by a steep hill.

2.06 Design speeds are reduced as terrain features become more pronounced. The costs of building a road to a specific design speed increase rapidly as the road goes from flat to rolling to mountainous terrain. The economic requirement that users' vehicle operating cost savings should equal or exceed the construction and maintenance costs of a road therefore prohibits application of the same design speed in different types of terrain and for different traffic volumes. Instead the road users pay a penalty in travel time, discomfort and increased vehicle operating costs so the government can build and maintain the road at a reasonable and justifiable cost.

2.07 This strategy is predicated on the following factors:

(i) access is not related to design speed;

(ii) traffic volumes on many feeder roads in developing countries are so low that vehicle operating costs are not a major decision parameter in overall project evaluation;

(iii) high speed operation is not desirable on feeder roads with mixed traffic including carts, bicycles or pedestrians;

(iv) feeder roads serve only local traffic whose operators are familiar with local safety conditions;

(v) time lost because of slower speeds over relatively short roads is insignificant compared to the slow loading and unloading procedures practised in many rural areas; and

(vi) traffic volume will increase only as the local transport needs increase, so national traffic volume projections are not applicable to feeder roads.

Any feeder road with the above features can be built at reduced costs if the design speed concept is waived.

2.08 **S-2: Concentrate on eliminating drainage problems.**

Unreliable access is often caused by improper drainage. Flooded water crossings are the most common deterrents to reliable access; however, poor road surface conditions and road side slope slides are also drainage related. Each geometric feature should be designed with its drainage implications in mind. The crown and/or cross slopes should be sufficient to remove standing water, which is important since the presence of water and traffic together is the primary cause of potholes and rutted surfaces. Longitudinal roadway slopes and side slopes should be erosion proofed. Roadway surfaces must be raised above the water table. Granular material will eliminate slippery earth road surfaces.

2.09 The road location itself should be drainage orientated, i.e., along watersheds wherever possible. Large cuts and fills disrupt the natural drainage patterns and the natural stability of undisturbed soil. Many older existing roads and tracks have reached an environmental equilibrium point; their horizontal and vertical alignment harmonize with the natural topography to encourage free drainage flow with a minimum of maintenance and their surfaces have been compacted to the point where they resist the disruptive action of traffic and erosion. To disturb such an environmental equilibrium often causes more problems than it solves.

2.10 **S-3: Use incremental improvements to meet changing access needs.**

Throughout the year, the majority of trips and transport in rural areas are largely local in character, relating to the population, its household and local activities. Travel activities include home-to-farm, home-to-market, home-to-social, and home-to-recreational trips, and trips to gather firewood or transport water. Most of these trips tend to be short in length and duration, and involve only small quantities of goods. Transport needs for agricultural purposes increase as development proceeds. These additional needs include input deliveries, product movements to both markets and consumers, equipment movements, and extension worker-travel. Travel for social and recreational purposes also grows significantly as increased economic activity provides additional discretionary spending capacity.

2.11 However, for the individual farmer, resident or small local community, transport demand and activity frequently increase slowly in response to development activities. The initial impact of improving transport infrastructure in developing areas is largely influenced by the degree of accessibility during key periods in the farming cycle. The ability to travel and transport without excessive effort, delays or cost when these agricultural needs arise, provides the motivation to participate in development activities. For some types of trips, timing and travel time are crucial, such as emergency health-related travel and the transport of perishable goods; however, the trip timing, travel time, and travel distance are seldom critical for the majority of the population.

2.12 The evolving increase in access needs can lead to a set of priorities for improving accessibility through RAI improvements:

(i) make passage certain;
(ii) improve the RAI's location where necessary once the access demands respond to development activities; and
(iii) consider improvements to alignment inadequacies only as further increases in access demand warrant them.

2.13 The most cost-effective approach to making passage certain is to improve the specific locations that limit accessibility. This activity, termed spot improvement, becomes the first step in a series of further upgrading activities to be undertaken if and when development increases access needs. Spot improvements are the most economical means of initially improving all rural access infrastructure because they:

(i) take advantage of existing suitable infrastructure sections;
(ii) concentrate resource expenditures on identified existing access constraints;
(iii) require minimal engineering input and drawings;
(iv) are frequently individually small operations; and
(v) can be easily tailored to fit specific transport aids.

The spot improvement, stage construction approach of responding to an area's increasing access needs both reduces the initial resource investment and limits future investments to the times and locations that ensure a favourable economic rate of return.

2.14 **S-4: Accept additional risk to reduce expensive overdesign.**

RAI improvements should be planned at the lowest practical level in terms of size, standard and cost. Planning should involve functional access rather than a predetermined level of engineering standards. Small culvert design and road surface thickness design are two areas in which the results of engineering calculations can sometimes be quite conservative.

2.15 Culverts are sized in accordance with a theoretical risk classified as a design storm or design flood frequency. Typical small culverts on a low-volume rural arterial road network are often designed for a 10 year storm, which means that the culvert has a one in ten chance of being too small to handle the run-off in any given year. The occurrence of such storms is based on a statistical analysis. If a one year storm design is used for a feeder road's minor drainage structure, the risk of flooding and of washouts will increase; culverts will overflow sooner and a few may need to be replaced with larger culverts. The exact number of additional floodings or replacements cannot be accurately predicted since flooding frequency is often based on unreliable statistical data. However, it is unlikely that the additional occasional flooding of minor drainage structures will impede access for very long. The cost of any required minor drainage structure replacement will be known once implemented; the excess cost of culverts that are oversized is never known. A true lowest cost drainage system will therefore require some minor drainage structure upgrading or protection.

2.16 At stream crossings and other places where flooding may have serious consequences such as flooding nearby houses, washing out the road, or endangering livestock or human life, flood risk must be reduced to an acceptable level by selecting a design storm frequency which reflects the risks involved. Where flooding presents an inconvenience with the possibility of minor damage, for instance in the selection of culvert sizes to provide cross drainage for water draining from the road surface, a one year storm may be an acceptable criteria. In many areas insufficient collection of rainfall intensity data prevents storm risk forecasting and judgement must prevail. The implementation of minor risk taking will be influenced by the adoption of the following strategy (S-5).

2.17 Formal road surface thickness design involves the prediction of the total traffic volume that will travel over the road during its design life. Such formal design is applicable only on roads which are to be provided with a permanent asphaltic or portland cement surface. The road component of the RAI is not likely to carry sufficient traffic to warrant the provision of such permanent surfaces. There may well be some particular circumstances where feeder roads require short lengths with a permenant surface such as on soft ground near approaches to river crossings, on steep grades where surface erosion makes it impossible to maintain a satisfactory gravel road surface, or through villages where the nuisance of dust from earth or gravel roads can be intolerable. In such situations, because of the light traffic it may be possible to use relatively light construction, adapting for this purpose the general methods of pavement design which have been evolved by the Highway Departments of almost every developing country to suit local circumstances. For gravel roads such methods of designing the thickness of the applied layer is neither possible nor necessary. At this rather crude level of highway engineering it is appropriate to surface feeder roads to some arbitrary practical minimum thickness when a gravel or other granular material surface is required to improve access as part of a development project.

***2.18* S-5: Adopt a 'review and correction' design and construction approach with a two year post construction correction period.**

The implication of risk taking (S-4) is that some of the works may prove to be inadequate for events as they actually materialize. Accordingly, the works may be damaged by environmental factors or traffic, or both. It is therefore appropriate to temper the advantages of aiming for minimum cost solutions by including time and funds after the initial construction is finished to identify and correct any deficiencies. This will ensure proper maintenance, which is particularly crucial for the strategies suggested by S-4 and S-5.

2.19 This is not to suggest that the original design should be casually undertaken or construction methods compromised. It is rather an acknowledgement that fine tuning a functional access infrastructure is less expensive than arbitrarily applying standard design features developed for a type of infrastructure intended to serve a different purpose. The adoption of the 'review and correction' approach serves several purposes:

(i) it weans design engineers from an overly conservative approach to solving problems whose solutions must be based on skimpy, incomplete data, by permitting them to use judgment without imposing the stigma of incompetency;
(ii) it assigns the funding and responsibility for delivering a functional access infrastructure to an organizational unit that has the resources to make improvements, rather than imposing post-construction improvements on an underfunded, underequipped maintenance organization; and
(iii) it guarantees that the appropriate accessibility and maintenance will be provided as is required, particularly during the critical first two years.

2.20 **S-6: Evaluate local off-road transport needs to determine an area's total transport demand.**

A local farmer's off-road transport needs may amount to more than 20 times the weight of his cash crop (Annexe 1, Technical Note 1.2). Project evaluatins that fail to identify total transport needs merely scratch the surface of the rural poor's transport demand. An improved feeder road may only satisfy 5% of the local farmer's transport needs and then only after the farmer transported his surplus to the road side.

2.21 Both total area production and weight of individual loads are important to the determination of a suitable transport service. A 50 kg sack on one's head may discourage long trips, but a truck with a 5 ton capacity may require an average volume of transport of about 70 sacks to be economically viable. Obviously an integrated transport service meeting the needs of the rural poor should address such disparities. As long as the small farmer's maximum load capacity and the trucker's minimum load needs are at such odds, one or the other's transport inefficiencies will reduce the farmer's income in rural development areas. Any transport study that limits its evaluation to produce transported over the road in trucks turns a blind eye to a major transport problem in development areas. In addition, the provision of suitable transport facilities to carry agricultural output to a road in sufficient quantities is often needed to induce private or government truckers to offer a reliable and economical forwarding service.

2.22 **S-7: Allow for small investments in appropriate transport aids to arrive at least cost solutions.**

The least cost solution to an area's total transport demand (S-6) cannot include building a feeder road to every farm. However, increased use of traditional transport aids will permit the farmer to carry a greater total load in the same time, or to carry the same total load in less time and with less effort. More importantly, from the farmer's viewpoint, the availability of a traditional transport aid and its appropriate infrastructure will provide for most of the farmer's transport needs, not just the need to transport his surplus products to the roadside. Therefore, the least-cost solution to improving access in areas where traditional transport aids are in short supply, is to make these aids available and to spot-improve the infrastructure that will accommodate them.

2.23 The need and justification for small-scale farmer credit in general is recognized in the preparation of agriculture and rural development projects. However, most of the projects which include small farmer credit ignore or overlook the possibility of providing credit for the provision of small-scale rural transport services and their necessary auxiliary repair facilities. Small farmers and small-scale entrepreneurs are often precluded from purchasing simple vehicles and transport aids because excessive demands by local banks deny the borrowers access to credit. Access to credit on reasonable terms for the purchase of transport aids is as important as good access on RAI.

2.24 **S-8: Consider as many alternative combinations as possible.**

Since different transport aids require different types of infrastructure, the infrastructure improvements should vary in accordance with sub-area transport demands. For example, a rubber plantation may justify a grid of tracks to allow tractor-trailer latex pickup at convenient loading stations. The same tracks can also provide access for the surrounding farmers to deliver vegetables to the plantation workers' villages via animal carts. Since many alternative transport aids can operate over the same type of infrastructure, the rubber workers can also reach their specific work locations by bicycle over the same track network.

2.25 Alternative combinations of transport aids and their appropriate infrastructure can also be used to reduce the initial investment in infrastructure. For example, a development project may ultimately provide a feeder road to a village and ox cart paths from the village to the fields. Until agricultural production warrants ox cart transport from the fields to the village, an ox cart track may suffice on the feeder road location. Such an approach reduces the initial infrastructure's investment and introduces ox cart technology at the village level before ox carts will be required in larger numbers.

2.26 **S-9: Retain as many feasible options as possible for further access improvements.**

While the highest priority should be given to immediately foreseeable development, RAI improvements should be planned to be flexibly adaptable to alternative future developments. Particular attention should be given to not foreclosing alternative short-term development options. In addition, foreclosure of long-term options should be minimized. Markets located at the edge of a village are more amenable to feeder road service than central markets. Motor vehicles require shallower fords than wading pedestrians, but foot logs require short spans that are also conducive to future culverts. Ridge routes require fewer drainage structures than water level locations. Longer routings that follow good soils may be more economically improved than more direct routes through areas of poor soils.

2.27 **S-10: Eliminate regulations inhibiting free entry into the transport industry.**

Any arbitrary restriction on the use of transport vehicles is a major deterrent to matching accessibility to actual transport demand. Yet many governments regulate and control the supply of transport services by the private sector by one or more of the following measures: (i) restricting entry to the transport industry; (ii) controlling the routes over, or areas within which services are provided or commodities may be carried; (iii) stipulating the days of the week during which services are provided; (iv) governing the fares and/or freight rates to be charged; (v) prohibiting the carriage of goods in passenger vehicles and vice versa; (vi) forbidding the owner of a vehicle to transport goods not produced by him; and (vii) restricting the choice of transport mode. Such measures may result in the absence of any form of service, or at best limited, irregular service with inefficient transport operations, and significant and unnecessarily high differences between farmgate prices and prices of agricultural products prevailing in cities.

2.28 **S-11: Promote local manufacturing facilities for appropriate transport aids.**

Governments should do more than just promote the use of appropriate transport aids. They should encourage the local manufacture or assembly of these aids so as to generate additional employment opportunities. Access to credit on reasonable terms and conditions is crucial to get local entrepreneurs started in the manufacture or assembly of transport aids. Servicing, repair and maintenance needs of these aids may also provide opportunities for local employment. In many countries, operation of for-hire transport services ranging from head-porters to motorcycles and mini-buses, are an important employment source. Governments can also use their own programmes to promote transport aids by having their employees use appropriate ones as demonstration 'projects'. For example, extension agents may be provided with bicycles rather than passenger cars or motorcycles.

2.29 **S-12: Increase the use of local labour and materials.**

The technology or work methods employed for constructing or improving RAI should be matched to the nature of the work, availability of local labour, size of the budget and method of maintaining the finished product. Small, scattered RAI improvements are particularly suited to low technology. Explicit attention should be paid to the possibility that other factors may have an undue influence on technology choice. For example, the existence of large equipment may be the basis for determining the technical scale and dimensions of the works, instead of the real transport demand and its infrastructure needs.

Designs, procurement procedures, and specification and bid documents can also encourage or discourage the most appropriate technology.

2.30 The use of local materials reduces the need for foreign exchange. Timber bridges and culverts can reduce steel imports where timber is available and not subject to insect attack. Using local materials can also reduce the need for depleting expensive, scarce natural resources. Gravel surfacing, for example, usually specifies very exacting material grading limits and properties to satisfy formal thickness design standards (Para 2.17). Relaxing these standards opens the possibility of using more abundant local granular material deposits for surfacing. Even when a feeder road must be upgraded to standard design criteria later because of increased traffic demands, this handbook suggests that in an economy with capital scarcity and uncertainty, discounted stage construction is more economical than initially building feeder roads to standard design criteria based on traffic predictions some years ahead.

2.31 **S-13: Adopt 'method' rather than 'performance' construction criteria.** The use of low technology construction methods should not imply relaxed construction standards. Workmanship requirements for mixing concrete, compacting backfill, placing earthworks in layers with controlled thicknesses, etc., must be maintained if the infrastructure itself is to be durable. However 'method' rather than 'performance' requirements are suitable criteria for many construction activities. The former requirements describe the type and quantity of work to be done and the method of carrying it out. For example, a concrete mix would be defined by indicating the proportion of each of its ingredients and its acceptable slump limits, i.e., the permitted range of height reduction of a fresh sample of the mixture. Performance criteria for the concrete mix, on the other hand, would stipulate that a properly cured sample of the concrete would develop a specific compressive strength in kilograms per square centimetre after a specific time period, while leaving the actual concrete production technique to the workers' skill and competence.

2.32 The generally small-scale and simple nature of RAI improvements, and the flexibility offered by the post-construction review and correction period (S-5) is well suited to the method criteria. Less academically qualified supervisors are able to complete method-specified work tasks satisfactorily. With the general shortage of professional qualified personnel in many developing countries, RAI improvements offer an opportunity for training and using para-professional or sub-professional technicians, supervisors, and foremen. This is particularly important since these improvements require a higher level of supervision than highway construction, due to minimal engineering design. RAI construction using method criteria may also be a suitable medium for developing small contractor capabilities for other small local construction works through provisions that:

(i) subcontract some activities within a project;
(ii) allocate some works to local groups or individuals to carry out under small contracts;

(iii) negotiate petty support contracts where force account operations are used; and
(iv) use labour under piece-work contracts.

2.33 **S-14: Adopt the performance budgeting concept.**

Performance budgeting assigns specific costs to specific task accomplishments. It forces engineers to develop definitions of work activities at the lowest level, thereby eliminating vague concepts of production goals. Specific definitions call for the determination of manpower, equipment and material needs and their associated costs for individual activities. These definitions help both to standardize work activities nationwide and to intelligently allocate appropriate funding. Unit costs can be readily developed by dividing the total task costs by the anticipated output. This formalized approach to unit costing is easily understood by both planners and field supervisors. It also permits estimates to be developed for activities that vary from the normal highway construction and maintenance tasks which constitute the data base for many unit cost estimates.

2.34 **S-15: Institute a workable system to monitor costs.**

In any situation where available financial resources are limited and unlikely to be increased, efficient and economic application of these funds is critical. Present cost-monitoring systems vary in their coverage, breakdown of costs, items included or excluded, etc. When cost classifications are inconsistent or incomplete, cost variations are difficult to trace or explain. For example, an equipment operator's pay may be treated as part of a project labour component, as part of the project equipment cost, or hidden in an equipment rental cost along with additional overhead charges levied by the equipment division. The impact of the operator's costs becomes even more difficult to analyze when he runs more than one piece of project equipment or is inactive for long periods of time.

2.35 Project expenditure recording systems should be simple, flexible, complete and adaptable for use by relatively poorly staffed local government agencies. Above all, no matter how simple or complex the recording system is, its accuracy will depend on the accuracy of the basic information collected at all levels. The most difficult problem is to get accurate reliable data and information from field personnel where the workers' education and analytical levels are usually minimal. The primary requirement of a costing system is to account for all expenditures and to relate these to the work carried out:

(i) by specific road or other RAI;
(ii) by type of work;
(iii) by task categories;
(iv) by inputs and outputs;
(v) by standardized classification headings; and
(vi) by consistent coverage of items.

Cost controls are often viewed as an unnecessary expense because they cannot be measured in cubic metres placed or kilometres completed. However, this is shortsighted since the lack of cost control, like overdesign, reduces the efficiencies the policy maker or planner is trying to achieve.

***2.36* S-16: Assess maintenance requirements in terms of necessary capabilities and funds while evaluating infrastructure improvements.**

Maintenance is critical to the spot improvement concept. It is therefore important to evaluate long-term maintenance requirements and available capacity and funds while assessing RAI improvements. The concurrent evaluation of construction and maintenance should consider the choice of construction technology, which is particularly important in countries with an unemployed or underemployed labour force and/or a scarcity of resources. Without the provision of maintenance funds and manpower on a long-term basis, any RAI improvement will be short-lived and development cannot be sustained.

***2.37* S-17: Modify existing institutions to meet new needs whenever possible rather than creating new ones.**

The implementation of a rural infrastructure and transport services programme involves new responsibilities at the central government level. Special project units, created specifically to execute infrastructure improvement needs for particular projects, work well only when there is a political commitment on the part of the central government to assign counterpart staff for training, to provide funds to cover recurrent costs, and to integrate project activities within an established agency's annual work programme. It may take some time for newly created organizational units to become effective since new relationships must be developed with the other agencies involved in the rural development process, with the beneficiary communities, and with the planning and budgeting authorities.

2.38 The creation of a separate branch within an existing sectoral agency rather than the introduction of special project units is often a more effective institutional arrangement for the implementation of RAI improvements. This approach can provide a channel for allocating funds to RAI improvements while permitting the development of new planning, construction, and maintenance methods. Such an organization enjoys the legitimacy and communications channels of the parent agency while becoming the focal point for institutional change. Its activities can be focussed on helping to develop local decentralized institutional capabilities for future RAI construction and maintenance without disturbing the parent agency's relationship with local communities in other issues.

***2.39* S-18: Promote local participation to strengthen the communities' institutional capacity.**

Beneficiary participation not only helps to ensure that project activities will meet immediate objectives, it also increases a rural community's capacity to identify and solve its own problems in other areas. Local people usually have an intimate knowledge of the existing local transport system, its problems, constraints and bottlenecks. They can usually identify locations for highest priority attention. Used judiciously, and with sufficient supervision to avoid poor workmanship, local participation can be helpful in avoiding unnecessary overdesign. Its prospects and incentives are complex and relate to:

(i) duration of required effort: short term efforts are more conducive to local participation;
(ii) scale of work: small-scale work is more conducive to local resource mobilization;
(iii) standards of workmanship: the lower and more basic the standards, the more the work will be perceived as related to local needs rather than the needs of transit users;
(iv) payment for local labour: recruitment goals are much easier to reach if some payment is made rather than by soliciting free labour; and
(v) timing of tasks: labour is more readily available if the RAI improvements are scheduled during slack periods in agricultural activities.

2.40 **S-19: Simplify contracting procedures whenever the possibility of developing small contractors exists.**

When small contractors are available, their use permits a government to base its force account capability on long-term and stable requirements. The peaks in any fluctuating workloads can be carried by the small contractors. The nature of RAI construction and maintenance activities offers an excellent opportunity to develop a small contractor capability. In view of the low technical and professional skills of many small contractors, it is important to keep the bidding documents simple. In many cases, standardized contracts at predetermined, pretested prices are the most suitable method for petty contracts, subcontracts, and piecework and taskwork contracts. Administrative practices and procedures should be streamlined to support small undercapitalized contractors by providing sufficient mobilization payments, rapid periodic progress payments, and by reducing retained fees as much as possible. Small contractors can also be encouraged by providing them with technical assistance through the supervision activity. If the supervising personnel use the method criteria (S-13) as a training tool and treat the small contractor as an asset rather than an adversary, the development of a local contracting capbility for the type of small public works projects common to rural developing areas is an obtainable objective.

2.41 **S-20: Promote the use of appropriate maintenance technology.**

Appropriate technology is defined as one that is both efficient and least costly (Para 1.25). One way of promoting the use of appropriate maintenance technology is to use an appropriate construction technology. For example, labour-based ditch construction results in ditches with steep sides and flat inverts which make them easy to maintain with labourers rather than with graders. Small ditches built by graders are usually 'V' shaped and are efficiently maintained by other graders. Their maintenance by labour rather than machine requires considerably more labour per metre to clean than trapezoidal ditches of the same capacity. 'V' ditches therefore may be more efficiently maintained by graders, but not necessarily at or below the cost of labour-based maintenance of the same ditch.

2.42 Other factors also affect the selection of maintenance technology. For example, a feeder road's granular surface can be maintained by labour even when built by machinery. However, as the traffic volume approaches ten vehicles per day, labourers begin to experience efficiency problems in corrugation removal. At this point in time it may be less costly to augment the labourers' efforts by periodically towing a drag over the road with a farm tractor than to replace the labour-based maintenance activities by a motor grader. Therefore, appropriate technology may change as development progresses, and that change may take place in phases. Technology that employs both men and machines, as in this example, is called intermediate technology in this publication.

***2.43* S-21: Assess the impact of inadequate storage facilities on agricultural production where reliable access is not provided.**

This handbook is based on the premise that resource investments in access should benefit as many of the rural population as possible within the constraints of available funds and economic return criteria. Under that premise, if better storage capability can help to spread the benefits of improved, general area accessibility to isolated sub-areas while meeting budgetary and economic constraints, storage should be evaluated as a substitute for uneconomical access improvements.

2.44 Improved storage has identifiable beneficiaries. The costs to obtain these benefits should be borne by their recipients. One of the goals of a development project is to improve the economic status of the local population so they can bear such costs in the future. Therefore, project storage components should focus on technology transfer relating to simple, low-cost storage facilities that will meet local needs, utilize local resources, and rely on local operational expertise.

***2.45* S-22: Simplify economic evaluation procedures when possible.**

Many governments are faced with the problem of how to reduce time and costs of appraising RAI components. This problem is particularly acute since most developing nations have limited qualified staff and a large number of kilometres of RAI to be improved or built. The use of simplified operational procedures for the appraisal of these components is therefore suggested. These procedures imply the determination of a simple economic return (ER), which is defined as an ER based on a portion of quantifiable benefits rather than all.

2.46 Simple ERs are underestimates of actual ERs. Their use is acceptable in situations where policy makers choose investments based on economic viability rather than the highest ER. If a simple ER of a proposed investment is equal to or higher than the prevailing opportunity cost of capital, the investment is economically viable. In addition to saving time for data collection and analysis, the application of simple ERs allows for the consideration of social and political criteria when making investment decisions. Consideration of these criteria is difficult if investment decisions are based on the principle of the highest ER for non-mutually exclusive investments.

2.47 Standard evaluation procedures call for the appraisal of interdependent RAI components as a group while requiring independent components to be evaluated individually. To further simplify the appraisal, Chapter 19 proposes to evaluate independent RAI components as a group when certain conditions are satisfied. It also suggests the employment of social criteria in situations where the determination of an ER is difficult or impossible with available data.

***2.48* S-23: Reduce formal design activities to reflect the reduced costs of appropriate construction technology.**

Most major highways are designed as 'one of a kind'. The heavy engineering involvement in designing construction details is warranted by high construction costs and heavy traffic volumes. The same design orientated approach to low-volume feeder road and other RAI construction can easily increase engineering costs to an unacceptable proportion of total costs. RAI engineering must therefore become basically a field evaluation activity. Fortunately, one experienced engineer can locate and assess a large number of spot improvement sites in a relatively short time, especially after consultation with local inhabitants.

2.49 Substantial specialized engineering construction design activities will still be required for major construction works such as bridge foundations. However, the adoption of a portfolio of standard designs for culverts, fords, ditch configurations, and typical cross sections, with their appropriate method specifications (S-13), will relieve the evaluating engineer of many details. Such an approach allows for supervision by junior personnel under the general guidance of a senior supervisor. Consequently, a considerable savings of the critical engineering resources often lacking in developing countries may be obtained.

II. TECHNOLOGY FOR IMPLEMENTING STRATEGIES

2.50 The strategies outlined in this chapter can be used to implement more than one policy (Para 2.04). The technical details and methodologies described in the rest of this presentation also often enhance the implementation of more than one strategy. For example, when policy makers choose to concentrate on eliminating drainage problems (S-2), they also promote the strategy of using incremental improvements to meet changing access needs (S-3). Consequently, many of the individual paragraphs that follow contain information that can apply to two or more strategies. Table 4 can be used to locate the pertinent paragraphs in Parts II and III and the salient technical notes in Part IV for each of the strategies described in this chapter.

III. CONCLUSIONS

2.51 Each individual policy may be considered as the apex of a pyramid. It is supported by one or more strategies which are implementation methods selected by the policy makers to effect the chosen policy. These strategies are in turn founded on a multitude of specific actions which a governmental agency may undertake to execute the specific strategies.

2.52 Specific actions of the nature described in the rest of this presentation are synergistic. They may help to promote several strategies which, in turn, may implement one or more policies. The role of the planner is to select the actions that implement the designated strategies with the least expenditure of the country's resources. The role of a country's technical staff is to execute the individual actions efficiently to further husband the country's limited resources. Planners and technical staff must therefore be aware of the national goals, and the policies and strategies chosen to reach them, in order to determine the appropriate investments which will satisfy, but not exceed, the country's developmental needs.

Table 4. Strategies and Their Technology

Strategy code	*Paragraph number*	*Technical notes*
S-1	4.02, 11.05, 12.02, 12.28–12.29, 13.03–13.05, 13.26	2.3–2.4, 4.01, 4.15, 5.01, 6.7
S-2	4.09, 6.03, 9.17–9.19, 12.01, 12.03–12.04, 12.06–12.19, 12.31–12.32, 13.05, 13.12, 13.16–13.18, 13.23, 15.02–15.05	1.1, 4.02, 4.04, 4.06–4.11, 5.01, 5.03–5.10, 6.2
S-3	3.02, 3.07, 4.14–4.15, 4.23, 6.02, 9.14, 12.17, 12.33, 13.02–13.26, 15.01, 15.06–15.08	2.4, 4.01, 4.11, 4.13, 5.01–5.10, 6.2–6.3
S-4	12.04, 12.06–12.10, 12.15, 15.05	4.03–4.04, 4.06–4.07
S-5	4.23, 12.10, 12.15, 13.06, 15.04–15.06	4.07
S-6	4.01, 4.04, 4.19, 5.04, 6.01, 9.20–9.23, 10.02–10.04, 10.09–10.12, 10.15–10.20, 11.08	1.2, 2.6–2.7, 2.9
S-7	3.06, 4.03, 4.22–4.23, 5.01–5.17, 9.02, 9.14–9.16, 9.21–9.23, 10.01–10.06, 10.09–10.20, 13.01	2.1–2.2, 2.4–2.8
S-8	3.08–3.16, 4.03, 4.05, 6.04, 6.12, 9.02, 9.13–9.14, 9.23–9.26, 10.03–10.05, 10.15–10.20, 11.02, 11.08, 12.33, 13.09, 18.01, 18.06, 18.12, 18.21	2.2–2.4, 2.9, 10.2
S-9	3.02, 3.07, 3.19, 6.30, 12.34, 13.12–13.13, 13.25–13.26	2.4, 4.11, 5.01–5.02, 6.2
S-10	3.04–3.05, 5.02, 10.12, 19.42	2.6–2.8
S-11	5.02, 5.06–5.07, 5.09, 5.12–5.13, 5.17, 10.13–10.14	2.6–2.8
S-12	4.12, 7.04, 7.09–7.15, 10.13–10.14, 13.11–13.13, 14.01–14.13, 14.27–14.29, 15.09–15.11, 15.15–15.24, 15.27–15.37, 16.02–16.05, 16.10, 16.16–16.20, 16.30–16.32, 16.39–16.42, 17.13–17.15, 17.20–17.21	2.7–2.8, 4.02–4.03, 4.05, 4.08, 4.11, 6.1–6.2, 6.5, 7.1–7.7, 9.03–9.05, 10.1–10.2

Table 4. *(contd.)*

Strategy code	*Paragraph number*	*Technical notes*
S-13	14.21, 14.34, 17.24–17.25	6.6, 9.08–9.09
S-14	6.13–6.20, 6.31, 7.17, 17.16–17.27	9.02, 9.06–9.10
S-15	6.15–6.20, 16.23, 17.01–17.35	7.7, 9.01–9.10
S-16	6.12, 7.04–7.23, 13.12, 15.01–15.37, 16.07, 17.26–17.27	5.01, 7.1–7.7, 9.10
S-17	7.05, 14.06, 16.08, 16.10–16.12, 16.15–16.16, 16.21–16.24	8.1
S-18	3.01, 3.18, 4.12, 5.03, 5.07, 5.12 5.17, 7.09–7.12, 8.11, 10.24–10.25, 14.41, 15.15–15.24, 16.02, 16.10, 16.14–16.20, 16.26–16.47, 18.11–18.12	2.7–2.8, 4.13, 7.2, 7.7, 10.1, 10.2
S-19	14.21–14.26, 15.23, 16.04–16.05	6.3, 7.2, 7.7
S-20	7.19, 14.01, 14.04, 15.09–15.11, 15.32–15.34	6.5, 7.1, 7.6–7.7
S-21	3.06, 3.14, 3.16, 8.01–8.11, 8.42–8.45, 18.01–18.20	2.7, 4.12, 10.1–10.2
S-22	3.15–3.16, 8.23–8.42, 16.02, 19.01–19.34, 19.47–19.50	11.1–11.2
S-23	12.20–12.25, 13.14–13.16, 13.19–13.20, 13.22, 14.11–14.12, 14.30–14.31, 14.41, 16.05	3.1, 4.01–4.04, 4.09, 4.11–4.15, 5.03–5.10, 6.3–6.4

PART II

CHAPTER 3

Planning Rural Access

3.01 Part II presents a series of guidelines for implementing the policies outlined in Part I. It draws heavily on Parts III and IV for the detailed technical considerations, approaches and evaluations necessary to improve access for the rural population living at the outer edges of a developing country's transport system. To the rural poor, transport problems begin at home and frequently end at the local marketplace, which may well be the terminus of the conventional road network. These people often have no concept of national, sectoral or programme objectives, policies and strategies that motivate the transport planner. However, the rural population are well aware of their own access problems and are often willing to accept minimal improvements, especially if they are induced to participate in some way in the decision-making process leading to the allocation of the limited resources available. In some cases a development project may lack any rudimentary existing RAI, such as a project in which immigrants are expected to settle in and improve an uninhabited area. The transportation planning task for such projects must develop a construction programme that is appropriate to the local environment and provides sufficient new transport infrastructure to accommodate the daily access needs of the immigrant population. The lack of an existing social structure within the new population often leads to a more ready acceptance of new techniques, both in transport and agriculture.

3.02 The project planning relationships for multi-sectoral programmes such as integrated rural development projects or rural roads projects complementing agricultural projects are important since such projects are the most likely vehicles for expanding rural access infrastructure. The close relationship between project components, their interdependencies, and their combined resource limitations, result in three planning tenets that form the basis for this book:

(i) planning rural development projects is a team effort rather than a competition in which each sector tries to accomplish a fixed independent goal;

(ii) planning access components requires sufficient flexibility to respond to changing needs discovered after the project has begun or whenever the scope of another project component is altered, increased or decreased; and

(iii) phasing access components must acknowledge the uncertainty of long-term projections by providing sufficient access only for the current development phase while retaining all reasonable options for the transport requirements of future phases.

3.03 Rural access planning differs from conventional highway planning in four significant aspects:

(i) using generalized predictions of national motor vehicle growth rates and usage is inappropriate for determining feeder road traffic volumes. Feeder roads, the highest level of infrastructure considered here, are defined as roads that link a defined area to the highway network. Their traffic volume and type will be related to activities and development within that area. By definition (Table I), these roads carry no through traffic, making generalized national trends irrelevant;

(ii) the concept of pavement design based on repetitions of a standard axle load now in general use as a means of coping with the wide spectrum of axle loads in the traffic using the through highway network cannot be readily applied to the design either of gravel roads or of the lightly constructed asphalt pavements which may sometimes be required on feeder roads;

(iii) rural access can be accomplished by a larger variety of transport means than normally considered in highway planning. Transport mode is a term used here to differentiate transport means by their power source, i.e., human, animal and mechanical. 'Transport aids' refer to individual transport means such as head baskets, bicycles, pack animals and trucks. Each transport means must be evaluated within the context of a specific rural access infrastructure network; and

(iv) local storage capability and capacity may modify the accessibility requirements of the rural access infrastructure. Therefore, investing in storage facilities at the local level may be more cost-effective than providing high level access through large investments in infrastructure.

I. PLANNING INTERACTIONS IN RURAL DEVELOPMENT PROJECTS.

3.04 Chapter I describes the linkages between national, sectoral and project objectives. Successful rural access planning results from the integrated planning of project components developed through insights into their interrelationships. Questions to be answered in the evaluation of a rural development or rural roads project proposed to increase agricultural output include the following:

- What level of access is required to eliminate transport bottlenecks hampering the flow of inputs and the evacuation of surpluses?
- What level of access is required for personal mobility for both economic and social activities?
- What transport capacity will be required?
- When will the major transport activities be needed?
- What transport modes, i.e., human, animal, or motor-powered means of transport, are currently in use?
- Is it desirable to introduce new transport modes or aids?
- What type of access infrastructure exists?

- What steps are desirable to incorporate local society in the planning, execution and care of the transport facilities?
- Have the farmers' views been considered?
- Are transport services reasonably competitive, or are they hampered by regulation?
- Are storage facilities appropriate and adequate?
- Are crop prices adequate to stimulate production; do they fluctuate during the year?
- What is the land tenure situation? and
- Is there access to credit, and what are its terms and conditions?

3.05 Answering these questions calls for examining not only the technical features of topography, soil and climate as they affect access infrastructure, but also present and future area activities requiring access. Cultivated and uncultivated areas available for future cultivation, storage methods and the current transport system all must be examined. Government-controlled crop prices, if set low or unclearly stated, or unresolved land tenure problems, may act as disincentives to farmers to produce more, resulting in low current and future transport requirements. In some countries, free entry into the provision of transport services is restricted, which may result in under-utilization of transport modes. Facilitating access to credit and introducing realistic and fair terms and conditions may enable farmers to procure transport aids. Any bottlenecks other than poor rural infrastructure and/or inadequate transport modes and aids must be examined before making transport-related investments. In addition, the impact of possible development activities, such as the introduction of improved seed, fertilizer, irrigation, extension activities, cottage industry and increased educational and social activity, must be weighed against transport requirements for the entire access area whether the activities are part of the original project or not. The integrated analysis should result in recommendations for an overall resource expenditure which will not only be beneficial to the area itself, but will also conform to national and sectoral objectives.

II. PLANNING THE TRANSPORT COMPONENT

3.06 Planning the transport component of a rural development or a rural roads project includes the evaluation of the three individual subcomponents that may alleviate access constraints:

(i) providing alternative transport modes and aids by either increasing the availability of existing aids, introducing modifications to them, or introducing new transport aids from another area, either domestic or foreign.
(ii) improving existing infrastructure, by either upgrading or expanding that infrastructure, or any combination thereof; and
(iii) providing local storage, by upgrading or increasing the local storage capability or by introducing new storage technology.

The transport planners' goal is to determine which combination of subcomponents or individual components will provide satisfactory accessibility with the least resource expenditure.

3.07 The following General Planning Chart depicts a planning approach that includes an evaluation of these three access subcomponents. More detailed planning charts for each subcomponent are presented in later chapters. For instance, Chapter 4 discusses the definition and evaluation of access. All planning must be phase- and time-oriented. Evaluations should be made using both the degree of access required to permit development to begin, and the accessibility needed if the project succeeds. Between these limits, accessibility can be upgraded through the stage construction techniques discussed in Chapter 13. The limits vary from the relative certainty of the access requirements at the beginning of the project to the higher-risk-inherent ones requiring the projecting of both future agricultural production and increased transport activity resulting from the successful project completion. The stage construction or incremental improvement approach should be oriented towards retaining as many future options as possible, while providing improved access and/or additional carrying capacity in the early stages when there are clear indications that these are required.

III. EVALUATING TRANSPORT SUBCOMPONENT INTERACTIONS

3.08 This section consists of a general discussion of the overall relationship of the planning activities shown on the General Planning Chart, hereafter referred to as the GPC. The rationale and procedures for approaching accessibility problems, alternative mode and transport aid selection, RAI improvements and maintenance, and storage problems will be outlined in later chapters.

3.09 The GPC and other charts included in this book are intended as guides. They attempt to address most of the more common facets that might require consideration in a comprehensive planning approach. However, many rural access problems require small diversified activities which do not warrant intensive professional resource investments. When the investment in data collection exceeds the value of any refinements expected from analyzing that data, a poor resource expenditure results. Common sense and experience will often indicate that some of the activities suggested in individual guides, including the GPC, do not apply to certain projects. This does not mean, however, that those activities will not be appropriate in other projects which may appear similar. Therefore, at the onset of project preparation, each project must be treated as being unique.

3.10 Conventional highway planning, for example, has worked well for many years using an abbreviated application of the GPC, namely the application of only Steps 1, 2, 10, and 12. In such planning, arbitrary assignment of all weather access to Step 1, motor vehicles to Step 2, standard highway design criteria to Step 10, and standardized unit cost figures to Step 12, has been, and will continue to be, a satisfactory approach. However, that same abbreviated approach will often result in overdesigned feeder roads when traffic volumes

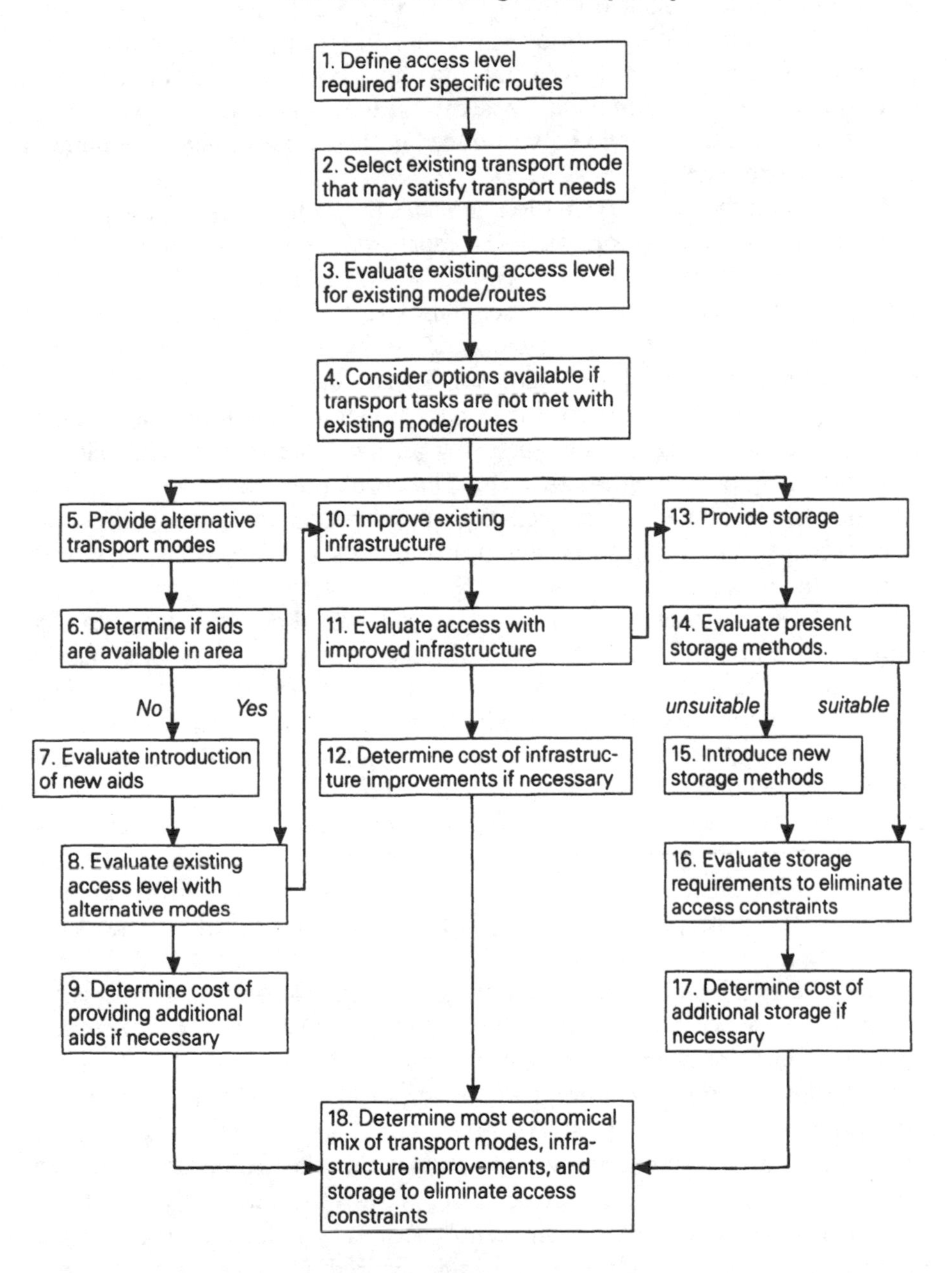

are low and include a mix of alternative transport aids.

3.11 The least time-consuming planning effort will occur when the first three steps of the GPC are followed in order, and the outcome of Step 3 indicates sufficient accessibility is available over existing infrastructure using an existing transport mode. While abandoning the access planning process at this point may not optimize the transport resource expenditure, its short-term financial implications can be very attractive. In this example, any sizeable expense to introduce improved transport aids, for instance, in a small project or for low traffic demands, would only be feasible if such an expense were shared within an ongoing nation- or region-wide transport aid improvement programme. However, if the transport aid improvements were simple enough so that the local population could accomplish them under the guidance of an extension agent, they should be introduced in the course of any development training programme.

3.12 When the Step 3 evaluation indicates Step 4 is necessary, the planner is faced with up to three options. The most frequently chosen option to be evaluated first is Step 10, improve existing infrastructure. If the planner determines in Step 11 that proposed infrastructure improvements provide satisfactory accessibility, a strong temptation sometimes exists to terminate the investigation after Step 12. Excluding the Steps 5, 6, 7, and 8 evaluation of the suitability of available alternative transport modes and their associated transport aids negates consideration of using an available resource in place of investing in improved infrastructure. The local population is unlikely to overlook the continued use of available transport aids on the new infrastructure, perhaps to the extent that the improvements add little if any economic value to the project.

3.13 In some cases, the solution is so obvious that many steps can be eliminated. For instance, if a rope bridge across a chasm satisfies access requirements, the construction of 10 km of paved road is not an economic alternative. In other cases, the cost of infrastructure improvements is so similar for families of alternative transport aids that it makes no difference which transport aid within that family the local population chooses. It would appear that recalculating the cost of improving infrastructure for two different transport aids would only be necessary if they required different levels of infrastructure, such as a track and a road, unless some unusual cost may be involved, such as a bridge which may be sufficient for some aids within a specific category but too narrow or weak for others.

3.14 Although Step 13, provide storage, is listed as an option, it really is not a means of improving access but rather a substitute investment for expensive infrastructure improvements. Its direct beneficiaries are cash crop farmers, although introducing new storage technology will also help subsistence farmers. While benefits from improving access are both economic and social, and accrue to a major segment of the population within the project area, benefits from providing storage facilities are realized only by the farmers who use those facilities to keep agricultural produce until access is feasible, after, for instance, the wet season, or until better prices can be secured. However, when the cost difference between providing minimum access and reliable access (Para 1.19), is substantial, local storage may be a viable option for reducing resource investments.

3.15 The rural transport planner's main problems occur when several options appear equally attractive, or when more than one combination of options is available. For example, if suitable access can be achieved either by minor infrastructure improvements and credit provision for an increased population of ox carts, or by a more expensive infrastructure improvement to permit better motor truck access, an economic evaluation is required. The evaluation, which should be undertaken only when the options appear equally viable, will need to include more data than is usually analyzed in a conventional transport economic analysis.

3.16 The evaluation possibilities increase if the accessibility provided by alternative rural infrastructure improvements and transport aids is about equal but expensive and/or insufficient, indicating the need of a storage subcomponent. Among the possible solutions are:

(i) substitute storage for all infrastructure improvement requirements if minimum access is already available;
(ii) provide enough storage to overcome access deficiencies after the infrastructure is upgraded for use by existing traditional transport aids; or
(iii) provide only enough storage to satisfy access needs after the infrastructure is upgraded for trucks.

A detailed economic evaluation will determine the least costly mix of these subcomponents.

IV. CONCLUSIONS

3.17 Rural access planning places the transport planner in a situation that is somewhat at variance from his role in planning specific links in a country's general network of through traffic roads built to standard specifications. The transport planner's approach must change from that of the major decision-maker to that of a member of a planning team considering not just a transport infrastructure investment, but an assortment of interacting resource investments which together form a fluid development planning activity where options abound.

3.18 The ultimate users of each rural access infrastructure network are identifiable people with specific needs, beliefs and methods; who are intimately aware of the transport needs of the specific development area. Since the local population will be among the beneficiaries of any development project, even when the project involves an immigration component, they should have some voice in the resource expenditures. This is particularly true when many of the resources such as transport aids, storage facilities and ongoing infrastructure maintenance are to be provided at least in part by these identifiable users rather than the government.

3.19 The transport planner must deal with two levels of options concurrently. The first level involves servicing other project components, and includes the type and timing of transport resources required to prevent the occurrence of any access constraints to the other development activities. Therefore, the planner must be prepared for unforeseen changes by retaining his resource expenditure options as long as possible through spot improvements and stage construction. The second level of options involves the manner in which the accessibility problems will be overcome within the project's transport component itself. These options must be assessed to determine the most economical mix of transport modes, transport aids, infrastructure improvements and storage capability. This second level of options is directly influenced by the first level of options involving the service of the other project components. This book concentrates on the second level, with the caveat that the decision-making process must always be tempered by circumstances imposed by the first level of options and by local considerations.

CHAPTER 4

Accessibility Evaluations

4.01 Access is the ability to enter into or exit from a location. To many, development area access implies entrance from the outside. However, this handbook considers access as the ability to:

(i) transport surpluses from a development area to markets foreign to that area;
(ii) receive inputs, goods and social services into a development area;
(iii) conduct trade within a development area;
(iv) move to and from villages, farms, fields, forests and water sources as required to transport the necessities of everyday life within a development area; and
(v) ensure personal mobility for economic and social activities.

4.02 All-weather access is defined in Para 1.19. Paragraph 9.26 indicates that such access normally requires relatively high infrastructure expenditures, especially for mechanized transport. This handbook is particularly concerned with the evaluation and implementation of projects that do not require all-weather access. However, the principle of reducing resource expenditures by not applying design speed criteria and by reducing other standards to a level suitable for the low traffic volumes and short-haul distances often encountered on rural access infrastructure, is equally valid for all-weather access. Evaluating need for all-weather access should begin by considering the possibility of not providing it. However, certain agricultural activities must have all-weather access. These include any programmes that necessitate movements on a year-long basis, such as milk production; and any crops that must be harvested and processed or consumed within a short period regardless of whether the rainy season has ended or not, such as premium coffee that must be harvested and washed at a specific point in its ripening cycle.

4.03 As defined in Para 1.19, reliable access is realized when the accessibility level is not a constraint to achieving the economic and social potential of a development area where substantial agricultural economic activity exists or is proposed. Accessibility is defined individually for each relevant transport mode and aid. It is measured in terms of the acceptable maximum duration of individual infrastructure closures and maximum accumulated closures per year. Therefore, the specific transport aids available or proposed are important in determining the needs for infrastructure improvements. Failure to recognize this interdependence can easily lead to uneconomical over-investment in rural access infrastructure (RAI)

4.04 Minimum access is defined in Para 1.19 as that access required to meet the local population's goods transport and personal travel demand in periods when economic activity is low. For example, cotton is the principal cash crop in an agricultural area in Tanzania [21] with an annual rainfall of approximately 1000 mm, three-quarters of which falls between December and May. Cotton is planted in February, as are the principal food crops maize, sorghum and rice. All cash crops can be harvested in the dry season, for example, maize and rice in June, cotton and sorghum in August. Therefore, all cash crops can be transported over dry-weather roads, indicating these roads represent reliable access for economic activity. The poor smallholders and landless work for the cotton producers during the peak growing season. Consequently, the smallholders plant more maize in June or December, all of which is consumed locally. This maize is harvested between March and June. April has the highest rainfall of the year, over 200 mm. Minimum access must permit the harvest and distribution of this maize within the area to allow the poorer segment of the population to survive. It is also required to permit the workers to travel to the larger farms to cultivate the cotton. However, the infrastructure required for this minimum access need not carry the volume of heavy traffic required for the export of the cash crops during the dry season.

4.05 The planner's task is to evaluate the access needs of each major infrastructure link and group of minor links to determine the accessibility level that will satisfy those needs. While this may seem an insurmountable undertaking, the options for improving access are limited, and individual options may have a significant impact on all accessibility levels. For example, the construction of a bridge to provide all-weather access for trucks also provides all-weather access for all other transport aids.

4.06 The following Access Evaluation Chart (AEC) outlines an approach to determining accessibility requirements. It can be implemented using the planning book method described in Para 11.04, and Technical Note 3.1. The purpose of determining accessibility requirements is to set the parameters for the ensuing evaluations required in the planning process. Determining the interruptions in transport that are acceptable for specific crops, and the quantity of transport for crops or goods requiring specific accessibility levels, defines the needs for transport aids, infrastructure improvements, and possible storage investments to be evaluated in subsequent steps. Accessibility levels by themselves do not define the type of rural access infrastructure required, but when combined with transport volumes, gives the planner the first clues to the form the transport network is likely to take.

I. EXPLANATION OF ACTIVITIES SHOWN ON THE ACCESS EVALUATION CHART (AEC)

4.07 All rural access infrastructure planning must be based on the development area's needs. These needs must be identified, located, and coordinated with each other so that the individual RAI links are properly located and provide suitable accessibility. Such planning can be carried out using the plan book system referenced in Para 4.06. The first step in preparing for any RAI planning activity is to acquire all available mapping and air photos of the

Access Evaluation Chart (AEC)

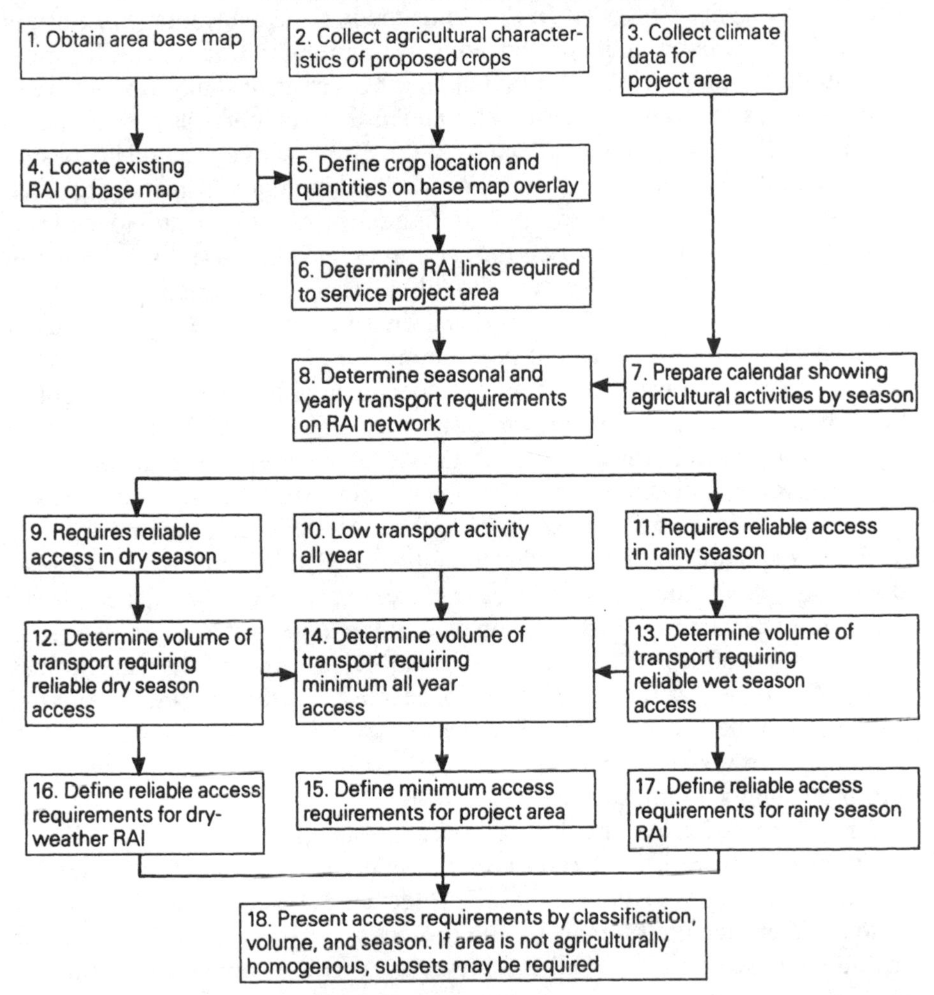

project area and develop a project base map.

4.08 The agricultural needs to be supported by the RAI require a comprehensive understanding of the crops to be transported. This agricultural data should be provided by the agricultural specialists on the project planning team. Step 2 crop data includes planting and harvesting seasons; amounts and timing of fertilizer, pesticide and fumicide applications; and anticipated agricultural production.

4.09 Since all-weather, reliable, and minimum access are functions, at least in part, of climatic conditions, Step 3 includes the collection of any rainfall data available for the development area. Climatic data includes rainy seasons, yearly and monthly amounts of precipitation, record monthly precipitation amounts, any rainfall intensity data available, average number of rainy days by month, or pockets of intense rainfall caused by topographical features. If such data are not recorded, local farmers are usually a good source of information about average and unusual occurrences.

4.10 All currently existing transport infrastructure should be recorded on the base map, since every effort should be made to take advantage of existing resource investments in the project area, as indicated in Chapter 13. Information about existing infrastructure should be checked against any available road inventories, air photos, and geographic and military mapping as part of Step 4.

4.11 All information developed from the Step 2 activity should be located on overlays to the base map in Step 5. This information should include the areas to be developed and the timing and phasing of the development. None of this information should be drawn on the base map itself because, as indicated in Technical Note 3.1, the base map should be reserved for existing conditions only. Otherwise, as plans change, critical links may inadvertently be dropped from the required improvement programme.

4.12 Step 6 involves the selection of specific RAI links that are thought to be critical to the development process. Once those links are chosen, their agricultural produce transport role in the development process can be evaluated. However, accessibility for extension workers, traders, etc. is not as easily determined, and the selection of individual tracks, trails and paths may be very difficult to quantify in the least developed areas. These lower RAI classifications will involve considerable local resource input; if not in their improvement, most certainly in their maintenance, if the improved RAI network is to be sustainable (Chapter 7). Therefore the initial selection of the trails, tracks and paths to be evaluated should incorporate the user's views. The initial decisions will be less subjective if they are based on the users' willingness to commit the necessary manpower and materials resources to assist in developing and maintaining an improved RAI network.

4.13 Step 6 further involves an initial evaluation of how the selected existing infrastructure satisfies the area's development needs. All major terrain constraints to access to and within the project area should be identified at this time to determine their impact on the project's feasibility. Step 6 should also include the identification of all specific access corridors not currently utilized by some potential users. This will allow evaluation of alternate routings to avoid physical barriers or to provide access where none exists. Once the proposed RAI network has been established on overlays, its specific terrain and soil conditions should be reviewed for probable access constraints.

4.14 During Step 6, the maximum use of existing resources, namely the existing RAI, should receive priority consideration. Chapter 13 highlights stage construction and spot improvements as the most cost-effective approach to improving access while conserving resources and maintaining future options. Step 6 determines where these principles may be applied. When a development project is planned to incrementally develop portions of a large area, those portions already served by existing RAI should receive consideration for the first development phase to limit investment while testing the assumptions about the development potential of the overall project.

4.15 The principle of providing the right level of access for the current development phase applies to new infrastructure as well. New RAI construction should only meet the immediate access needs while retaining as many options as possible, including later spot improvements, stage construction and possible abandonment for a higher level of access infrastructure in a future development phase. A simple method for determining the cost effectiveness of deferring ultimate improvements in favour of incremental improvements in the first phase of a rural development project is described in Technical Note 5.01.

4.16 Before any complete evaluation of the access requirements of any specific RAI links can be made, the seasonal amounts and types of transport on those links must be estimated. Since climatic conditions represent a major access constraint, the agricultural activities required for each crop must be superimposed on a rainy/dry season calendar in Step 7.

4.17 Step 8 involves allocating the transport activities determined from Steps 2 and 5 to the RAI network determined in Step 6, using the calendar developed in Step 7 to determine the seasonal transport demands. This permits an evaluation of transport needs and the anticipated infrastructure conditions within the same time-frame. Such an evaluation should permit classification of project sub-area or individual RAI links into one of three categories:

(i) requiring reliable access in the dry season (Step 9);
(ii) having low transport activity all year (Step 10); or
(iii) requiring reliable access in the rainy season (Step 11).

4.18 Reliable access is determined by the needs for agricultural transport. These needs must be satisfied for development to occur. Steps 12 and/or 13 include the evaluation of transport volumes requiring reliable access by specific season. The determination of seasonal volumes is necessary because accessibility is transport-aid-specific. For example, donkeys cannot carry as much as they can pull, but as a pack animal they may be able to transport over wet sand or on a narrow trail where a donkey cart could not operate.

4.19 Minimum access must be provided during seasons of reduced economic activities so the local population can survive and carry out the day-to-day activities that generate the requirement for reliable access. Step 14 therefore involves determining the transport needs when economic transport requirements are low. This step is based on the assumption that reliable access automatically provides minimum access. In cases where this is not true, for example when a cableway provides reliable access for fruit from orchards on a mountainside to a paved road, obviously minimum access must be determined independently. Minimum access requirements will govern when the transport activity is low all year. They will usually be a factor when reliable access is only required in the dry season; however, they may be completely dominated by reliable access requirements occurring during the rainy season.

4.20 The definition of both reliable and minimum access requirements in Steps 16, 17 and 18 is the most critical activity in access planning. Chapter 9, Sections I and II, should be reviewed before determining accessibility levels. Suitable economic or reliable access is both site- and crop-specific. Certain generalities are sometimes made, as for example, milk collections require all-weather access to prevent souring. Cassava, on the other hand, can be harvested at any time between 12 and 18 months after planting, inferring that its harvest may be delayed until the dry season with no damage to the tuber in the ground. Once dug, cassava must be cleaned and processed or consumed within a relatively short time. Therefore cassava-growing areas that supply large amounts of the staple food for urban areas may well require some degree of access during the rainy season to prevent urban food shortages. In such a case, marketing considerations rather than the agricultural characteristics of the cassava plant may determine the accessibility for cassava-producing areas.

4.21 The cassava example illustrates the dangers of generalization. No other generalizations to assist in determining accessibility are therefore offered in this chapter. The planner must approach accessibility evaluations with an open mind, and make determinations on a case-by-case basis, while resisting the temptation to select all-weather access as the easy solution. Each major agricultural activity must also be evaluated separately and in the context of reasonableness. For instance, if farmers individually use small quantities of fertilizer during the rainy season, it may be more cost-effective to distribute sacks of fertilizer before the rainy season. While this procedure requires the farmer to store one or two sacks of fertilizer at home, it is economically more viable than increasing accessibility solely to provide fertilizer to small users.

4.22 Step 18 involves the documentation of the project's accessibility requirements. Only RAI classifications need be listed when all individual links within a classification share the same requirements. Unique requirements necessitate listing individual links. These listings, which include estimated seasonal volumes for each crop and other material requiring transport, serve as input to the transport aid selection described in Chapter 5, and the infrastructure evaluations described in Chapter 6. The accessibility criteria also affect the maintenance evaluations, and determine the necessity for storage evaluations outlined in Chapters 7 and 8, respectively.

4.23 Large projects warrant access documentation by phases. When developing such documentation, the following principles apply:

(i) phases are selected so the lowest access costs are incurred in each phase. The early phases should consist of areas where existing infrastructure is currently suitable for development demands or can be improved to that state at low cost. No new infrastructure should be built unless it is expected to have a high utilization rate during that phase;
(ii) further phases should include increased accessibility only when transport volumes become significant and losses from access constraints can be estimated and evaluated. The 'review and correction period' concept, (S-5), allows for correcting errors in underestimating the resource commitment to achieve suitable access on individual RAI links. There is no such reprieve for over-investing in access improvements; and
(iii) access for different transport modes can be considered for the same link in different phases; for example, access for animal carts should be considered if truck access is not required until a later phase.

II. EXAMPLE OF A SEASONAL CROP CALENDAR

4.24 The following example shows the development and use of a crop calendar to assist in determining accessibility requirements. The data are not complete but are typical of the information that might be available. The planner must use judgement and proxy data from other areas to fill in the missing information with assumptions that seem reasonable.

4.25 In this example, our project involves the construction of three coffee wash stations to increase the value of a portion of several existing coffee cooperatives' output. This is in line with (i) the national objective to increase foreign exchange earnings from coffee exports, and (ii) the agricultural sector objective to improve the income of the members of the coffee-growing cooperatives. To realize these objectives, any constraint to the movement of the coffee from the bush to the processing plants must be removed, since the coffee must be washed within five hours of being picked in order to retain its quality. The coffee must be harvested when it reaches a specific point in its ripening process regardless of whether the rains have ended or not, since it is intended as a premium export commodity. The remainder of the coffee crop is either picked or allowed to fall from the bush during a longer time span; and is dried and stored in sacks by the individual cooperative members for domestic consumption or sale throughout the year.

4.26 It is further assumed that three different areas, A, B and C, all prime coffee producers, will be involved in this project. These areas also produce maize and sorghum, chiefly for local consumption. The following tabulation of project data: Table 5, Project Activities by Season, consists of two parts. Part I contains information supplied by the agricultural planners to provide general data on the agricultural cycle in the three project areas. Part II contains specific information from the country's Meteorology Department.

4.27 Evaluating the rainfall data in Part II, we find that in the 7-month main rains period, as identified by the agricultural planners in Part I, 84% of the yearly rainfall occurs in area A, 82% in area B and 84% in area C. However, if the rainy season is defined by including months when rainfall exceeds 100 mm, area A's rainy season includes 6 months, April to September, with 79% of the yearly rainfall; area B rainy season remains at 7 months but begins a month earlier, continues from March to September, and includes 84% of the yearly rainfall; and area C rainy season includes 9 months, March through October, including 94% of the yearly rainfall. Additional data is available from another area with a rainfall of 1050 mm per year and a five-month rainy season averaging 100 to 200 mm of rain per month. These data indicate that 75% of the rain falls in the rainy season, and that the average rainfall per rainy day is 12 mm/day during the rainy season and 11 mm/day during the dry season. This information may be somewhat relevant to area A but becomes completely unreliable for areas B and C because the annual rainfall in area B is 43% greater than that in the sample area while area C's annual rainfall is 130% greater. When we compare the elevations of the three areas, we find that we cannot factor rainfall by altitude.

4.28 The average annual rainfall and the amount of rain that falls during the rainy season, no matter how it is measured, is approximately twice as much in area C as it is in area A. The highest average monthly rainfall in area C is almost 65% more than in either area A or B. During the sorghum sowing season, area B receives 11% more rain than area A, but area C receives 86% more than area A and 67% more than area B. During the first month of harvesting coffee and maize, area C receives almost three times as much rain as area A an almost two and one-half times the rainfall of area B. During the second month of the three-month coffee harvest, which is also the last month of the maize harvest, and the first month of the two-month sorghum harvest, area C averages over two and one-half times as much rainfall as area A and over three times that of area B.

4.29 Before final design of the RAI network in these three areas is undertaken, more detailed information about the rainfall would be helpful, but the following general conclusions can be drawn from the data presented above:

(i) traction will be a major access constraint because the earth is red coffee soil. The area adjacent to the coffee wash stations can be serviced by foot or donkey, thereby reducing traction requirements. However, to expand the service area to include sufficient coffee production to operate the stations economically, while observing the required five-hour time-frame for both picking and transport, motorized transport such as tractors and trailers will be necessary. Their routes will require well-drained granular-surfaced tracks to provide reliable access to the coffee wash stations;

(ii) storage facilities are required to keep the washed coffee. However, if the access from the stations to the paved highway system includes existing earth roads, the storage in area C must be considerably increased over that required in areas A and B to account for the additional transport delays from the higher rainfall amounts. Therefore the cost of gravelling the roads from the wash stations in area C should be compared to the additional storage costs;

(iii) minimum access includes delivery of unwashed coffee, maize and sorghum to the cooperatives for redistribution among all of their members. That delivery takes place in the dry season in areas A and B, but begins in the rainy season in area C. Minimum access must account for this since distribution to the cooperative members is based on the timely receipt of the new crop; and

(iv) since these coffee wash stations are erected to process a diverted portion of an existing crop, they should be able to operate at capacity shortly after opening. This indicates a stabilized producing area. Therefore the inputs may determine the minimum access requirements for part of the infrastructure. This development project is likely to be a single phase operation with actual accessibility levels easily identifiable.

4.30 This example shows the impact of data collection and application on the accessibility determinations for development projects. Each of the three areas in the example can be considered the same in all respects except rainfall amounts. Because of this single difference, access costs will differ. If the agricultural calendar had been used without considering the actual rainfall amounts and timing, reliable access would require surfacing all coffee transport

Table 5: Project Activities by Season

Part I – Project phasing

Months

. J . F . M . A . M . J . J . A . S . O . N . D .

Main rains

Coffee activities

Planting (in-filling)

Main flowering

Fungicide spraying

Picking

Stumping

Maize activities

Planting

Harvesting

Sorghum activities

Planting

Harvesting

Inputs received

Delivery of agricultural produce to cooperative

Part II – Climatic data

Location	Monthly average rainfall (mm)												Av. annual rainfall	Altitude (m)
Project area A	28	29	77	199	108	129	234	164	184	69	51	19	1291	1842
Project area B	37	53	107	151	140	235	227	215	186	84	43	25	1502	1700
Project area C	39	46	116	166	164	376	372	368	386	203	130	38	2420	2002

routes to the wash stations and between the stations and the highway system. The maize harvest would be carried in the rainy season but the sorghum harvest could, when necessary, be carried on dry-weather routes. Detailed data indicate that while deliveries to the wash stations require all-weather access for some transport modes, the cost of providing appropriate accessibility to the main highway system and for grain harvesting can vary considerably between the three project areas.

4.31 If such a coffee wash station programme were to be financed in tranches so that the areas were developed one at a time, area A should be developed first to determine the economic viability of washing coffee. Had the evaluation been limited to the project phasing calendar, thereby allocating the same portion of funding for access to each area, the expenditures would have been excessive for areas A and B, insufficient for area C, or inappropriate for any individual area.

III. CONCLUSIONS

4.32 The sole purpose of rural access infrastructure is to provide accesibility from and throughout rural areas. Accessibility is measured as the lack of constraint to travel and transport. All-weather infrastructure provides 100% accessibility, but in many cases such access is not an economically viable option for low transport volumes. Reliable access provides sufficient accessibility to eliminate constraints on the economic development of rural areas. As such it is both site- and development-activity-specific. Minimum access provides sufficient accessibility for the rural population to sustain their economic and social activities in periods of low economic activities in otherwise economically active areas, and throughout the year in areas with low economic activity.

4.33 Access is provided by a network of rural access infrastructure. Different accessibility requirements apply to individual links or groups of links within that network. A procedure such as the one outlined in the access evaluation chart (AEC) must be followed in planning agricultural and rural development projects to identify: (i) the location of the important and sustainable links in the RAI network, and (ii) the seasonal transport demands on these links. A key tool in determining the seasonal transport demand is a calendar showing the relationship between the agricultural activities and the rainy season/s. Agricultural activities determine if reliable or minimum access must be provided by specific RAI links. Once the accessibility, i.e., the acceptable interruptions, of the various links is determined, the infrastructure requirements can be evaluated as described in Chapters 6, 7 and 8. The seasonal transport demand becomes a key factor in determining the type of transport aid required, which in turn dictates the infrastructure type, i.e., path, trail, track or road. The accessibility of different transport aids over the same section of infrastructure varies.

4.34 Accessibility requirements increase as development grows. The most economical resource expenditures in access improvement occur when accessibility and development increase in parallel. Therefore, determining accessibility by phases which respond to increased development activities, preserves transport resource expenditure options until their validity is assured, even if development takes longer or falls short of original estimates. This concept of accessibility is the keystone to an understanding of this handbook.

CHAPTER 5

Transport Aid Evaluations

5.01 Present transport aid technology has been developed within the context of assisting the rural population in their daily tasks. As a single sub-component in a development project, transport aids cannot be evaluated in isolation, but must interact with other possible transport sub-components to maximize the potential use of resource investments as described in Chapter 3. Transport aids include equipment to assist in human porterage activities such as shoulder or pickul bars; wheeled transport aids propelled by humans, such as wheel barrows and bicycles; animal transport aids such as pack animals, and animal-powered aids including sledges and carts; and motorized transport aids such as motorcycles and tractors and their associated aids like trailers.

5.02 Unless transport aids can be easily produced in the project area, as for example the making of a simple animal sledge, the successful introduction of new transport aids necessitates an involvement of the national government as outlined in Chapter 10. The planning of transport aids must assume a two-tiered approach, a country-wide assessment and a project assessment. The country-wide assessment involves an evaluation of the potential for domestic production, required imports of parts, impact of required imports on the balance of payments, necessary technical capabilities, etc. Such an assessment is normally too time-consuming and costly for a specific project activity. The costs for the country-wide assessment and the start-up costs for introducing new or modified transport aids should be accounted for in some other manner than being pro-rated against the units produced for the initial development project. The following Transport Aid Selection Chart shows the division of country-wide and project planning activities.

I. EXPLANATION OF ACTIVITIES SHOWN ON THE TRANSPORT AID SELECTION CHART (TASC)

5.03 The activities of this chart involve discussions with farmers, private transporters and other transport aid users like merchants and persons having to carry firewood and water. Some of the activities are technically orientated, involving discussions to collect facts such as costs, carrying capacities, etc. Other activities involve user interface, collecting opinions and personal experiences. The following paragraphs briefly outline the rationale for including each step in the chart. More specific technical details about transport aid selection can be found in Chapter 10 and Annex 2. For example, Technical Note 2.1 is a list of transport aids; Technical Note 2.3 contains suggested infrastructure requirements for transport aids; and Technical Note 2.7 outlines both the constraints to transport aid usage and possible actions that can be taken to overcome individual constraints.

Transport Aid Selection Chart (TASC)

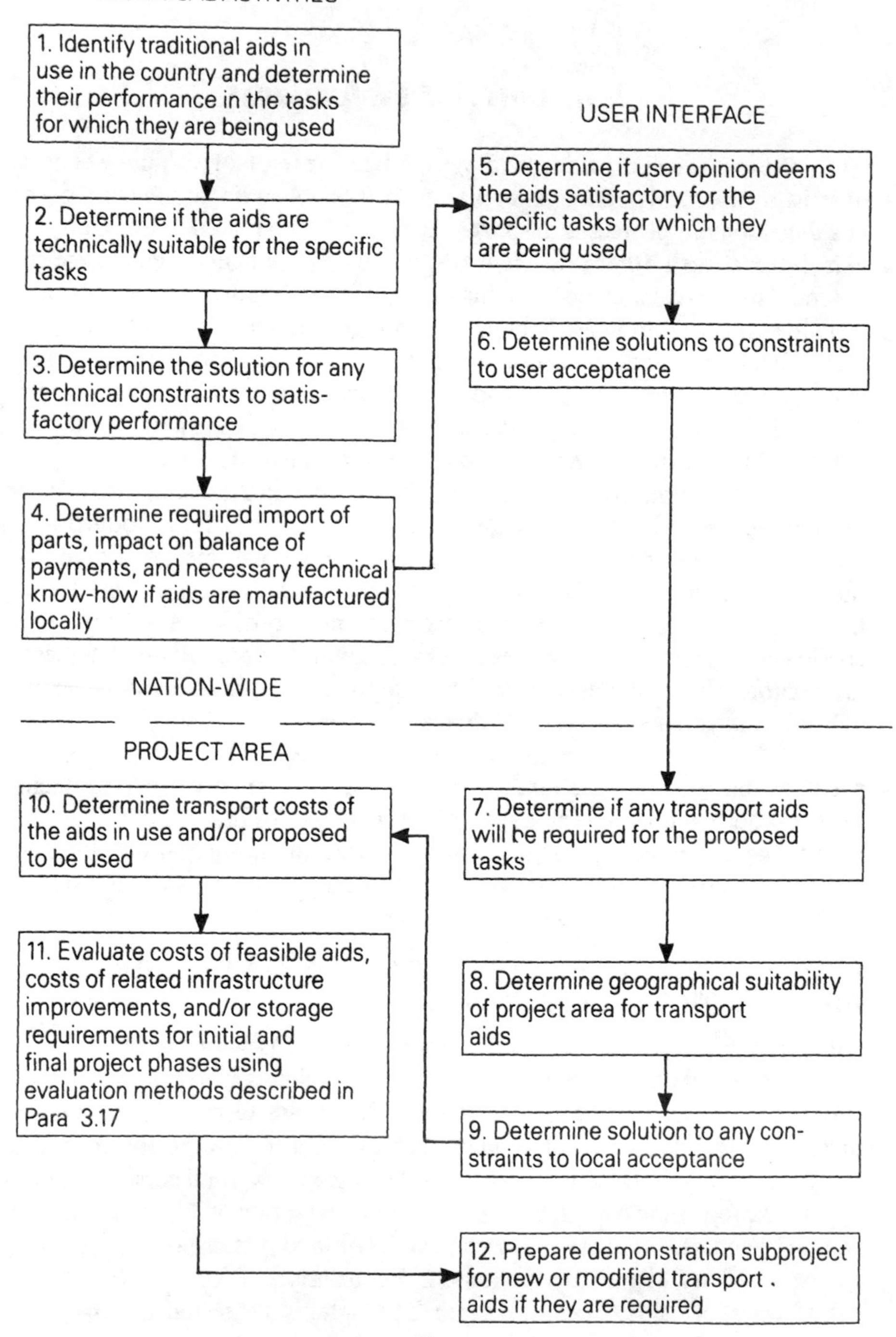

5.04 Step 1 of the TASC identifies existing transport aids which provide the most feasible means of expanding the zone of access beyond the conventional zone of influence of existing or proposed roads. The tasks of such transport aids may be to provide access for development material or crop evacuation from small areas. Other uses can include commerce between cash crop farmers and subsistence farmers, perhaps located beyond the fringe of current development areas, who wish to enter the cash economy by marketing vegetable garden produce to the cash crop farmers concentrating on crops for export. However, these existing transport aids may not currently be in country-wide use.

5.05 Topography, environmental conditions, soils, tractive power, capacity and infrastructure constraints, as well as supply problems may be hampering the increased usage existing transport aids and are therefore considered in Step 2. Any constraints that can be identified as area-specific, for example gradients which are too steep for a specific aid to climb or descend, should be catalogued for later project evaluation.

5.06 Solutions to technical constraints, the subject of Step 3, can involve redesigning the existing aids to reduce the constraints, for example, by replacing wooden block bearings with roller bearings; or by classifying the aid for use in specific conditions as noted in Step 2. If neither solution will satisfy the anticipated task requirement, a transport aid not currently used in the country may be the solution. For instance, if a pack animal cannot carry enough sugar cane over level firm ground, an animal cart may be evaluated even if carts are not currently used in country. The subsequent steps in the TASC should still be followed in sequence to evaluate a new aid.

5.07 Step 4 addresses the question of how to sustain extended use of transport aids. If the aids, or parts of the aids, are to be imported, a supply system for the imported material must be developed and its impact on the balance of payments evaluated. Local artisans must be trained in the production, repair, and maintenance of the transport aids. A local credit system may be required for the manufacture, repair and acquisition of locally produced transport aids [30].

5.08 While a technical assessment should identify any weakness in a transport aid, it does not identify the user's perceived imperfections in existing or proposed aids, nor does it identify any sociological or economic constraints as called for in Step 5. This step follows Step 4 rather than accompanying Step 2 for the following reasons:

(i) it eliminates bias in the Step 2 activities:
(ii) it serves as a check to the conclusions reached in Step 2 activities;
(iii) it permits the investigator to discuss with the users improvements identified during the Step 3 activities; and
(iv) it provides a feedback on the evaluation of any new transport aids identified during Step 3 activities.

5.09 In Step 6, the results of the Step 5 activities are used to modify the conclusions developed in Step 3, and any affected evaluations from Step 4. A comprehensive approach to the need, practicability, and procedures for introducing or modifying transport aids can now be assembled, packaged, and presented to governmental policy-makers to overcome any institutional constraints. A firm central government policy should be elicited to allocate the required resources for introducing specific transport aids to appropriate development projects.

5.10 In Step 7, the short-term and long-term access requirements for a specific project, developed using the AEC technique described in Chapter 4, indicate the transport tasks involved. Comparing the list of possible transport aids developed previously to the project transport task requirements will indicate a choice of transport aids that may satisfy these needs.

5.11 Using the previously developed terrain and environment constraint list for the specifically proposed transport aids (Step 2), Step 8 calls for an investigation of the project area to decide whether aids so far identified are suitable for this area. Some constraints present on the existing infrastructure, such as width, are not disqualifiers. However, topography indicating that grades on specific infrastructure are excessive for a given aid, eliminates that aid from further consideration. During the same investigation, the present use of any transport aid in the project area should be determined and the local population interviewed concerning the use of both their current transport aids and any transport aids likely to be proposed as a project subcomponent.

5.12 Step 9 addresses constraints to a transport aid's local acceptance in a specific area. For example, constraints can be social, in the form of local tribal, sexist or religious beliefs; economic, in the sense that the local population believe they cannot afford the aid; or technical, such as a local disease which makes livestock raising chancy, or a shortage of livestock feed. Most, if not all, of these local constraints may have already been considered during Step 5 and 6 activities, in which case Step 9 examines ways of how to overcome them. For instance, training activities, provision of credit, opening supply lines, providing veterinary services, and introducing alternative cropping patterns for fodder may overcome problems related to the local acceptance of transport aids.

5.13 Step 10 calls for the determination of the economic costs of each transport aid to be used. These costs vary from area to area in response to such things as the cost of credit, the cost of fuel or animal feed, and delivery and maintenance costs. The economic costs includes both the fixed and variable costs of the aid, including any tractive force required. Fixed costs or overhead costs include the purchase amount amortized over the aid's economic life, which may be different for an ox-cart and its oxen; the interest on that capital investment; routine maintenance and repair costs; and any insurance costs. Variable costs are considered as usage-dependent and include the costs of spare parts, extra tyres, labour costs for repairs and maintenance required because of usage, fuel and lubrication costs, and operator costs.

5.14 Fixed costs are based on an assumed usage rate since high usage implies a short lifetime with a higher yearly amortization. However, some variable costs for mechanical locomotion, like fuel, oil and grease, become predominantly fixed costs for animal locomotion, since animals consume some food daily regardless of any economical activity. Consequently, each aid must be evaluated within the context of its probable transport usage. This is especially true when multipurpose locomotion, such as tractors and draught animals, is being evaluated.

5.15 Motorized transport costs tend to increase per ton-km when long loading times are involved, and when repair time and cost are increased because of poor infrastructure. Thus transport costs on the specific network being evaluated must be properly prorated between the actual time spent and costs incurred while using that infrastructure and the time and cost in other activities such as long distance truck hauling on better roads.

5.16 The total resource investment in transport includes not only the economic costs of transport aids but also the economic costs of building and maintaining an infrastructure that provides sufficient access to accomplish the transport task. Storage availability may be a determining factor during periods of economic activity within a specific project area. Step 11 calls for an economic evaluation of total resource investment alternatives. Minimum resource investment for a project's access component is achieved by finding the most effective combination of transport aids, infrastructure needs and storage capacity to relieve access constraints on the other project components (Chapter 19, Section II).

5.17 If the previous steps have determined that a transport aid not currently being used is suitable to the transport needs of the local population, it is logical to question why it is not already in use in the area. The answer should already be known, having been uncovered during the evaluation of the technical and user constraints. The reason that the transport aid is now thought suitable is because solutions have been found to these previous constraints. However, the local population must also believe that these solutions have been found. The most suitable means of convincing the local population, and reassuring the policy-makers, that a specific transport aid can and will both serve a need and be beneficial to the local population is to test its suitability in the field, as suggested in Step 12 of the TASC. Only when the needed education, training, demonstration and promotional activities are formally addressed and satisfactorily achieved will a new or improved transport aid be accepted and used by the local population. Demonstration plots have already proven their worth in agricultural projects financed by the World Bank. There is no reason to believe that a demonstration component to familiarize users with the advantages of transport aids would not be equally successful.

II. EXAMPLE EVALUATION OF A TRANSPORT AID

5.18 The following example was adapted from a study report, Farming with Work Oxen in Sierra Leone [64]. The currency has been converted to 1982 US Dollars. This example is illustrative only and is presented to delineate a range of possible evaluation considerations, not to suggest valid current costs for ox-cart usage in any country.

5.19 Farmers in Northern Province, Sierra Leone, have been using oxen for cultivation and transport for some time. Ox-ploughs are imported across the Guinea border. About 90% of Sierra Leone's 1981 herd of 333,000 cattle were located in Northern Province. Consequently ox-carts can be considered as a traditional transport aid which has limited in-country use.

5.20 In 1981 cattle ownership was restricted to about 5% of Sierra Leone's farmers. The study [64] was made to determine the feasibility of increasing the agricultural use of oxen by training non-cattle-owning farmers in animal-oriented agricultural practices, including the use of ox-carts. Approximately 34,000 head of cattle, including trained oxen, are imported from Guinea each year. Therefore quite large ox-training schemes, whereby the farmers would learn to work with trained oxen and learn how to train their own oxen, could be contemplated without unduly disrupting the existing cattle trade. Such work-oxen projects already underway in Sierra Leone have convincingly demonstrated that farmers who had not previously owned cattle are now successful in farming with oxen.

5.21 While this example will concentrate on the transport aspects of the study, the report also presented the following order of cost comparison of primary cultivation, i.e., land preparation by hand, oxen and tractors in Sierra Leone. The foreign exchange component of ox cultivation is for farm implements.

System	*Total Cost per Hectare*	*Foreign Exchange per Hectare*
Hand	$ 197	$ 0
Oxen (restricted use)	134	16
Oxen (medium use)	110	20
Tractor	149	134

5.22 The local oxen are castrated bulls of indigenous Ndama cattle. They are small (160–300 kg), stocky, short-legged, muscular, and considered capable of an average traction equivalent of about 14% of their body weight as compared with 10–12% for many other breeds. Ndama cattle are well adapted to the Sierra Leone environment and, in particular, are tolerant of trypanosomiasis, transmitted by the tsetse fly. They appear to be more resistant to certain other blood parasites and to streptothricosis than some other West African breeds. Their tolerance to disease, however, is significantly reduced by too much work or by underfeeding.

5.23 Oxen feed, in Sierra Leone, consists of grazing and rice straw, plus 200 g of a salt/mineral mix per work day. The recommended oxen supplemental feeding of 1500 g of cereals per day for light work; 500 g per hour for medium work; or 750 g per hour for heavy work, is not practised in Sierra Leone because of the likely social barriers to the feeding of oxen with rice or cereals destined for human consumption. Instead the oxen work at cultivation tasks for only four to five hours a morning for five days a week, and are consequently able to maintain their weight by grazing, thus consuming only a free, renewable, energy source.

5.24 Since carting generally involves periods of waiting during loading and unloading, it is difficult to determine the actual hours worked, but the report [64] indicates that three hours of pulling during a six-hour period is probably not excessive when the oxen are in good condition. However, the animals will become more tired if the load is not well balanced, or if the work involves pulling on rough or muddy ground or up slopes.

5.25 The oxen are trained when their weight reaches approximately 160 kg (three to four years old). With adequate grazing, the oxen generally grow during their working lives, reaching about 300 kg at the age of eight years. They are then sold at a profit for slaughter, and younger oxen are trained to take their place. Oxen therefore not only contribute valuable manure to the farms during their work life, but when sold also contribute to the national supply of animal protein.

5.26 The oxen are used in pairs, providing tractive force through a carved double neck yoke tied to their horns, rather than a shoulder yoke. The neck yoke is said to provide better control, restrict head movement and accompanying danger from the horns, reduce skin sores and problems associated with rubbed skin, and is suitable to the small, humpless Ndama oxen with their strong necks. A pair of oxen can pull a cart with a load of 800 kg and the operator on a reasonable track.

5.27 Two types of carts were evaluated in the study report [64]. One cart type consists of an angle-iron frame with a wooden platform and wooden removable sides. It is mounted on a ready-made axle with pneumatic tyres imported from Senegal, although it could be mounted on an old car axle. It has a box capacity of about half a cubic metre. The second cart type is made from old car axles with angle-iron struts attached to a wooden-framed platform of the same size. It is heavier and less durable than the first type, but less costly. Wooden-wheeled carts were still in an experimental stage in Sierra Leone at the time of the report. The annual cost of the less expensive second-type cart is estimated as follows:

Initial Cost		$320
Depreciation Time[1]		10 years
Depreciation/Year		$ 32
Interest/Year		$ 19
Maintenance/Year		$ 12
TOTAL YEAR COST [2]	$ 63	

5.28 The annual budget for maintaining a pair of oxen is estimated as follows:

Item	Annual Cost	Notes
FIXED COSTS		
Two oxen purchased at $646 and sold at $1121		Bought at 170 kg each and sold at 295 kg each for $1.90 per kg

[1] Depreciation in this example is assumed to be caused by the elements rather than by usage.
[2] 12% interest on mean value. Total yearly cost is sufficient to repay interest and capital on a 12% loan over the full period of depreciation.

Item	Annual cost	Notes
5-year amortization	– $95	Profit resulting from weight increase averaged over 5 years
Interest – 12% on mean value of oxen	$106	0.12 (646 + 1121)/2
Insurance (risk of accidental death)	$71	8% of mean value as a 1 in 12 risk of death
Grazing costs	$197	$0.54 per day for a year to cover grazing supervision
Ropes and reins	$27	Locally made ropes
Housing	$16	Simple shelter only
Animal health costs	$35	Tick, stray and would oil, etc., no veterinary services available
Total annual overhead	$357	
VARIABLE COSTS		
Ox handlers	$4.20	2 men at $2.10 per day
Supplementary salt	*$0.20*	0.2 kg salt mix per working day
Total daily cost	$4.40	
Total hourly cost	$0.74	Assuming 6 hours work, 3 hours pulling.

5.29 Using the above figures, the impact of usage on the cost per ton-km of ox cart haulage in Sierra Leone is as follows:

Ox Cart Ton-Km Costs by Different Annual Workloads, Assuming Oxen are Used for Nothing but Transport

Days worked per year	50	75	100	150	200	250
Ox-cart costs	$63	63	63	63	63	63
Oxen fixed costs	$357	357	357	357	357	357
Daily costs at $4.40	$220	330	440	660	880	1100
TOTAL COSTS	$640	750	860	1080	1300	1520
Daily cost	$12.80	10.00	8.60	7.20	6.50	6.08
Hourly costs (6hrs/day)	$2.13	1.67	1.43	1.20	1.08	1.01
Cost per ton-km[1]	$1.07	0.83	0.72	0.60	0.54	0.51

5.30 Using the same figures from Paras 5.27 and 5.28 but assuming the oxen work full-time, which according to Para 5.23 is 5 days per week for 52 weeks or 260 days per year, the costs per transportation per ton-km become the following:

Ox Cart Ton-km Costs by Different Annual Transport Workloads, Assuming Oxen Work 260 Days per Year

Days transporting per year	50	75	100	150	200	250
Ox-cart costs	$63	63	63	63	63	63

[1] Assuming 3 hours × 0.8 ton × 5 km, loaded both ways.

Ox Cart Ton-km Costs by Different Annual Transport Workloads, Assuming Oxen Work 260 Days per year

Prorated oxen fixed costs	$69	103	137	206	275	343
Daily costs at $4.40	$220	330	440	660	880	1100
TOTAL COSTS	$352	496	640	929	1218	1506
Daily cost	$7.04	6.61	6.40	6.19	6.09	6.02
Hourly cost	$1.73	1.10	1.07	1.03	1.02	1.00
Cost per ton-km[1]	$0.59	0.55	0.53	0.52	0.51	0.50

5.31 A similar analysis should be made during project evaluation for each transport aid being considered for specific transport tasks. The analysis should evaluate the full economic implications of the aid's impact on the target population, not just the ton-km transport costs. In this example, the oxen can be used for farming which has a definite economic impact on the farmer's other activities. Other economic impacts include the financial burden the transport aid imposes on the farmer. The economic fixed costs may not all be financial costs if the farmer's family supervise the grazing oxen, or if another person accepts produce in payment for supervision. The economic variable costs are not true financial costs from the farmer's viewpoint if he and/or other family members handle the oxen while they pull the cart. The cart and oxen may even provide cash income if the farmer hauls other people's goods while driving his cart.

5.32 Many studies of motor vehicle operating costs [1, 8, 20] present methodologies for calculating motorized transport costs. Paragraphs 5.29 and 5.30 show cost comparisons of ton-km costs vary by usage, both daily and yearly. Motor vehicle costs are more sensitive to reduced usage than are the ox-cart costs shown here since a truck's fixed costs constitute a much larger proportion of its total ton-km costs. The variable costs in this ox-cart example exceed the fixed costs when the oxen are used solely for transport for 100 days or more. When the oxen are also used for agricultural purposes, the variable costs constitute almost two-thirds of the total costs even when the ox-cart is used only one day a week for transport purposes.

III. CONCLUSIONS

5.33 The selection procedure for identifying appropriate transport aids for specific transport tasks must begin with a nationwide evaluation to:

(i) collect data about various aids;
(ii) identify and solve any technical, user or institution constraints to the use of those aids;
(iii) determine the resource investments required to introduce transport aids, or modifications to transport aids, in areas where they are not currently used; and
(iv) develop a national policy favourable to introducing or encouraging transport aid use.

[1] Assuming 3 hours × 0.8 ton × 5 km, loaded both ways.

If no transport aid is available in-country to satisfactorily carry out a specific transport task, aids used in other countries may be substituted in the first step in the above procedure.

5.34 Project transport aid selection procedures apply the results of nation-wide evaluations to the solution of specific project transport tasks. The nation-wide findings must be modified, or fine-tuned, through a further evaluation of project-specific technical and user constraints. Transport aids should be subjected to a project economic evaluation not only to determine their economic viability, but also to obtain the most effective investment package, possibly consisting of transport aids, their related transport infrastructure costs, and storage facilities if access proves costly.

CHAPTER 6

Infrastructure Evaluations

6.01 Rural access infrastructure (RAI) consists of paths, trails, tracks and feeder roads. This infrastructure carries an assortment of different transport aids as described in Technical Note 2.1, Annexe 2. Each aid has its own infrastructure requirements, as indicated in Technical Note 2.3. The evaluation of different links in the infrastructure network requires a degree of investigation which depends on:

(i) the types of transport aids anticipated on that link, such as headbaskets, pack animals, and/or tractors with trailers;
(ii) the level of accessibility required as described in Chapter 3; and
(iii) the amount of resource investment anticipated for that link, varying from trails and paths requiring little improvement to feeder roads that may be substantially upgraded at some time during the project.

6.02 The specific RAI links requiring evaluation are determined by a method similar to the plan book procedure described in Para 11.04, whereby each link is assigned a specific role in the RAI network. In standard highway planning, the preliminary plans, specifications, and estimates (PS & E) are based on an evaluation of the specific features of an acceptable alignment before the detailed plans are developed. Under the stage construction/spot improvement concept described in Chapter 13, the significant spot improvement activities to eliminate access constraints must be individually identified, but in homogenous areas their costs may be derived from representative improvement costs in a comparable PS & E exercise. Spot improvement activities are often expressed in aggregate terms such as providing 1½ pipe culverts per km, or one 65-metre long ford every three km, or granular surface for 30% of the total length of the existing earth roads. These aggregate quantities must be developed from specific link evaluations based not only on the physical nature of project location, but also on the various accessibility levels required throughout the project area. Only then can representative costs be developed. For example, the costing criteria of 1½ 60-cm diameter pipe culverts per km might be based on an inventory of the total number of those culverts required for a project involving two 10-km feeder roads, however one 10-km link may require 20 culverts distributed evenly along its entire length, while the other 10-km link may require only ten culverts, all located within the first 3 km of that link.

6.03 Chapter 12 describes the technical features which influence accessibility levels of rural access infrastructure. They include operational constraints caused by (i) geometrics such as steep grades and insufficient widths, (ii) soils which soften, expand or get slippery in the presence of moisture, (iii) climatic features such as wind and rain which cause erosion, and (iv) inadequate drainage facilities which either present physical barriers to transport or affect the infrastructure's soils. Technical Note 4.11 offers several options for the staged construction of water-crossings to provide various accessibility levels. Part IV also includes several technical notes (Technical Notes 5.03 to 5.10) to assist technical personnel in identifying the possible causes and remedial actions for the more common drainage-related problems that may be ameliorated by spot improvements. Such drainage problems are often the overriding cause of access constraints. Technical Note 5.01 offers guidance for determining when spot improvements are preferred to complete reconstruction of feeder roads and tracks.

6.04 The planner's task is to both select and evaluate the critical links which comprise the RAI network. Selection is a joint planning exercise involving inputs from project area inhabitants and from agricultural and transportation personnel to assure the project's transport component services will enhance other project activities. Evaluation is primarily the function of the transport planner. The following Infrastructure Improvement Chart outlines an approach to analyzing the infrastructure costs of satisfying predetermined levels of accessibility using pre-evaluated transport aids. Aids which are unable to satisfy speed or distance constraints should be eliminated during the access level determination activities. When evaluating a project with different short-term and long-term access needs, the possibility of expanding the use of the same transport aids as the needs increase should be considered. This will reduce resource investments in introducing new aids (Para 2.25).

I. EXPLANATION OF ACTIVITIES SHOWN ON THE INFRASTRUCTURE IMPROVEMENT CHART (IIC)

6.05 A combination of inputs must be evaluated before determining the proper resource investment for a specific rural access infrastructure network. The individual network links, existing and/or new, must first be identified. This activity, Step 1 in the IIC, evolves from the general planning of a rural or agricultural development project, (Chapter 3), and is a response to other project component needs. However, restrictions to locating or improving transport infrastructure may in turn influence the possible development alternatives. One feature of the plan book system (Para 11.04), is that all existing infrastructure is shown on a base map while new infrastructure and existing infrastructure improvements are shown on transparent overlays. Even when this system is not used, a project base map must be developed to determine infrastructure link requirements and locations.

6.06 The actual transport demand on these individual links must be forecast as indicated in Step 2 of the IIC. The demand criteria includes both the quantity and timing of the transport needs. The sensitivity of this demand, i.e., the accessibility requirements, must be defined in terms of acceptable constraints imposed not only by the weather but also by physical limitations to the size and type of possible carriers. Step 2 is outlined in greater detail in Chapter 4.

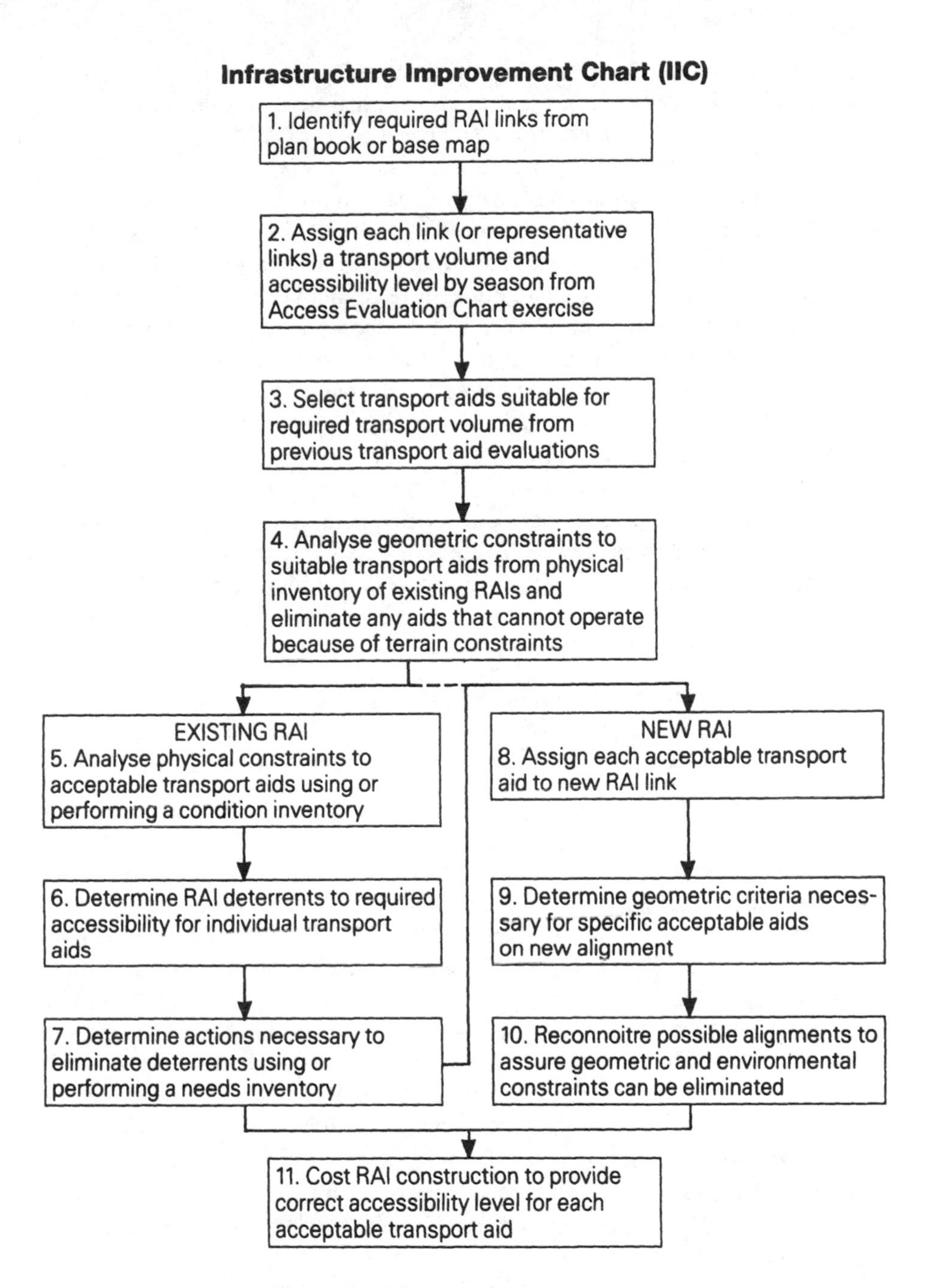

6.07 Chapter 5 indicates that any general evaluation of possible carriers, or transport aids, should be undertaken nation-wide, independent from and prior to rural access infrastructure evaluations on a project-by-project basis, to avoid the cost of constantly reinventing the wheel. The third input to the IIC evaluation, Step 3, is the determination of which of these transport aids may satisfactorily meet the project demand criteria identified in Step 2. The methodology is further outlined in Chapter 5.

6.08 Step 4 is also outlined in the Chapter 5 description of transport aid selection. It includes the project area technical activities and user interface steps outlined in the transport aid selection chart (TASC) in Chapter 5. It is included here as the fourth input required before any detailed analysis can be made of project-specific RAI links in development project evaluations.

6.09 Steps 1 to 4 therefore represent a compilation of the data necessary before an intelligent approach can be made to an analysis of the costs of a specific RAI network for an individual development project. The same data is the basis for analyzing alternative network schemes and project phasing. Once this data is available, it can be used either to determine the costs for (i) spot-improving existing infrastructure by applying Steps 5, 6, 7 and 11, (ii) constructing a new RAI network by applying Steps 8, 9, 10, and 11, or (iii) increasing access through the improvement of existing infrastructure and the construction of any missing network links by following Steps 5 to 11. Item (iii) is the most likely scenario for agricultural and rural development projects.

6.10 The difference between evaluating existing rural access infrastructure and new RAI is in the basic approach. Steps 5 to 7 involve evaluating existing RAI to determine access constraints. This evaluation is undertaken by conducting some form of:

(i) a physical inventory, Technical Note 4.12;
(ii) a condition inventory, Technical Notes 4.13 and 4.14 and
(iii) a needs inventory, Technical Notes 4.13 and 4.14.

The form and details of each inventory depend on the type and use to be made of the individual links. Para 12.20 classifies these inventories into three levels. The lowest inventory level, trails and paths, that may be in need of some improvement, can be presented in narrative form rather than in the formalized manner outlined in Technical Notes 4.12 through 4.14. This narrative format is outlined in Para 12.24. The needs inventory for existing RAI is based on the infrastructure improvements required to achieve the accessibility determined in Step 2 for each of the transport aids that survived the Step 4 analysis.

6.11 The evaluation of access improvement by constructing new RAI begins with the assignment of the transport aids which survived the Step 4 analysis to specific non-existing RAI links identified in Step 1. The Step 4 criteria must still be met, i.e., the aid must be able to function on any suitable existing infrastructure in the project area so it can be used without restriction within its operating range. Otherwise, the aid may not be acceptable to the local population. Each candidate transport aid should undergo the Steps 8–10 procedure, although some aids may later be eliminated when the combined resource investment for individual aids and their required infrastructure improvements are evaluated.

6.12 The construction cost of RAI is often a substantial portion of the transport component of agricultural and rural development projects. However, construction costs alone do not determine the most economic investment. The investment for high volume through traffic rural roads is determined by three factors: construction costs, maintenance costs, and users' benefits. The investment for rural access infrastructure is determined by up to five factors: construction, maintenance, transport and storage costs, and economic benefits. In neither case should construction costs alone be the determining factor. Therefore, the alternative construction costs associated with each of the viable transport aids should be estimated and retained for further evaluation. As indicated in Para 3.13, several viable transport aids may meet the accessibility requirements of a specific project as a result of the same infrastructure improvement. Consequently, the number of construction estimates in Step 11 is limited even when several individual transport aids are being considered.

6.13 The major problem in estimating the cost of rural access infrastructure improvements for rural and agricultural developments is not the number of options to be considered. Para 6.12 indicates that a minimal number of options exists. The problems encountered in estimating individual, dispersed, and small construction activities lie more in correctly allocating costs to these activities than to the number of activities or alternatives considered. Costing RAI construction using existing unit prices, derived from highway department construction experience will result in unrealistic costs (Para 17.03). The estimating procedure must begin with a new engineering determination of the output that can be expected from the construction technology chosen. That output should become the basis for a new set of unit costs to apply to specific rural access infrastructure construction activities. Furthermore, the resources mobilized for some of the infrastructure improvements, e.g., paths and trails, should include a self-help or community component which cannot be properly evaluated using standard unit costs.

6.14 The process of developing new unit costs is obviously lengthy and complex. However, once a performance budgeting procedure (Paras 17.23–17.26) is in place additional unit price modifications can be simply and logically introduced. The procedures outlined throughout this handbook constitute a technology for developing access in rural areas that may be new to many countries. As in any new technology, the provision of rural infrastructure and transport services will require an investment in effort and learning in the beginning. In the long term, that investment should pay rich dividends.

II. ESTIMATING IMPROVEMENT COSTS

6.15 Since costing RAI construction is a key element in the planning process, this section discusses estimating in greater detail. Section III gives an example of the estimating procedure for determining the costs of improvements. Chapter 14 details the options available for RAI construction and describes procedures used in determining the appropriate technology to be applied. Chapter 17 outlines the procedures used to make cost evaluations and to implement cost controls. Cost evaluations include determining the unit cost of each activity, be it undertaken by labour, animals or equipment. Examples in Chapter 17 are amplified in Technical Note 9.02, which describes the derivation of different cost elements, and Technical Note 9.06, which includes sample calculations for the use of an individual piece of equipment, namely a self-propelled 10-ton steel-wheeled roller. These calculations determine the unit cost for a particular work task such as the compaction of a cubic metre of embankment or surfacing.

6.16 Such unit costs have their greatest application in standard construction procedures. When these activities have been repeated over several projects, the average output and its average unit cost for individual units of equipment of labour crews doing specific tasks can be determined with some accuracy. These unit costs for specific activities remain valid only under the conditions for which they were determined. For example, a front-end loader's output is calculated as the amount of material it can lift from a stockpile and dump into a truck in the course of an hour or a day. These calculations assume a truck will be available to receive the material when the loader picks it up. This is termed equipment balance. A similar balance is assumed when labourers load material into a trailer. The labourers' efficiency is calculated from their ability to dig, lift and load a specific number of shovel-loads per day, assuming they have some place to put the material.

6.17 Once such an activity changes, its unit cost is no longer valid. For example, if the labourers must dig, lift and then walk five metres before dumping the material out of the shovel, the activity has been changed to include a degree of transport, and the labourers' excavation productivity is reduced. This added activity can be, and usually is, calculated when the task is defined for a specific project.

6.18 The balance between work tasks must also be determined whenever an individual task is altered. This balance is sometimes neglected. A common example of ignoring balance between work tasks is seen in road gravelling operations when a fixed number of trucks is assigned to a gravelling crew with no regard to the distance from the pit to the job site. In such a situation the maximum output of a front-end loader occurs when a truck is always available to fill, while the maximum output of the truck fleet occurs when every truck is loaded without waiting. Since the dumping location varies for each load as the spreading operation moves nearer or farther away from the pit, the output of the combination of the front-end loader and thr truck fleet will peak infrequently. When the spreading is farther away the loader will have to wait for the trucks, which are then said to control, and when it is closer the trucks will have to wait for the loader.

6.19 In real life, construction operations are even more complex. The above gravelling operation may include equipment to spread, water and compact the gravel surface. Obviously all the equipment cannot work at maximum output since they all have different production rates, and all depend on the production of the other equipment in the operation. Usually only one activity will be undertaken at calculated efficiency. That activity is then said to control. All other equipment will work below calculated efficiency and the unit cost of the operation will be higher than the unit cost of its individual components working at their peak efficiencies. In the above example, the gravel operation may be controlled by the production capability of the following equipment:

(i) the front-end loader in the pit;
(ii) the truck fleet hauling the gravel;
(iii) the bulldozer(s) spreading the gravel;
(iv) the grader(s) shaping the gravel surface;
(v) the water truck(s) supplying water for compaction; or
(vi) the compactor(s).

6.20 Whenever a new project is undertaken, the equipment fleet, labour force, and animal herd must be balanced for that operation. Perfect balance is never achieved, nor can it be expected, especially in any operation involving variable travel distances. However, proper balancing will lower the overall cost of the operation and will permit realistic estimating, i.e., the cost of the operation will be higher than the sum of the cost of the individual activity rates at maximum production. This is especially true for spot improvements, stage construction, and low-cost rural infrastructure construction in general.

III. EXAMPLE OF BALANCING A SURFACING OPERATION

6.21 The following example is included to illustrate the variables encountered in balancing a specific operation. The principles are valid for other operations using labour, animals, or equipment. Labour and animal activities are more suitable to balancing because each unit performs less work than a piece of heavy equipment, so the optimum output for the entire operation can be more closely approached.

6.22 An equipment-based example has been chosen because equipment outputs are high enough so that imbalances introduce substantial variations in unit costs. The example ignores some of the practicalities of RAI construction by assuming an unlimited number of trucks would be available, again to emphasize the variations in unit costs. Personnel costs such as supervisors and labourers, and service equipment costs such as pickup trucks, fuel trucks and lubrication trucks, are ignored to simplify the example. A three-year project was selected so no piece of equipment could be written off against the project. The practice of assuming that equipment life and project life are the same sometimes leads to serious simplifications. For instance, if a piece of heavy equipment has an anticipated 10,000 hour useful life, a brigade is sometimes formed assuming the life of all its equipment will be expended in five years. The published hourly production of the selected piece of equipment is used to forecast the five year production of the entire brigade. The total cost of the brigade is then divided by the anticipated output of the selected piece of equipment to determine the brigade's unit production cost.

6.23 In this example a sandy granular surface is to be applied as a single 30 cm loose layer which is compacted to a 25 cm surface, 5.5. metres wide. The granular surfacing requirements are 1800 loose cubic metres (LCM) per kilometre. The heavy equipment has an 8-year clock life operating at 80% efficiency for 200 days per year. Trucks have a five-year life. Spare parts used during the three years are anticipated to equal 40% of the purchase price of all units. Therefore the daily equipment costs are as follows:

	140 hp front-end loader ($)	4 M³ dump truck ($)	140 hp bulldozer ($)	135 hp grader ($)	6000 litre water truck ($)	86 hp compactor ($)
Total equipment cost on site	127,400	48,000	100,000	85,000	61,700	68,500
Cost/day	79.60	48.00	62.50	53.10	61.70	42.80
Cost spare parts on site	51,000	19,200	40,000	34,000	24,700	27,400
Cost/day	85.00	32.00	66.70	56.70	41.20	45.70
POL/day[1]	37.50	21.70	94.70	41.00	27.90	60.70
Total cost per day	200.10	101.70	223.90	150.80	130.80	149.20

6.24 The equipment output evaluation is as follows:

(i) *Front-end loader*. Heaped capacity of bucket is 2.1 cubic metres × 0.95 (load factor) or 2.0 Loose Cubic Metres (LCM). The loose material weighs 1700 kg/metre. One bucket-load therefore weighs 3400 kg. The loaded bucket capacity of the loader is 4100 kg, so the loader is properly sized for the pit operation. The cycle time to load, dump and manoeuver is 0.4 minutes per bucket. The maximum daily output at 80% efficiency is (8 hr × 60 min × 80)/0.4 min. or 960 bucket-loads, or 1920 LCM.

(ii) *4 LCM dump truck*. Two bucket-loads will fill the truck. This is an acceptable ratio of truck to loader capacity. The truck rating indicates a design payload of 7000 kg. Since two bucket-loads weigh 6800 kg, the trucks will not be overloaded. The loading time is 2 × 0.4 min or 0.8 min per truck. At 80% efficiency, the loader can load one truck per minute or 480 trucks per day. Assuming dump and turnaround time at two minutes and a travel speed of 15 kph loaded and 40 kph empty, the following table represents round trip time for trucks at 80% efficiency.

Table 6: Dump Truck Round Trips for Various Distances

4-cubic metre capacity

One-way trip (km)	*Round trip* time (min)	*No. of trips* per day[1])	*LCM*/day *(4/trip)*	Trucks required *for F.E.L. control*[2]
1.0	10.4	46	184	11
1.5	13.8	34	136	14
2.0	17.3	27	108	18
2.5	20.7	23	92	21
3.0	24.1	19	76	26
3.5	27.6	17	68	29
4.0	31.0	15	60	32
4.5	34.4	13	52	37
5.0	37.9	12	48	40

[1] POL costs include fuel, lubricants, filters, and grease.

Table 6. *(contd.)*

6-cubic metre capacity trucks (alternative evaluation)				
One-way trip (km)	*Round trip* time (min)	*No. of trips* per day[1]	*LCM/day* *(4/trip)*	Trucks required *for F.E.L. control*[2]
1.0	10.9	44	264	8
1.5	14.3	33	198	10
2.0	17.8	26	156	13
2.5	21.0	22	132	15
3.0	24.6	19	114	17
3.5	28.1	17	102	19
4.0	31.9	15	90	22
4.5	34.9	13	78	25
5.0	38.4	12	72	27

(iii) *Bulldozer*. The bulldozer selected can push in excess of 12 tons of material. One 4 LCM truckload weighs 6.8 tons (the 6 LCM truckload weighs 10.2 tons) so the bulldozer is large enough. Each truckload is dumped at 4.5 metre intervals along the lane being gravelled. The bulldozer has the capacity to push 400 LCM per hour. Correction facors are 1.20 for loose stack pile, 0.75 for average operator, and 0.80 for job efficiency. Therefore actual quantity becomes:

$$8 \text{ hr} \times 400 \text{ LCM} \times 1.20 \times 0.75 \times 0.80 = 2300 \text{ LCM/day}.$$

(iv) *Grader*. At 2.5 km/hr the grader can make one round trip over 1 km of road grading one side each way in 50 minutes, including turnaround time. Assuming two passes are required, an operator efficiency of 0.75, and a job efficiency of 0.80, it would take less than three hours to grade 1800 LCM. The grader is also used to maintain the road while the gravel is being hauled. The grader can be ignored as a control for production when only one front-end loader is working in the pit.

(v) *Water truck*. The amount of water to be added to reach optimum moisture content for compaction varies due to weather and other causes. In this example, assume water requirements to be 50 litres per LCM. Since the truck holds 6000 litres, each truckful can service 120 LCM. If the truck takes 10

Table 7: Water Truck Round Trips for Various Distances

One-way trip (km)	*Round trip* time (min)	*No. of trips* per day[1]	*LCM/day* *(120/trip)*	Trucks required *for F.E.L. control*[2]
1.0	29.5	16	1920	1
3.0	38.5	12	1440	2
5.0	47.5	10	1200	2
7.0	56.5	8	960	2
12.0	79.0	6	720	3
15.0	92.0	5	600	4

[1] No truck starts a trip if it cannot finish within an 8-hour day.
[2] Front-end loader can produce 1920 LCM/day.

minutes to load and 10 minutes to discharge its load, and travels at 25 kph full and 50 kph empty, Table 7, which includes an 80% efficiency factor, applies.

(vi) *Compactor*. If the vibratory roller works in increments of approximately 100 metres, each pass will require 1.25 minutes at 80 m/min. Stopping and starting requires 0.5 minutes. The roller width is 1.9 m, therefore three passes theoretically cover the road width. However, because of the required allowance for overlap, four passes are required to roll the material one time. Assuming the road requires six rollings over the total width to achieve sufficient compaction, the roller must make 24 passes of 1.75 minutes each, allowing for stopping and starting. Factoring in 0.75 for operator efficiency and a job efficiency of 0.80, the total time to roll 100 metres of road is: $(24 \times 1.75)/(0.80 \times 0.75) = 70$ minutes. One hundred metres contains 180 LCM, therefore, the roller can compact 1230 LCM per day.

6.25 The calculated production of the heavy equipment and the unit costs per LCM are as follows:

Front-end loader,	1920 LCM at $0.14
Bulldozer,	2300 LCM at $0.097
Grader,	Consider as fixed cost per day
Compactor,	1230 LCM at $0.121

6.26 The calculated production of the trucks requires an assumed distance. The following table gives the costs by distance for both 4 LCM and 6 LCM trucks. The daily cost for a 6 LCM truck is assumed to be $130.

Table 8: Unit Costs for Various Haul Distances

	4 LCM Truck		*6 LCM Truck*	
One way trip (km)	*Total trips*	*Unit cost ($)*	*Total trips*	*Unit cost ($)*
1.0	184	0.553	264	0.492
1.5	136	0.748	198	0.657
2.0	108	0.942	156	0.833
2.5	92	1.105	132	0.985
3.0	76	1.338	114	1.140
3.5	68	1.500	102	1.275
4.0	60	1.695	90	1.444
4.5	52	1.956	78	1.667
5.0	48	2.119	72	1.106

This table indicates that as long as the 6 LCM Truck costs $145 or less per day, it is more efficient; however, in this evaluation the 4 LCM truck is considered the standard truck in use in the country road work.

6.27 Now the cost of the entire operation will be evaluated under these given conditions:

(i) gravel pits are 10 km apart; and
(ii) water is located no more than 12 km from any point on the road.

Evaluation 1 – Front-end loader (F.E.L.) controls. Production is 1920 LMC:

Equipment required	*Item cost*	*Total cost*
1 F.E.L.	at $200 =	$ 200
40 dump trucks	at 102 =	4080
1 bulldozer	at 224 =	224
1 grader	at 151 =	151
3 water trucks	at 131 =	393
2 compacors	at 149 =	298
	Daily cost =	$5346
	Unit cost =	$2.78/LCM

Evaluation 2 – Compactor controls. Production is 1230 LCM:

Equipment required	*Item cost*	*Total cost*
1 F.E.L.	at $200 =	$ 200
26 dump trucks	at 102 =	2652
1 bulldozer	at 224 =	224
1 grader	at 151 =	151
2 water trucks	at 131 =	262
1 compacor	at 149 =	149
	Daily cost =	$3638
	Unit cost =	$2.96/LCM

Evaluation 3 – Average haul distance controls. Average haul distance is 2.5 km. For half of the distance, the front end loader will control, on the other half the trucks will control. Output is 1920 LCM for 2⅓ days, while the front-end loader controls; the total time to gravel 5 km is six days for an average daily output of 1500 LCM.

Equipment required	*Item cost*	*Total cost*
1 F.E.L.	at $200 =	$ 200
21 dump trucks	at 102 =	2142
1 bulldozer	at 224 =	224
1 grader	at 151 =	151
3 water trucks	at 131 =	393
2 compacors	at 149 =	298
	Daily cost =	$3408
	Unit cost =	$2.27/LCM

Evaluation 4 – Arbitrary number of trucks controls. Number selected is 14. For the first 1.5 km from the pit, the front-end loader will control, thereby requiring a full complement of equipment. The total time to gravel 5 km will be eight days for an average daily output of 1125 LCM.

Equipment required	*Item cost*	*Total cost*
1 F.E.L.	at $200 =	$ 200
14 dump trucks	at 102 =	1428
1 bulldozer	at 224 =	224
1 grader	at 151 =	151
3 water trucks	at 131 =	393
2 compactors	at 149 =	298
	Daily cost =	$2694
	Unit cost =	$2.39/LCM

6.28 In this particular example, the most economical choice of the four evaluations made is the average haul distance choice. The most expensive choice is to let the compactor control, which is 30% higher. It should be noted that the difference between the unit cost of the front-end loader control and the average haul control is 22%, but in a given length of time the front-end loader control will produce 28% more work. Therefore the entire project cost must be evaluated because the time-dependent costs of support such as supervision, lodging, clerical and other non-producing staff may be high enough to warrant front-end loader control. The most common error made in such evaluations is to apply the front-end loader control to derive output and use that value to determine the number of trucks for the average haul balance. If the output is not corrected for the time period when the trucks control, the unit costs are deceptively low. More importantly, a brigade formed with a fixed investment for buying the equipment and paying the crew will produce less than anticipated if their output is based on this type of calculation. Costs determined by any theoretical approach are usually on the low side because factors not considered tend to delay production rather than increase it.

IV. CONCLUSIONS

6.29 Infrastructure evaluations are one facet in the orderly investigation of relationships that jointly eliminate transport constraints in rural and agricultural development programmes. Before any detailed infrastructure costs estimates are prepared, decisions affecting the infrastructure's use must be made. Access requirements and transport aid availability and use must be determined. The infrastructure evaluation chart included in this chapter outlines all the activities to be performed in conjunction with such an evaluation. More detailed information about determining accessibility is included in Chapter 4. Transport aid investigations are described in Chapter 5. The purpose of the infrastructure evaluation activities described in this chapter is to find the cost of the various infrastructure options that will satisfy the accessibility requirements using the feasible transport aids.

6.30 The infrastructure to be evaluated may exist or be proposed. In either case, the improvements to be priced must satisfy the requirement of eliminating access constraints at minimum cost while retaining as many options for future improvements as possible. New construction is as amenable to staged construction as are existing paths, trails, tracks and feeder roads.

6.31 Each proposed transport aid represents a separate option for satisfying access requirements. The access requirements themselves may be twofold, reliable access (Para 4.03), and minimum access (Para 4.04). However, the variety of infrastructure improvements is limited. A single improvement may provide suitable access to all the feasible transport aids. Many transport aids function equally well on the same type of infrastructure. Therefore, the major problem in infrastructure evaluation is often not the number of options that must be evaluated, but the evaluation technique used to determine realistic costs for those options. However, the savings realized by providing only the necessary level of access amply justify an evaluation using the recommended cost estimating procedures.

6.32 When infrastructure improvements are part of a multi-investment package that also proposes transport aids and/or additional on-farm storage facilities, the cost of these project components must be evaluated together. The total investment in these components represents the cost of improved access. A combined economic evaluation not only helps to limit the number of transport aids requiring different infrastructure criteria, but also determines the most inexpensive way to satisfy access requirements. However, the only way to determine the best investment mix is to estimate the infrastructure costs for every viable option.

CHAPTER 7

Maintenance Evaluations

7.01 Rural access infrastructure (RAI) maintenance can be defined as that process by which (i) the investment in the infrastructure is protected, and (ii) the infrastructure is kept in a condition appropriate to its current use. This definition is an oversimplification of the method of determining expenditures for maintaining high-volume highways. The level of highway maintenance has a direct, calculable impact on vehicle operating costs that significantly influences the economics of highway investments. However, when considering tracks, trails and paths, it is safe to assume that there is a fixed cost on each link to combat climatic wear and an often negligible variable cost dependent on traffic volume.

7.02 The different forms and purposes of road maintenance are defined in Chapter 15. Track, trail and path maintenance take similar forms and have the same purposes. The institutional allocation of responsibilities for routine, periodic and emergency road maintenance varies from country to country as discussed in Chapter 16 and its companion Technical Note 8.1. As a general rule it will be desirable to build up a local maintenance organization which can cope with those RAI maintenance tasks which can be done by hand labour, and to draw on the resources of central or regional highway departments for those tasks which involve complicated and heavy tasks requiring special skills in design and the use of mechanical equipment. In effect this means that a local maintenance organization can be responsible for the routine maintenance, and perhaps for minor emergency repairs, of low-volume feeder roads and for the complete upkeep of tracks, trails and paths. Chapter 15 recommends (i) that maintenance of paths and trails always be the responsibility of the users, and (ii) that a written agreement be made specifying community participation in the maintenance of tracks and some feeder roads in return for the government's investment in their initial construction or improvement.

7.03 Periodic maintenance and spot improvements on feeder roads normally, but not always, involve the use of heavy equipment and are therefore best allocated to a central or regional highway authority. Periodic maintenance is only undertaken after the role of the infrastructure is reassessed. Normally, paths and trails require very little periodic maintenance. If the area development is progressing as anticipated, additional access improvements such as spot maintenance is needed. These activities require an additional capital investment over and above that available to the decentralized maintenance organizations suitable for a RAI network.

7.04 RAI routine maintenance consists of small but frequent, dispersed activities. A concrete plan for routine maintenance must be developed at the same time as the plan for RAI improvements. The plan must be based on the likely local capacity to care for the facility. It will have more chance of succeeding if it provides incentives to encourage this care. In some cases the RAI network must be reduced to a level at which maintenance resources are available. Therefore, any maintenance programme planning should begin by identifying who benefits from RAI. It should attempt to establish the closest possible linkage between the beneficiaries and the suppliers of maintenance resources. Local, regional or central government maintenance involvement should be a last resort, except in providing unplanned or emergency maintenance caused by natural disasters such as unexpected floods and landslides.

7.05 Chapter 15 describes many variations for, and approaches to, setting up feasible maintenance programmes. Technical Note 8.1 Annexe 8, describes organizational possibilities for road maintenance decentralization. The following Maintenance Planning Chart (MPC) presents a logical approach to developing a maintenance strategy and programme when used in conjunction with the technical considerations outlined in Chapter 15. The chart also includes the option of abandoning all or any part of a rural access infrastructure development programme that lacks identifiable maintenance resources.

I. EXPLANATION OF ACTIVITIES SHOWN ON THE MAINTENANCE PLANNING CHART (MPC)

7.06 The activities in this chart are based on four premises:

(i) all maintenance activities should be as decentralized as possible;
(ii) different links within an RAI network can be maintained by different organizations or combinations of organizations;
(iii) maintenance is only sustainable if the resources for its activities are sustainable; and
(iv) if a long term maintenance capability cannot be developed, the rural access infrastructure should not be built or improved.

7.07 Without an assessment of the amount and type of resource expenditures required to maintain a RAI network, no maintenance strategy can be developed. Step 1 of the MPC includes such an assessment. Technical Note 7.6 itemizes the considerations that must be evaluated when planning a RAI maintenance programme. Each infrastructure tier in a RAI network, i.e., paths, trails, tracks and roads, has its own maintenance requirements. Some tiers, such as paths, may be evaluated as a group. Other links, such as roads, may require an individual evaluation since increasing traffic may limit the available maintenance options.

7.08 The Step 1 assessment should cover the lifetime maintenance needs of the RAI network. Those links which may experience significant traffic increases as development progresses, should be evaluated by development phases. For example, an earth road with less than ten vehicles per day may be maintained by labour (Technical Note 7.3). That labour may be provided by the local government under the lengthman system, which involves a contract between each labourer and the local government (Technical Note 7.2). As the traffic increases, some form of mechanized maintenance such as dragging will be

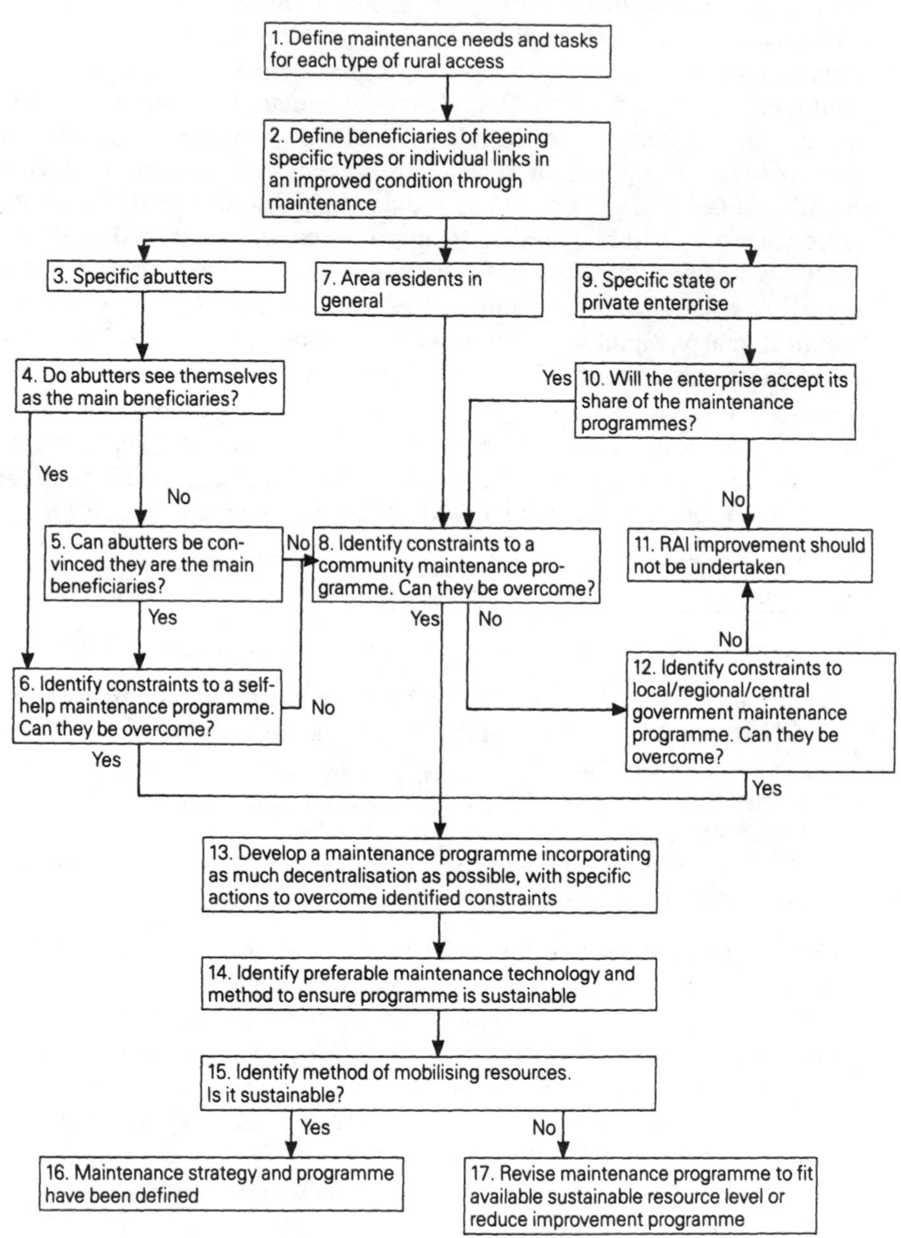

required. At that time, the regional government may provide a tractor and drag to complement the local government lengthmen. Thus, the maintenance resources for this single link eventually are provided from two sources, each with a limited capability. The Step 1 assessment must consider these probabilities and include an evaluation of any existing maintenance activities in the project area for possible expansion and/or improvement.

7.09 The ultimate solution to effective RAI maintenance is to capitalize on sustainable local resource bases. The beneficiaries of RAI network maintenance activities are the people most motivated to sustain these activities. Step 2 in the MPC includes the identification of these beneficiaries. The purpose of decentralizing the maintenance activities to the greatest extent possible is to take advantage of both motivation and available resources. Three groups of beneficiaries may be identified: (i) people who live adjacent to the specific RAI links, (ii) people who live in the area served by portions of the RAI network, and (iii) specific businesses, farms, plantations, mines, etc., privately-owned or government-controlled, which benefit from portions of the network.

7.10 The most decentralized maintenance source is the abutters to the RAI links. The term abutter is defined as a person who lives immediately adjacent to a road; however, in the MPC the people who live in small settlements straddling any RAI links are all considered abutters. Abutters to heavily travelled highways are usually not the main beneficiaries of the highway, nor do they have the resources to maintain a highway. However, abutters to paths and trails are frequently the principal beneficiaries of those links. Many exceptions occur, such as paths to a village well or trails between a settlement and a school. Therefore, the abutters' perception of their status as beneficiaries is important. When the local residents have a history of self-help, or when they will accept a self-help role in improving the infrastructure, they are likely to view themselves as the main beneficiaries of any maintenance activity. If the infrastructure link serves more than one group or settlement, the intermediate abutters are unlikely to consider themselves as beneficiaries unless the settlements are closely connected by family, tribal, or religious ties. Steps 4 and 5 are included to determine if the abutters will participate in self-help maintenance activities and to encourage that activity where possible.

7.11 Any constraints to accomplishing the proposed maintenance strategy being evaluated, whether it is a self-help programme (Step 6), a community maintenance programme (Step 8), or a government maintenance programme (Step 12), must be identified. Specific actions to overcome these constraints must be determined and implemented as an integral part of the final maintenance programme. Technical Note 7.7 identifies existing or potential constraints to successful implementation of maintenance activities and potential methods for overcoming them. Adopting any options or combination of options presented in the MPC is predicated on satisfying the maintenance needs free of any constraints.

7.12 Rural access infrastructure most often benefits more than just its abutters, e.g., the path to the village well or the school house trail. Feeder roads, some tracks, and longer trails often benefit the area residents in general (Step 7). Roads servicing through traffic, as noted in Para 3.03, are not considered as rural access infrastructure in this handbook. Therefore, the MPC addresses only RAI links in some way beneficial to local area residents. Maintenance of these links may be delegated to an entity termed 'community' herein. Para 15.15 identifies several organizations which may provide maintenance on the community level. These organizations may take the form of a legally constituted association for community development, a village or tribal leadership group, a cooperative, a religious group, a non-profit organization, or a merchants' association. 'Communities' may contract with governmental organizations to provide RAI maintenance, in which case the 'community' warrants satisfactory maintenance and some degree of supervision in return for fixed payments. If a community utilizes the lengthman system the contract or sub-contract is drawn between the 'community' organization and the individual rather than between the local, regional or central government and the individual.

7.13 When no community organization exists, or if the maintenance tasks are beyond the resources of the community, two possibilities exist: develop or strengthen the community organizational capability, or shift the maintenance responsibility to a local/regional/central government organization (Chapter 16). Local communities are sometimes so resource-poor that they cannot survive any additional infrastructure maintenance demands. When the Step 8 evaluation leads to this conclusion, the risk of losing the proposed resource investment in improving the RAI must be carefully evaluated. Such evaluations usually lead to a transfer of maintenance responsibilities to a formal local, regional, or central government organization. The actual maintenance activities may remain the same. For instance, local government may enter into a lengthman contract instead of the community doing so. Only the responsibility shifts since the maintenance requirements are independent of the organizational structure.

7.14 In some situations, a state or private enterprise such as a plantation or mine, may be located in a development area. More frequently, development projects may be designed specifically to create a state enterprise, or support a private enterprise, such as a rubber, palm, sugar or cocoa plantation. One objective of such projects is to encourage worker migration to the project area. In such cases, certain sections of the RAI network, usually feeder roads, but sometimes tracks used during the harvest season, most directly benefit the enterprise. Four maintenance scenarios are possible when such enterprises are located in a development area: (i) the enterprise pays its own maintenance crews; (ii) the enterprise supports a 'community' organization financially or becomes a 'community' organization, undertaking maintenance in return for a fixed fee from an existing government organization; (iii) an existing government organization provides maintenance for the appropriate roads; or (iv) the maintenance problem is never addressed. Step 10 in the MPC is intended to focus on the selection of the appropriate maintenance scenario for development projects which include such enterprises.

7.15 The Step 10 evaluation should subject the enterprise's maintenance programme to the same tests for constraints as are applied to any other maintenance programme if the first scenario is proposed. When the second scenario is being considered, the community maintenance programme must be tested to assure the added burden will not overtax the local capabilities investigated in the Step 8 evaluation. If the community concept is unworkable, the evaluation must next consider local/regional/central government maintenance programmes (Para 7.13). The third scenario bypasses the community option and immediately allocates maintenance to a governmental agency. This third scenario is the most commonly used approach. It is based on the premise that governmental road agencies already have a viable maintenance capability. This premise is addressed in Para 7.17. Unfortunately, the fourth scenario, deferring the maintenance question, also occurs in some development projects, either intentionally or unintentionally. This presentation recommends that no RAI should be improved or built until its maintenance problems have been solved. Without adequate maintenance, such investments have no long-terms benefits. Failure to properly maintain RAI links to large enterprises in a development area jeopardizes both the investment in the infrastructure and investment in the enterprise itself.

7.16 The Step 12 evaluation embraces all governmental road maintenance agencies for simplicity. Under the maintenance decentralization concept recommended in this book (Chapter 16), local government agencies should be investigated first, followed by regional and central government agencies. The evaluation procedure (Para 7.11) is the same in each case.

7.17 Three factors complicate assignment of RAI maintenance to government agencies:

(i) they have no special expertise in maintaining paths and trails, and may lack experience in the technology best suited for many RAI links, i.e., labour-based maintenance;
(ii) they assign a low priority to maintaining tracks and earth roads; and
(iii) they cannot increase their maintenance capacity without an increased budget, but unless they have adopted some method of performance budgeting they may not be able to determine or justify their increased budget needs.

The MPC attempts to reduce these complications by systematically:

(i) evaluating the maintenance needs for paths and trails and assigning their maintenance responsibility to abutters where possible, or else to 'community' groups;
(ii) evaluating the maintenance needs of tracks and low-volume roads, and, wherever possible, assigning their maintenance responsibility to motivated 'community' groups; and
(iii) finally, evaluating the needs and determining the resource requirements for the remainder of the RAI network which is most similar to the roads government agencies normally maintain.

If no combination of governmental agencies has the resources to carry out the government's share of the maintenance programme as developed in Step 12, the RAI improvement should not be undertaken.

7.18 Step 13 consolidates the various links of the RAI network under one overall maintenance programme. It allocates specific maintenance tasks to the groups best suited to execute them; and incorporates specific actions, such as supplying hand tools, to overcome previously determined constraints. The overall programme may include a self-help component from Step 6, a community maintenance component from Step 8, and a government component from

Step 12, or other combinations excluding one or two of the components. The government component may be further partitioned into activities undertaken by various government agencies from one or more levels of government. Many participants lighten individual resource commitments but may create administrative problems.

7.19 Until the maintenance assignments are made, the maintenance technology cannot be completely defined. Certain links require specific technologies, e.g., paths and trails must be maintained by labour. However, roads or tracks may be maintained by labour or machinery. For example, a community may not be able to sustain a mechanized maintenance effort on an earth road while the regional government would only maintain the same surface with equipment. These decisions not only affect the sustainability of the maintenance effort, but also affect the choice of construction technology (Technical Note 7.1). The Step 14 evaluation should result in a compilation of all the proposed maintenance activities with their technology options, if any, and the responsible organizations.

7.20 Mobilization of resources is the key to continued maintenance. As the final maintenance programme is developed (Step 13), resources play an important role in Steps 6, 8, and 12. The methods of mobilizing these resources for each element are evaluated as a constraint to each activity. Once the maintenance programme has been decided, the resource requirements should be reevaluated as Step 15, to ascertain no double-counting has occurred. The resources must also be evaluated against the requirements of the other development activities to determine if any conflicting demands for resources will be generated as development proceeds. For example, the manpower pool may diminish as more labour is required for increasing agricultural activities. If the resources for the programme as planned can be mobilized and sustained, the proper maintenance strategy has been defined. The maintenance programme, including any activities to develop the required resources, should provide adequate RAI maintenance (Step 16). If not, the programme must be revised in Step 17 to accommodate the available sustainable resource level, either by identifying additional available resources or by reducing the RAI network. Otherwise, the RAI improvement programme will not be a viable investment and should not be undertaken.

II. CONCLUSIONS

7.21 Maintenance considerations are not listed in the general planning chart in Chapter 3. Maintenance is an integral part of the resource expenditure to provide improved infrastructure, not an alternative means of upgrading an RAI network. The resource requirements for maintenance will vary depending on (i) the choice of transport modes and aids, (ii) the type of infrastructure selected, and (iii) the traffic volume and seasonality. However, once the transport demands are determined and the infrastructure network decided, maintenance activities and resource investments become a necessity, not an option.

7.22 Insufficient maintenance, not only on rural roads, but on all segments of the transport network, has long been a major problem in developing countries. It is unrealistic to assume that the already over-extended maintenance capabilities of these countries' highway organizations can readily be increased to include the network of paths, trails and tracks proposed for improvement as rural access infrastructure in this handbook. Experience has shown that without proper planning and execution of a well-defined maintenance programme, any resource investment in infrastructure improvement is short-lived. The maintenance planning chart in this chapter attempts to reduce these problems by identifying new maintenance resources at the local level and developing a decentralized maintenance programme whereby the beneficiaries of improved maintenance relieve the government of some of the maintenance responsibilities.

7.23 The maintenance planning activity outlined in this chapter should be undertaken as soon as the infrastructure evaluations (Chapter 6) are completed. If an acceptable maintenance strategy is defined as Step 16 of the MPC, its costs should be determined and added to the construction costs estimated for the infrastructure as noted in Para 6.12. The sum of the construction and maintenance costs represent the resource investment for the RAI necessary to provide the level of accessibility determined in Chapter 4. If resources are inadequate to sustain the maintenance programme developed from the MPC, then (i) the amount of infrastructure to be improved as determined in Chapter 6 must be reduced, or (ii) the type of infrastructure to be provided as identified in Chapter 5 must be modified, or (iii) the accessibility level determined in Chapter 4 must be lowered. Each choice necessitates a re-evaluation of all succeeding determinations. If none of the choices results in sufficient access to economically remove the constraints to other development activities, the development project itself should be re-evaluated to determine its economic feasibility with reduced access.

CHAPTER 8

Storage Evaluations

8.01 As indicated in Para 3.03 (iv) local storage capability and capacity can alter the level of accessibility requirements, and therefore the costs, of specific RAI links. Para 18.06 indicates that local storage facilities also dampen periods of maximum transport demand and consolidate local transport needs. However, it should be noted that storage is not a means of improving access. It is a substitute investment for expensive infrastructure construction and/or maintenance; consequently, the benefits derived from improved storage are not as extensive as those from improved access.

8.02 The evaluation of storage as an access proxy is in many ways similar to the evaluation of transport aids. Traditional storage methods are practised throughout every developing country, but may be localized by area, social group or tribe. Therefore, improved storage technology may have its source within the same country. New storage technology may also be imported from abroad, but because storage is so widely practised, imported storage technology is much more difficult to introduce than are improvements to local storage practices gleaned from the more innovative in-country farmers. Examples of improved in-country storage technology may be as simple as placing an existing basket granary on a foundation to reduce moisture from the ground, or placing rat guards on existing maize crib legs. Imported technology may include the local manufacture and distribution of metal storage bins for in-house use.

8.03 Sometimes storage methods will vary within a single village where the present inhabitants continue the storage practices common to their diversified backgrounds. A study in one such village [21] indicated that each tribal group tended to believe its storage methods were the only correct ones. Storage technology transfer between groups of different heritages had not occurred. No common evaluation base was available, even when the local inhabitants worked side by side and accepted each other's presence, since their storage activities were confined to their own household compounds.

8.04 Any physical survey of indigenous or foreign storage technology is beyond the purview of transport planners. The transport planner's initial approach to any storage evaluation is to search the government agencies, quasi-agencies, and relevant donor or non-profit organizations for any repository of on-farm or local storage data. If none is available, the transport planner will have to rely on the undocumented knowledge available to the agricultural planners assigned to development activities and extension programmes. The following Storage Improvement Chart (SIC) is based on the assumptions that the transport planner has been able to locate a source of storage technology and the government is willing to provide technical expertise in dealing with the local farming communities during the storage substitution evaluation.

Storage Improvement Chart (SIC)

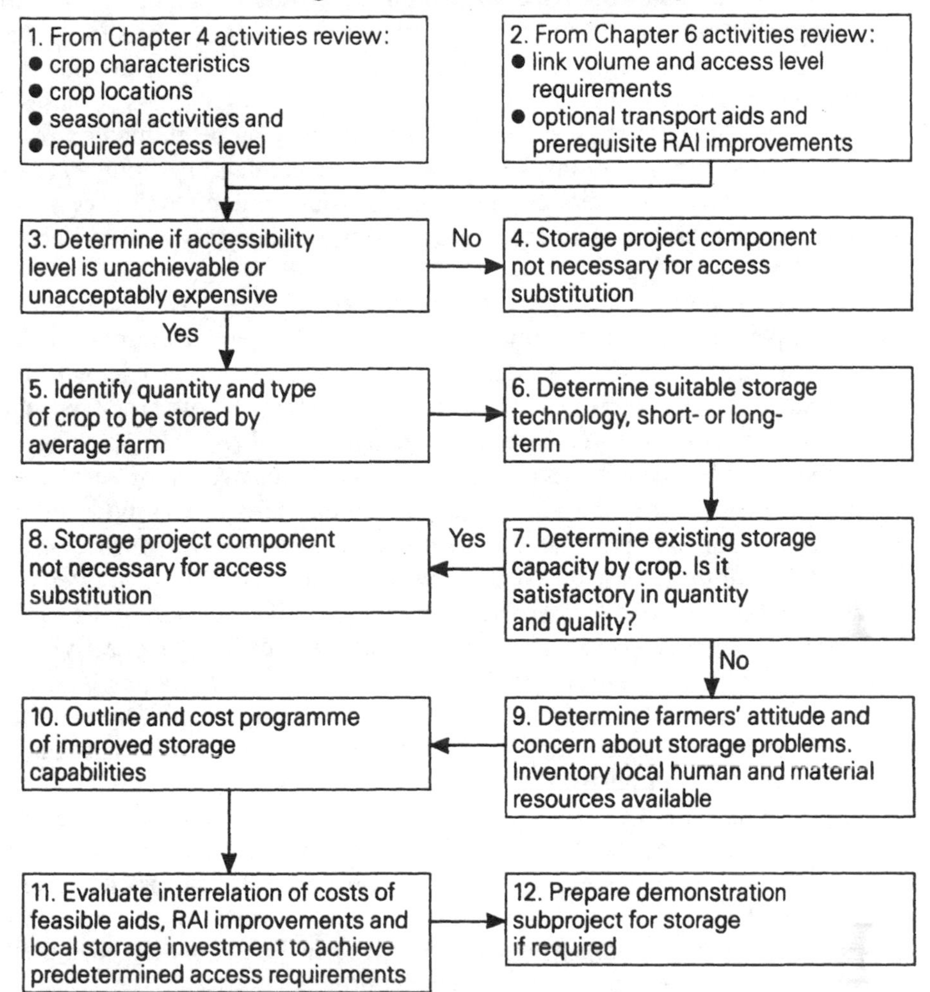

I. EXPLANATION OF ACTIVITIES SHOWN ON THE STORAGE IMPROVEMENT CHART (SIC)

8.05 The first three steps of the SIC outline the derivation of the data used to determine if a storage component will be required in any areas of a proposed development project. Crop information from Step 1 will be used in Step 5 to evaluate the types and quantities of produce to be stored if accessibility is not suitable. Chapter 6 activities, reviewed in Step 2, identify the choices for providing proper accessibility to specific areas within the development project. The third step uses the costs developed in the evaluations of infrastructure link costs determined in Chapters 6 (construction costs) and 7 (maintenance costs) to determine if the required accessibility level is unacceptably expensive.

8.06 The actual determination of an unacceptable cost for proper accessibility can only be found during the final economic review of the project. If a project proves economically viable, the access costs are often assumed to be acceptable even when an unnecessarily high level of access has been included.

However, in the evaluation methods outlined in this book, the goal is to find the least resource cost for providing the proper level of access. Step 4 is intended to eliminate from further study those links that clearly provide the required level of accessibility. Any link requiring all-weather access because of specific crop characteristics, Paras 18.02 and 18,03, may be eliminated from further storage considerations. In addition, when eliminating a physical obstruction to minimum access automatically provides reliable link access, no further storage evaluations for access substitution are necessary for the affected links.

8.07 After the RAI links providing proper accessibility have been eliminated, the planner can concentrate on evaluating storage substitution in the problem areas, if any. Step 5 identifies the specific storage problems to be evaluated. These problems will vary by link, depending on both crop type and farm size. On-farm storage facilities are relatively small (Para 18.18). Therefore, they may be considered as standardized units for evaluation purposes. There is some latitude in the size of common storage units such as mud silos or bamboo cribs, allowing the farmers to make adjustments to their individual needs. Major differences in on-farm storage requirements can be accounted for by assuming the affected farmers will build more of the same size units when storage requirements increase substantially because of development activities.

8.08 The type of storage technology selected for cash crops depends on the crop type and the length of shipping delays. Short delays in shipping some crops may require no more than supplying additional containers, usually sacks, and providing credit to purchase them. However, encouraging farmers to plant additional crops usually requires long-term improved or increased storage facilities to store produce. The farmer then has the ability to: (i) store the subsistence portion of his crops, (ii) sell the portion of his subsistence crops previously retained to account for spoilage, and (iii) retain the option to sell his crops at a later date to obtain a higher price. Step 6 in the SIC should be limited to an evaluation of the storage technology requirements in the areas of concern since these are the areas where storage will become a proxy for access. However, improved storage technology may also benefit small farmers in other areas of the project.

8.09 Steps 1 through 6 considerations are office activities. The first on-site investigation of the possibility of substituting storage for improved access occurs in Step 7. This investigation may indicate suitable storage technology and capacity exists (Step 8). If not, the Step 9 field evaluation should be undertaken by agricultural personnel. This field evaluation is analyzed as an office activity in Step 10. A programme for improving local storage facilities is determined from this analysis. Its costs are calculated in cooperation with the transport planner's agricultural counterparts.

8.10 Step 11 includes an economic evaluation of the entire transport package which may consist of resource investments in (i) transport aids, (ii) infrastructure improvements and/or construction, (iii) maintenance activities for both transport aids and infrastructure, (iv) storage improvements, and (v) demonstration components.

8.11 Storage improvements at both the on-farm and village levels are suitable to extension agent implementation. Improved technology can be demonstrated either at the local farm-level or on demonstration plots used for other extension activities. As with transport aids, improved storage technology transfer will only occur when the local population believes that better storage will serve a need and be beneficial to them. Therefore, Step 12 is included to formally promote and demonstrate improved storage techniques and to educate and train the target population in their construction, maintenance, and use.

II. EXAMPLE OF STORAGE CONSIDERATIONS

8.12 Since many transportation planners have not dealt with on-farm storage, the following data have been abstracted from two studies, one in Tanzania [21], the other in India [9]. These data are presented to identify typical types of storage problems and possible approaches to their solutions. The examples are project-specific and intended to provide only a sampling of problems and possible answers.

8.13 Crop losses are usually expressed as percentages of the crop. Some percentages express theoretical values, for example, drought losses are expressed as a percentage of an area's hoped-for yield. Predator losses in the field before, during, and after harvest are also very difficult to measure in real terms. On the other hand, actual storage losses once the crop is harvested and placed in a storage facility, are amenable to scientific measure, although such methods are seldom used in developing countries for on-farm storage. Local storage demand depends on the farmer's perception of the size of each loss factor and his ability to prevent that portion of the total loss. For instance, if a farmer suffers large losses from rat infestation but believes such losses are inevitable both in the field and in his storage facility, he will not be motivated to improve or increase his storage capability until he is convinced he can reduce rodent losses. Furthermore, the rodent losses may be so large that they mask insect losses, so the farmer may never consider protecting his stores against insects. Obviously, education is the first step in reducing crop losses in such situations.

8.14 Predator crop losses in the field are thought to peak in the period when the crop has been harvested and is drying in the field before storage. In fact, predator losses before the crop is normally dry enough to be harvested can be substantial. Animal pests in Tanzania, especially wild pigs and monkeys, were reported [21] as major threats to unharvested maize, sorghum and rice. Wild pigs can easily finish half a standing crop during the night even if night guards are stationed. Monkeys feed during the day, as do birds who also destroy considerable sorghum and rice. To combat these problems, the farmers may station family members in the field while the crop matures. They often harvest their grains wet to reduce predator losses prior to harvest. Long-term grain storage usually occurs after the harvested crops are threshed (separated), winnowed (cleaned) and dried. However, the Tanzanian report [21] indicated many farmers stored their grain wet as soon as it was harvested to reduce post-harvest field losses.

8.15 Cribs are a common method for storing grain outside before it is husked and dried. Inside storage methods for storing undried grain include the Tanzanian dari [21] which consists of a platform of thin poles about head height above the hearthstones of the kitchen area in mud-walled houses with thatched roofs. Dari platforms hold up to half a ton of unshelled maize. The heat and cooking smoke dries the maize, which has been picked moist, to about 12% moisture content within a month. The dari's advantages are its efficiency in drying and its ease and low cost of construction; its deficiencies are its vulnerability to rat attack, its inaccessibility for cleaning and inspection, and its limited storage capacity. It has three other indirect health implications [21]: (i) rat droppings and contaminated broken grains can easily fall from the dari directly into the family cooking pot, (ii) it requires cooking to be done indoors, which means the house is almost constantly filled with smoke, an environmental contaminant of uncertain long-term health consequences for the family, and (iii) even the safest insecticidal dusts cannot be applied freely to the stored grain because the dust will fall through the platform into the family food. Furthermore, unhusked maize is more hospitable to insects and insecticide applications are less effective on unshelled grains.

8.16 An improved version of the dari is the dungu [21]. This is essentially a small storehouse on stilts about 1.75 metres above the ground, with a thatched roof. It employs the same drying principle, a cooking fire below the floor. Its capacity is larger, an oblong dungu 1.5 by 3 metres by 1 metre high has a storage capacity of 1.3 tons of unshelled maize or 1.6 tons of head sorghum. It is better ventilated than the dari, since the air can circulate underneath and around the sides, which are usually made from sorghum stalks, reeds or bamboo. However, the cooking fire is less effective in drying the grain because of the cross-winds. Access to the dungu is generally by ladder through a half metre square door in one end, making inspection easier. In addition, the dungu may be cleaned thoroughly before placing fresh grain. The old roof is removed and the new grain placed before constructing a new roof. Removing the old thatch eliminates the previous season's rat population and allows the grain to be properly stacked in the dungu, a technique thought to reduce insect damage. If the dungu is located far enough from the house roof and overhanging trees (Para 8.21) and rat guards (rodent baffles) are placed on the stilts, its resistance to rodents is considerably enhanced. However, insecticides and droppings can present the same problems as with the dari while the fire is used for cooking.

8.17 The Tanzanian villagers in this study [21] rejected airtight storage, i.e., long-term storage, as being too expensive and time-consuming to build; being unreliable, i.e., not easy to keep airtight; and requiring better drying techniques than were available, although Para 8.15 indicates that maize is dry enough after a month to husk and store. Another study in India [9] concentrated on long-term farm-level storage improvements for airtight storage methods in conjunction with a Save Grain Campaign extension programme. This second study evaluated losses, e.g., weight, quality, nutritional and germination, occurring during the post-harvest period. The study [9] contends that long-term storage is of particular importance because: (i) given the seasonality of the production of cereal and root crops, the heavy dependency of the lesser developed countries on these dietary staples requires the crops to be stored for longer periods, and (ii) the concentration of these countries in the tropical and sub-tropical belt makes stored food more susceptible to deterioration as the climate is more conducive to the growth of insects and micro-organisms – the two most important biological agents responsible for loss of food value.

8.18 A brief summary of parts of this study [9] is included here. Several traditional storage methods for rice paddy exist in the study area. Once the paddy has been cut, usually by hand, it is spread in small bundles for two to seven days to dry. The moisture content of the paddy at harvesting is 16% to 20%. After drying, the paddy may be stacked in a circular structure, i.e., a pile called a kuppa, of up to 3 metres high and 10 metres in diameter. Paddy is placed with the stalks outward to prevent pilferage, if bundles of stalks are pulled out the grain will be shed. The stack is waterproof but is vulnerable to rodent and insect attack over time. The paddy moisture content stabilizes at between 13% and 14%. Kuppas are considered as temporary storage to allow the fields to dry so the bullock carts can remove the paddy.

8.19 After threshing, cleaning and drying, the rice may be put into the final store immediately or may remain for varying periods in bags or in bulk inside a room or on a verandah until the final stores has been prepared to receive the grain. The final storage structures in use in the study area fall into several categories, most being used for subsistence storage:

(i) underground pits used for home consumption storage, cylindrical, rectangular or cup-shaped. The pits must be sealed. Construction is inexpensive and labour-based. This storage method changes the taste of the rice. Pits may only provide short-term storage in areas where the water table rises during the rains. Their usual capacity is half to two tons;

(ii) basket stores constructed of woven strips of bamboo or other materials. Baskets may be used indoors or outdoors. They are sometimes plastered with a mixture of clay and animal dung. Baskets are vulnerable to rodent attack unless they are placed on a platform, which is seldom done indoors because of lack of headroom. Usual capacity is one-half to two and one-half tons;

(iii) circular structures woven from paddy straw rope. These structures are commonly used outside, although they may be used indoors. They may have a coat of mud/dung. Woven rope structures are used only for one season, the rope being unwound as stored grain supply decreases. Such structures are vulnerable to insects, rodents, and moisture;

(iv) attached rooms or separate buildings used for grain storage. These structures are constructed of timber and mud plaster, or bricks and mortar, with earth, wood, or stone floors, and tiled or thatched roofs. Doors may be sealed with removable wood slats. The structures are vulnerable to rodents and insects and often have leaky roofs;

(v) bins constructed of stone slabs, cement plaster, or timber and mud, within the dwelling. Such bins are usually located in a room corner and may have a lid that can be plastered shut. Stone slab bins are rodent-proof, easy to clean, and can be sealed for fumigation. Timber and mud bins are easily breeched by rodents; and

(vi) small-scale storage containers. Pots and jars holding 75 to 240 kg of rice often have earthen covers sealed down with a mud/dung paste. Bags are considered short-term, small-scale storage and are placed on wooden benches or a platform within the house to provide moisture protection. While piles of bags are relatively insect-proof, individual bags are subject to attack. Bags are vulnerable to rodent attack but damage from rodents is easy to see on individual bags.

8.20 The study [9] determined that the average paddy weight loss for the traditional storage methods described above was 4.26%. These loss estimates are quantitative losses. Qualitative losses, from fungi and insects which prevent the use of grain as seed and reduce the nutritional value of the stored produce, were not measured, but are thought to be substantial in the partially damaged but still consumed grain. It also determined that the losses for an improved outdoor bamboo woven basket (Para 8.21) was 1.02%. The average paddy weight loss in a metal bin (Para 8.22) was 0.62%. Using these findings, the report calculated the benefits of using improved storage as outlined in Paras 8.24 to 8.30. Paddy storage losses in traditional storage facilities are normally smaller than other grain storage losses. Consequently, the benefits for storing maize or sorghum in improved facilities should exceed those derived for paddy storage.

8.21 Improvements to the outdoor bamboo woven basket of the type described in Para 8.19 (ii) were made to reduce or prevent rodent access and soil moisture migration. The basic features of the improvements were: (i) constructing a wooden platform on wooden legs and providing metal cones (rodent guards) for the wooden legs, or (ii) where timber was in short supply, constructing a flat stone or concrete platform on a narrow central stone or brick column, the platform overhang being too wide for rodents to pass. Since rats cannot jump higher than 0.5 m, the bottom of the cones were placed above that height, as were the bottoms of the stone or concrete platforms when the second method was used. The overhang of either the cones or the stone or cement base was at least 225 mm in every direction to prevent rats from climbing over the overhang. Care must be taken to locate the structure at least 60 cm to 90 cm from any stationary objects like house roofs, trees, etc., to prevent rats from jumping onto the thatched roof covering the basket. Rats can jump horizontally approximately 20 cm. The roof overhang generally protrudes 30 cm beyond the sides of the woven basket.

8.22 The 787.5 kg capacity metal bins evaluated in the study [9] were made domestically from galvanized sheet metal. It was assumed that metal bins can replace any of the indoor storage methods described in Para 8.19, while the improved basket would only be used outdoors because of the lack of indoor headroom. Under the right conditions either improved storage unit could be used both indoors and outdoors. The baskets can of course be made in various sizes (up to 2½ tons); however, the metal bins were locally manufactured in only two capacities, approximately 0.38 tons and 0.79 tons.

8.23 The study [9] used two evaluation methods to determine the benefits of improved storage: (i) a financial cost-benefit analysis, and (ii) a social cost-benefit analysis. The principles of the financial evaluation are outlined in the next paragraphs.

8.24 The costs of the average improvements to the basket granaries were determined excluding the basket itself. Improvement costs varied throughout the area, depending on: (i) the costs of available materials, i.e., the usage of timber, stone, or cement bases, (ii) the transport costs of those materials, and (iii) the differences in labour costs, partly due to the differences in time when the improvements were carried out. The costs of these improvements were averaged out for the project area and converted to the cost to store one quintal (100 kg) of paddy. The costs of the baskets themselves were consistent throughout the area. These costs, converted to 1982 US dollars, were:

Improvements	= $1.05 per quintal
Basic basket	= $0.74 per quintal
Total initial investment = $1.79 per quintal	

The basket has to be freshly mud-coated very carefully each year. This cost was estimated at $0.06 per quintal. Yearly fumigation costs for both the improved basket granary and the metal bin were estimated at $0.15 per quintal based on using one 6 cc ampoule of EBD (ethylene dibromide, a common fumigant) per quintal per year. Therefore, the total recurring costs for improved basket granary storage were $0.06 + 0.15 = $0.21 per quintal per year.

8.25 The cost for the metal bin with a capacity of 787.5 kg was $50.47 delivered to the nearest town. The transportation of the metal bin was assumed to average 10 km from the nearest town to the villages being studied. The transport cost by bullock cart was therefore calculated as $1.11 and the total initial investment cost per quintal was (50.47 + 1.11) (100/787.5) = $6.55, as compared to the total initial investment cost of the improved basket of $1.79 per quintal (Para 8.24). Recurring costs for the metal bin consist solely of the yearly fumigation costs of $0.15 per quintal (Para 8.24).

8.26 The recurring costs of the improved storage must be discounted to determine their present net value using the technique explained in Technical Note 5.1 for routine road maintenance costs. The criteria for discounting costs determined in the report [9] were: (i) an average life of both units, the improved basket and the metal bin, of 15 years, (ii) a scrap value of 20% of the present market value, and (iii) a present worth factor based on an 11% discount rate. The present worth factor extends from year zero (discount factor 1.000) to year 14. The discount factors for years 1 to 10 may be taken from Table 1, Technical Note 5.1. The discount factor of the remaining years, 11 to 14, may be calculated using the following formula:

Discount factor $= 1/(1 + p)^n$,

where,

p = the discount rate as a decimal,
n = the year for which the discount applies.

The discount factor for the fourteenth year therefore becomes $1/(1.11)^{14}$ or 0.232. The sum of the discount factors for the recurring costs becomes:

1.000 + 0.901 + 0.812 + 0.731 + 0.659 + 0.593 + 0.535 + 0.482 + 0.434 + 0.391 + 0.352 + 0.317 + 0.286 + 0.258 + 0.232 = 7.983

The present value of the recurring costs becomes:

Improved basket:	$0.21 × 7.983 = $1.68
Metal bin:	$0.15 × 7.983 = $1.20

8.27 The farmer's benefit from adopting improved storage practices is the increase in grain availability later in the season. This grain can be sold at the increased price prevalent at that time, the income representing the financial value of the farmer's benefit. That financial benefit can be expressed as follows:

$$Q \times L/100 \times (100 + P)/100 \times B$$

where,

Q = quintals of grain stored,
L = the percentage of loss saved,
P = the percentage increase in the price during storage, and
B = the basic price at harvest time.

8.28 In the present example, the savings (L) for the improved storage can be found using the figures in Para 8.20. The savings becomes:

Improved basket:	4.26 − 1.02 = 3.24%
Metal bin:	4.26 − 0.62 = 3.64%

The seasonal price increase averaged 30% over the base price B of $14.80 per quintal, so the farmer's financial gain for each quintal of paddy stored in the improved basket would be:

$$1 \times 3.24/100 \times 130/100 \times 14.80 = \$0.62$$

and his savings for each quintal of paddy stored in the metal bin would be:

$$1 \times 3.64/100 \times 130/100 \times 14.80 = \$0.70$$

8.29 These savings also accrue over the 15-year life of the storage structures. However, the first return occurs in year one although the first recurring expense, i.e., mud plaster and/or fumigation, occurred in year zero. Therefore, the discount factor for year zero (1.000) is deleted from the summation in Para 8.24 and the discount factor for year 15 (0.209) is substituted. The discounted benefits per quintal are as follows:

Improved basket:	\$0.62 (7.983 – 1.000 + 0.209) = \$4.46
Metal bin:	\$0.70 (7.983 – 1.000 + 0.209) = \$5.03

8.30 The following table, using the figures generated in Paras 8.24 to 8.29, outlines the cost-benefit ratio of the two improved storage structures being evaluted.

Table 9: Financial Cost-benefit Ratio of Improved Storage

	Structure	
	Improved basket (\$)	*Metal bin* (\$)
Initial cost, \$/quintal	1.79	6.55
Present value, recurring costs	1.68	1.20
Total cost	3.47	7.75
Discounted value of scrap[1]	0.07	0.27
Net cost	3.40	7.48
Benefits	4.46	5.03
Ratio	1:1.31	1:0.67

8.31 The preceding evaluation indicates that improving the woven basket granary, a traditional storage structure, is cost-effective using the prices in this example. Therefore, it may be possible to convince the local farmers of the merits of improved woven basket granary storage if accessibility is a continuing constraint. Introducing metal bins, a new storage technology, is not cost-effective, and unlikely to be acceptable to the local farmers over time.

8.32 Each of these examples illustrates different storage-related problems. In the Tanzanian example, the farmers have serious problems in preventing losses prior to storing their crop. Their traditional storage techniques reflect the need for both drying and storing their harvest in the same structure to reduce these pre-storage losses. Storage improvements consisted of providing an increased storage capacity and reducing rodent and insect losses while retaining the traditional storage technique. The Tanzanian farmers in this example have not yet accepted the more advanced long-term airtight storage methods illustrated in the Indian example. This latter example illustrates the type of improvements possible even when traditional storage structures are long-term oriented. The Indian example also illustrates the economic evaluation methodology and proves that, in this example at least, storage improvements are cost effective.

8.33 However, neither of these examples represents the substitution of storage for access. Research for this presentation failed to uncover any such evaluations. Para 9.21 indicates that even farm families below subsistence level sell produce immediately after harvest and later buy like amounts if storage facilities are unavailable. Starting with such a scenario, an order of magnitude

[1] The scrap value (Para 8.26) was discounted and subtracted from the cost figures to determine the net cost.

cost example can be developed using the typical Benin farm family of eight described in Technical Note 1.2.

8.34 The following storage assumptions have been made to simplify this example:

(i) the farm family of eight can store 5 quintals (Para 8.15) in their house. This is the only storage available under the no-storage scenario;
(ii) short-term storage consists of outdoor basic basket granaries costing $0.74 per quintal (Para 8.24) which can provide storage through the rainy season but cannot store grain long enough to take advantage in post harvest price increases;
(iii) long-term storage consists of outdoor improved basket granaries costing $1.79 per quintal (Para 8.24) which can store grain for a full year. The per quintal cost of all basket granaries is insensitive to size as was the actual case in the Indian example. Therefore the farmer can build one or several granaries with little magnitude of scale savings;
(iv) the rise in farm gate prices for stored material is 30% after all of the short term storage produce has been purchased (Para 8.28). For simplicity all stored produce will be considered to have a base cost of $14.80 per quintal (Para 8.28);
(v) the farmer will produce the surplus shown in Technical Note 1.2 as soon as he is assured of a market;
(vi) in the no-storage scenario the farmer will buy back produce at the current farm gate price, i.e. no middle man profits or penalties for small purchases are included; and
(vii) the discounted future benefits for reduced losses (Para 8.29) are ignored to simplify the evaluation.

8.35 Two accessibility levels will be considered:

(i) Reliable access is sufficient access to eliminate losses of any surplus crops temporarily stored in the field awaiting sale for export.
(ii) Minimum access is a lower accessibility level that introduces a substantial risk that crops stored in the field will deteriorate and become unsaleable for export before they can be transported. However minimum access assures transport of crops within the short-term storage time frame specified in Para 8.34(ii).

8.36 Three storage scenarios will be considered for the farm family described in Technical Note 1.2. Each scenario includes a total production of 52.5 quintals of foodstuff per year, of which 36.2 quintals is required for family consumption and 16.3 quintals is saleable surplus. To simplify these scenarios, short-term storage will be considered as effective for six months. However any produce that is sold must be purchased back at the 30% premium price, i.e. all surplus that is placed in short-term storage is sold as soon as possible to meet current financial needs while those farmers with long-term storage capabilities will not sell any produce unless they receive the 30% premium paid when produce becomes scarce.

8.37 The first scenario considers only the use of in-house storage. There is no capital investment for this 5 quintals storage. The farmer sells 47.5 quintals at $14.80 for a total of $703 and then repurchases his remaining subsistence needs, 31.2 quintals at $19.24, throughout the year, at a total cost of $600. The farmer contributes 16.3 quintals of surplus to the economy and receives a $103 income. This scenario is only possible when reliable access is provided.

8.38 The second scenario involves the construction of traditional storage facilities and assumes no in-house storage. The farmer with reliable access can store one half (18.1 quintals) of his subsistence requirements by constructing a basket granary at a cost of $0.74 per quintal or $13. He sells 34.4 quintals for $509 and repurchases 18.1 quintals for $348; he contributes 16.3 quintals of surplus to the economy and receives a $148 income. The farmer with minimum access must store his entire production in a 52.5 quintal granary costing $39. He also sells 34.4 quintals for $509 and repurchases 18.1 quintals for $348. He contributes 16.3 quintals of surplus to the economy and receives a $122 income. If the basket granaries last more than one year the costs must be prorated, but for the sake of clarity all storage facilities in this example are evaluated as having a one year life.

8.39 The third scenario involves the construction of improved storage facilities. The farmer with reliable access has the choice of selling his surplus immediately for $14.80 per quintal or later for $19.24 per quintal, a difference of $4.40 per quintal. Since the investment in improved storage in this example is $1.79 per quintal, one would assume he would elect to store the surplus. This difference in price increases when the storage facility life span is realistically evaluated. The farmer with minimum access has no choice but to store his entire production. The cost of storing 52.5 quintals is $94. The sale of the surplus crop of 16.3 quintals at the premium price grosses $314. The farmer therefore contributes 16.3 quintals of surplus to the economy and receives a $220 income.

8.40 The following summation represents the financial impact on the farm family in these scenarios when a new storage facility must be acquired every year. A more realistic economic evaluation assigning a multi-year life to the storage facilities will increase the later year financial benefits from additional storage. In each case 16.2 quintals of surplus crops are added to the economy.

Access type	*Storage type*	*Storage structure cost ($)*	*Buy-back cost ($)*	*Total cost ($)*	*Gross income ($)*	*One-year financial gain ($)*
reliable	in-house	0	600	600	703	103
minimum	traditional	39	348	387	509	122
reliable	traditional	13	348	361	509	148[1]
minimum	improved	94	0	94	314	220
reliable	improved	94	0	94	314	220[2]

[1] The difference between $122 and $148 in this example, i.e., $26, represents the value to the farmer of improving access from minimum to reliable in areas where traditional storage is practiced. This value, properly modified, becomes an input to the economic evaluation of the costs for improving accessibility.

[2] This example is based on a very narrow evaluation of financial gains from storage and does not take into account the degree of difficulty of marketing, or social benefits derived from reliable access.

8.41 An evaluation of the above scenarios leads to the following conclusions, which are example specific:

(i) providing minimum access when no on-farm or village storage is available or proposed will not encourage the farmer to produce a surplus since crop evacuation is not assured;
(ii) providing minimum access in combination with traditional storage ensures crop evacuation thereby encouraging the farmer to produce a surplus for sale as soon as access is practical, however the farmer cannot take advantage of premium prices and remains dependent on the market place for part of his food supply;
(iii) providing reliable access allows the farmer to produce and market a surplus crop without investing in storage facilities, but lack of storage forces him to rely heavily on the market place for his food supply.
(iv) providing reliable access in combination with traditional storage enhances the farmer's ability to market his surplus and reduces, but does not eliminate, his dependency on the market place for his food supply;
(v) providing either minimum or reliable access with improved storage encourages a farmer to produce and market a surplus crop because he is able to take advantage of price variations while remaining free of the open market variations for his own food supply; and
(vi) providing improved storage may be an attractive financial alternative to improving access above minimum standards.

8.42 Since increased access improves other social services which receive no benefits from improving storage facilities, any transport investment in storage should only be contemplated if it results in reducing overall transport resource expenditures. However, other sectoral objectives may be achieved by introducing storage facilities that show a favourable cost benefit ratio. Since planning rural and agricultural development projects is a team endeavour (Para 3.02), the results of all transport-motivated analyses of storage facilities should be shared with the other planners involved in the development project.

III. CONCLUSIONS

8.43 The ideal storage facility should [9]:

(i) afford maximum possible protection against insect, rodent and bird pests;
(ii) afford protection against excessive moisture and temperatures favourable for both insect and mould development;
(iii) allow adequate ventilation, yet be capable of being made reasonably airtight for fumigation;
(iv) provide for convenient loading and unloading;
(v) give protection against fire and theft;
(vi) allow facilities for inspecting the stored material; and
(vii) facilitate cleaning.

Of these, the most important requirements are the protection against insects, rodents, birds and moisture. The ultimate criterion for providing any of the above storage characteristics must be cost-effectiveness if the storage sub-component is to meet the same sustainability requirements outlined previously for transport aid usage and infrastructure improvements.

8.44 Local storage considerations are a concern to transport planners if a high investment is required to eliminate reliable access constraints to sections of agricultural and rural development projects. However, storage substitutions for access do not provide many of the social benefits induced by reliable access. Storage should therefore be considered as a last resort solution for the problem of providing reliable access, although the technology itself may still be of interest and value to the local farmers.

8.45 Data must be gathered to evaluate the possibility of introducing, improving and expanding the use of transport aids and to estimate the costs of their necessary RAI improvements. This data becomes the base information for evaluating the possibilities of including a storage component as a proxy for reliable access. However, considerable additional input from agricultural experts is also necessary to make a realistic evaluation and estimate of any storage sub-component of a transport investment. The Storage Improvement Chart included in this chapter outlines a possible approach to the evaluation of a storage sub-component.

PART III

CHAPTER 9

Accessibility

9.01 Part III presents detailed guidelines and recommendations for preparing, designing, constructing, and maintaining rural access infrastructure (RAI). It serves to expand the strategies developed in Part II, which, in turn, are methods of implementing the policies outlined in Part I. Each major factor affecting the selection of a RAI network is the subject of a chapter in Part III. Specific technical features concerning some of the subject matter in Part III are further amplified in the Annexes which make up Part IV. Accessibility is the first subject dealt with in Part III.

9.02 Accessibility, as used in this handbook, is a measure of the ability of a particular mode of transport such as a truck, minibus, bicycle, bullock cart or pedestrian, to reach a specific destination over a specific route. The differentiation of accessibility by mode is necessary because this handbook addresses several transport modes that may have different degrees of accessibility over the same route. This chapter distinguishes three principal levels of accessibility, i.e., all-weather, reliable and minimum access.

I. DEFINITIONS

***9.03* Trafficability**

is a road-related term used to quantify the number of days a low-volume earth road is likely to be impassable each year because of slippery road conditions, loose sand or excessive rutting or roughness [49]. Since trafficability applies particularly to motorized transport and motorable roads, the trafficability concept is generally included under the term Accessibility in this presentation.

***9.04* Level of Service**

is a term often found in highway literature. It is used in relation to motorable roads and highways and may mean one of two things. It is sometimes used in the context of highway capacity to describe levels of traffic congestion during peak hours of travel. At other times, it is used to define a serviceability index based on pavement condition for the purpose of projecting the remaining service life of the road surface.

9.05 Neither of these meanings of Level of Service is directly relevant to the term Accessibility as used in this book. Congestion is not normally an issue on the infrastructure under consideration herein; and as long as the various travel ways permit the passage of the relevant travel modes, surface roughness is not a major factor. Assured passage, not the speed of travel, is paramount. Travel speed becomes important for long distance routes but is less relevant to shorter local trips [51]. However, for some items, such as fragile produce like tomatoes and bananas, there is the potential for damage due to surface roughness; while the collection of palm kernels for manufacturing palm oil requires both reasonable transport speed and a relatively smooth surface. In such cases, the 'Level of Service' is a factor insofar as it will influence the proposed surface maintenance strategy.

***9.06* Level of Access**

is often used in reports about low-volume rural roads as a hybrid term implying the definition given above for Accessibility. Frequently such reports also further define the level of Access required for low-volume rural roads as All-Weather Access. The term 'Level of Access' is not used in this report because of its close association with the concept of all-weather motor vehicle access.

***9.07* All-Weather Access**

implies that a facility can be travelled 24 hours a day, 365 days a year. To meet this criterion the route must have a surface that provides strength and traction for the transport mode under consideration as well as waterway crossings that do not inhibit movement during or after storms [81].

***9.08* Reliable Access**

infers sufficient access for evacuating produce whenever necessary. This modification to the all-weather access concept can substantially reduce access infrastructure costs since it accepts intermittent flooding at fords and other weather-related closures which may otherwise be expensive to eliminate.

***9.09* Minimum Access**

implies the lowest level of accessibility acceptable in areas where people live and farm. It is necessary in areas populated by subsistence farmers and landless rural labourers so they may satisfy their daily needs and social activities. Minimum access is acceptable during periods of relatively low economic activities, but must be complimented by reliable access during periods of anticipated high economic activity, before development will occur. Consequently, in areas where colonization is a factor in the development process, care must be taken to provide both reliable and minimum access to attract and sustain the new immigrants. Minimum access problems may be reduced by locating development facilities near to the users, i.e., planting wood lots, digging wells, and building small dispensaries, schools, extension stations, etc.

II. IMPACT OF ACCESSIBILITY

9.10 Each project involving low-volume feeder roads needs to be assessed individually to determine the relationship between rural road accessibility, rainfall, drainage conditions, and engineering properties of the subgrade and pavement materials. An Indonesian study [45] indicates that road closures of one day or less have little economic impact on the population. In such cases travel delays are seldom crucial, while fresh produce can be either stored or perhaps delivered to another market. Of course, storage would not be an option in specific cases where an agricultural product, such as milk, is highly perishable unless a relatively high cost specialized storage facility is used.

9.11 As interruptions to access increase in length, impacts become more severe and more widely felt. Disruptions in rural road accessibility lasting several days begin to have a significant impact on weekly markets, distribution of non-agricultural products, social and administrative functions, extension activities and storage costs, as well as on opportunities lost to firms operating agricultural-related industries. When road closures extend over several weeks, closure-induced costs increase considerably depending on when the disruption occurs. If it occurs in the planting or harvesting period or when insecticides are required, there can be a substantial negative impact on the area's agricultural economy. Similarly, the impact on non-agricultural trade and social functions is proportionally greater when the closure continues for several weeks, because the population begins to make longer-term adjustments to the lower level of economic activity [45].

9.12 When rural road closures of several weeks are repeated annually, there is frequently a reduction in land use intensity or a change in the type of land use, leading to a real loss in area income. If the rural road closure period is extended to several months, the community costs assume very large proportions, as the area's characteristics revert to those similar to an area without access [45].

9.13 Short term road access problems may be relieved by alternative detour routes which, if available, at least permit urgent trips to be made, albeit over a longer distance. In many situations, using alternative transport modes (Chapter 10), which can operate under conditions which would temporarily deny motor vehicle access, will lessen the impact of short term feeder road closures.

III. ACCESSIBILITY BY VARIOUS TRANSPORT MODES

9.14 Each segment of RAI, defined in this book as feeder roads, tracks, trails and paths, provides access for one or more transport modes. The level of accessibility for each mode on a particular route is different, for example, a burdened pedestrian can traverse a road that is too muddy for truck traffic. It is often possible, on a route used by several different modes, to improve the accessibility for one mode at less cost than improving the accessibility for all modes using the route. For instance, a foot-log placed across a small stream next to a motorable ford which is frequently impassable during the rainy season can eliminate pedestrian delay without improving motor vehicle accessibility.

9.15 Access may also be prevented by physical obstructions. A trail through a wooded area may prohibit motor vehicle access if the trail width between trees is too narrow to allow a vehicle to pass, even though the soil and drainage conditions are suitable for periodic low-volume motorized traffic. Clearing the trees next to the trail may provide access for four-wheel drive vehicles. Physical constraints to access are permanent obstructions to specific transport modes which, when removed, increase accessibility for the affected transport mode to the point at which environmental constraints control. In the terminology of this presentation, the removal of physical obstructions often results in a change in the classification of any infrastructure not already identified as a road. For example, a trail would be reclassified as a track if it was widened to permit access by some means of motor vehicle traffic during part of the year.

IV. DETERMINATION OF ACCESSIBILITY

9.16 Physical restrictions to access can be readily identified by travelling the route in question using the various transport modes under investigation. Local residents are usually aware of previous attempts at using various means of transport and can identify trouble-spots for each such mode and the potential limit of its penetration into the area.

9.17 Determining the number of days per year that a route will be impassable due to environmental constraints, and the length of each individual closure, involves a complex and normally impractical analytical process which results in statistical values that do not accurately forecast the yearly variations in rainfall intensity and frequency. For the relatively small areas of the numerous rainfall catchments involved, rainfall and runoff records are rarely available. Such records would be needed to determine the frequency of flooding in rivers and streams, and the probable length of time fords will be impassable after storms of any selected frequency. In most cases flows in small streams rise and fall rapidly because of short intense storms. The duration of individual closures is usually well less than half a day although such closures may be frequent throughout the rainy season. Local historical knowledge and information can be an invaluable substitute for analytical procedures, and are often reliable.

9.18 Usually large rivers are not affected by short heavy downpours. Due to the larger size of their catchment areas and the higher capacity of their tributaries, they require larger, more extended rainfalls than do smaller streams to cause significant changes in their water depth. River closures, therefore, tend to be less frequent, but of longer duration, than small stream closures. Consequently, river fords may be closed for several days at a time after prolonged rainfall. An exception is the larger, dry river beds in hot deserts, which are subject to flash floods from infrequent but sometimes severe thunderstorms. These storms cause high velocity flows and high rates of runoff that result in a rapid but short-term rise in such river crossings [70].

9.19 The number of days an earth road is impassable because of weather-affected surface conditions such as softness or slipperiness is even more difficult to predict by analytical processes than stream flow problems. Road softness frequently lasts for some time after the surface has become saturated, while slipperiness usually occurs immediately on surface wetting and lasts only a short time after the rain ends. Road surface conditions are affected by many factors, including rainfall intensity and duration, soil type, moisture content at the time of the rain, hours of sunshine following rainstorms, time of day of the rainstorms, type of vehicles using the roadway, drainage characteristics of the road, wind speed and direction, shading by roadside trees, and other variables [26]. Technical Note 1.1, Rainfall and Slippery Soils, Annexe 1, Part IV, describes the tentative findings of a study of the rainfall amounts that cause slippery soils in Kenya. Similar information is no doubt available for other locations. However, there is no objective analytical approach that will provide specific solutions to this type of problem in all areas although terrain classification and evaluation can serve as a guide. Again, local historical knowledge and information would be the most reliable basis for planning and investment decisions.

V. FACTORS INFLUENCING NEED FOR ACCESSIBILITY

9.20 The most significant transport needs of the rural population are those relating to farming households. Transport is required for both crop production, e.g., the localized movement of seeds or plants, fodder, fertilizer, insecticide, agricultural implements, harvested crops and household needs such as the collection of firewood and water. This book calls all of these transport movements, 'on-farm activities' to differentiate them from the farm families' other transport needs. Transport distances for on-farm activities are generally relatively short, approximately 1–2 km for agricultural activities, increasing to as much as 13 km for firewood and water; and loads are small, ranging from 15–150 kg [23]. Off-farm agricultural transport can be divided into two elements: farm to roadside, and roadside to market. The limits of off-road transport are sometimes, but not universally, considered as between half a day's and one day's walk [29].

9.21 It does not necessarily follow that because a farm family is at or below the subsistence level, they do not market any of their produce. They sometimes must sell produce to raise cash after the harvest to pay debts, to supplement their diet, or because they lack storage facilities to store harvested crops. In these cases, they might later have to repurchase some of this same produce, often by borrowing again against their next crop. Small plots of 1–2 ha are capable of generating yearly transport demands of several tons [29]. One example of such a situation in Benin is described in Technical Note 1.2, Evaluation of a Farm Family's Transport Requirements. In this example, the family produces 1650 kg of surplus from five different crops, but their on-farm transport demands are more than 20 times this figure. Normally this type of transport is ignored when analyzing transport demand, or when making decisions about how scarce resources should be used to improve the standard of living and increase the productivity of the rural poor. Various transport means of the types described in the next chapter, operating on a variety of RAI types, may prove to be particularly relevant to providing accessibility for this type of transport demand.

9.22 Rural activities, both social and economic, involve the movement of people as well as materials. A study of feeder road characteristics in Ghana [57] indicates that the greatest percentage by weight of motor vehicle loads moved along the roads surveyed were passengers. They accounted for over 80% of the total carriage of goods and people when the average passenger weight was assumed to be 59 kg. Ideally, these people and materials should be able to move in either direction, into or out of the area under study, at will. However, the big financial investments required to provide all-weather accessibility, mean that such investments can normally only be justified where large, valuable agricultural crops and large local populations are involved. Smaller agricultural surpluses and smaller populations often cannot justify all-weather motorable roads into remote areas, nor even to the boundaries of the areas of access of existing all-weather motorable roads. These areas can only be served by less costly branches to the all-weather trunk route. Such branches can serve their function quite satifactorily even though they might provide reduced accessibility for motorized truck traffic. For example, an earth road might provide accessibility for smaller transport means such as minibuses or four-wheel drive vehicles. Other types of infrastructure might provide accessibility for alternative modes of transport that will cater for the travel and transport needs of the area, such as: tracks for tractors, trailers and carts; trails for pack and saddle animals; and paths for human porterage.

9.23 Economic development of an area is based on the population having the opportunity both to carry larger quantities of produce to and from markets, and to expand their access to more distant marketplaces. In the absence of all-weather motorable roads, such opportunities and improved access can be provided by: (i) introducing alternative transport modes into areas where they are presently not used, (ii) improving infrastructures for alternative transport modes, both in areas where the modes are newly introduced as well as where they presently exist, and (iii) communal or cooperative use of transport to make the more expensive aids available to the poorer farmers. Alternative modes of transport, combined with a local storage capability, can transform an otherwise isolated subsistence farming community into an improved, economically productive unit even when the costs of all-weather roads can not be economically justified.

VI. CONCLUSIONS

9.24 Accessibility is a term used in this report to indicate the ability of different types of transport to travel over specific routes. Accessibility constraints can be weather-related seasonal closures or permanent physical obstructions. Economic and social activities expand only if the population in an area perceives access to the area as reliable; however, the lack of all-weather road access is not always detrimental to the economic and social activity of an area if interruptions are infrequent and short-lived. Specific outputs such as milk and rubber production require constant access, while other outputs such as soft fruit and palm kernels require a well-maintained road surface. However, in the majority of cases in the more remote rural areas, the speed and/or discomfort of access over low-volume transport infrastructure is less important than the fact that useable access exists.

9.25 In order to achieve complete benefit from areas within the physical area of access of an all-weather motorable road, all such areas will need a level of accessibility appropriate to the area's economic and social activity potential. This accessibility can be provided with other than motorable all-weather roads if the access is suited to the alternative modes of transport available, and if storage facilities, markets and social services are within the range of these alternative modes of transport.

9.26 Although the relevance of levels of accessibility for each mode for a particular area will differ according to the levels of economic and social activity of the area, there would appear to be several critical plateaus in these relationships. The most critical aspect of an interruption to movement is the length of the individual closures, while the total accumulated length of closure in a year will also have an impact. It would appear that the relevant plateaus in the lengths of individual closures are as follows:

(i) All-weather access allows no closures, permitting travel by any mode at any time. It normally requires high cost, high design features such as: relatively flat grades; paved or at least gravel surfacing; full culverting of all streams; vented fords or bridges at short river crossings; ferrys at long river crossings where bridges are uneconomical or interfere with navigation; and high embankments across flood-prone areas. Subcategories of this level can be defined according to particular modes, for example, all-weather pedestrian access, all-weather pack animal access or all-weather motorcycle access. The relatively high cost of providing all-weather motor vehicle access would rarely be justified in rural areas except on major feeder roads carrying large volumes of traffic.

(ii) Reliable access substitutes, for all practical purposes, for the provision of a 'no closures' level of motor vehicle access in areas where all-weather access is not a viable goal. The lesser level of 'reliable access' needs to be defined on the basis that accessibility is not a constraint to the achievement of the economic and social potential of the area. For each particular area, this will need to be defined in terms of the particular modes necessary and feasible to perform the economic and social activities of the area. For each mode, the acceptable maximum length of individual closures, and the maximum accumulated closures for that mode per year, need to be determined, with major emphasis on the duration and frequency of the closures during periods of major economic activity.

(iii) Minimum access applies to areas of relatively high economic activity during the periods when the transport demands of that economic activity are low. Equally important, it also applies to populated areas beyond the area of proposed immediate development but within the zones of access of the rural roads servicing economic development activities. Within these areas, the population should be provided with sufficient access to perform, albeit with some difficulty, the various economic and social activities necessary to sustain their well being. As with reliable access, the specific minimum levels of accessibility to be aimed for will need to be determined in the context of particular cases and the current policies and strategies of the government concerned.

CHAPTER 10

Transport Aids

10.01 Transport aids, as referred to in this report, are any devices used to carry or facilitate the movement of goods or people. They may be motorized, animal-powered, or human-powered. The term *'conventional' transport aids* refers to those motor vehicles commonly used for transportation in industrialized countries. Such vehicles may be useful for long-haul and medium-haul transport on primary and seconfary roads in developing countries. However, they are not particularly suited to short distance transport on low quality access infrastructure due to their size and heavy axle loads, and to the low-volume transport requirements in rural communities [2]. Pickup trucks, minibuses and jeeps, are not considered as conventional transport aids in this handbook. These vehicles are well suited for use in many developing areas. The term *'alternative' transport aids* is used to refer to all traditional transport aids that are being used effectively in various countries throughout the world.

I. TRANSPORT AIDS' DESIGNS AND FEATURES

10.02 Traditional aids are those transport means adopted by rural farmers and residents in developing countries to help with their local level transport needs in the absence of conventional transport aids. Many of these traditional aids, such as pack animals and animal carts, are known, if not used, worldwide; others such as the South Korean chee-kee, Figure 1, a highly efficient traditional load-carrying backpack frame, are limited to specific geographical areas although their local use is extensive [10]. Figures 3 to 5 and Figures 40 to 55 of Technical Note 2.1 show a representative cross-section of various traditional transport aids. Technical Note 2.1, Transport Aids – Excluding Conventional Motor Vehicles, presents a comprehensive list of the various traditional aids used throughout the world.

10.03 Different modes of transport serve characteristically different movement demands and are seldom in direct competition [29]. Their applicability to different tasks and the way they are used in performing these tasks depends on a number of factors such as: locomotive power, load capacity in weight and physical dimensions, average operating speed and practical operating range. Similarly, their infrastructure requirements will vary in terms of grades that can be negotiated, travel way or tread widths required, clearing heights and widths required, bridge widths required, travel surface conditions and their ability to negotiate water crossings.

10.04 Technical Note 2.2, Features of Transport Aids, includes the load capacity, average speed, maximum trip length, and load width for each of a wide variety of traditional transport aids, as reported in various identified sources. Technical Note 2.3, Suggested Infrastructure Requirements for Transport Aids, presents suggested tread widths, clearing widths, minimum bridge widths, maximum and sustained grades, ford depths and surface requirements for a more complete list of specific transport aids. The research for this text included reviews and discussions of these various suggested features to produce a consolidated list of transport aids and the suggested criteria for the infrastructure to accommodate them. Technical Notes 2.3 and 2.4 (Para 10.05) are presented as general indicators of acceptable criteria and must be modified to suit the actual circumstances found for specific field conditions and transport aids.

10.05 The tread width/cleared width combinations in Technical Note 2.3 tend to fall into several fairly well-segregated categories – tread widths of less than 0.5 metre with a cleared passing width between 1.5 and 2.0 metres; tread widths of 0.6 to 1.0 metre with a cleared passing width between 2.5 and 3.0 metres; and tread widths of approximately 2.0 metres with a cleared passing width between 4.0 and 5.0 metres. In order to define a classification of paths, trails, and tracks for use in the rest of this handbook, these width requirements were combined with the other physical or geometric requirements for the types of transport aids listed in Technical Note 2.3 that can be accommodated within

Fig. 1 *South Korean chee-kee*

the specific width criteria. Technical Note 2.4, Suggested Criteria for Various Types of Infrastructure, was derived in this manner. It lists general geometric criteria for paths, trails, and tracks. The transport aids that can operate over each type of infrastructure, and therefore on all of the higher infrastructure classifications, are listed under the appropriate classification. Any modifications to the infrastructure's general geometric criteria necessary to permit a specific transport aid to operate on that class of infrastructure are listed in parentheses on the transport aid line of Technical Note 2.4 under the appropriate geometric criteria column. For example, although a bicycle can be operated on a path, the allowed sustained grade for the path must be reduced from 20% to 4% to accommodate bicycle usage.

10.06 Using the same type of tread width/passing width analysis, single-lane roads are considered, in this handbook, as requiring a tread width of 3.0 to 4.0 metres with passing bays at suitable intervals (Technical Note 4.01). The recommended total width of the road at passing bays is 6.0 metres (Figure 2). Two lane roads need a tread width of 5.5 to 6.0 metres. The tread width of earth and gravel roads is considered as the full platform width.

10.07 Surfacing for either width can be either soil or gravel, depending on the characteristics of the available soil. Some Asian countries pave all feeder roads in flood plain areas. In these areas they often pave one lane on a two lane platform and permit passing at any location. This handbook recommends a sealed surface width of 3.5 metres on a 5.5 to 6.0 metre platform in such cases (Figure 2) when the full width of the platform is capable of carrying passing vehicles.

10.08 The following rules of thumb for feeder roads are suggested:

(i) Single lane roads with passing bays for traffic of less than 50 vehicles per day (vpd). This figure is based on traffic volume only. Variations in construction costs may alter the vpd breakpoint between single and two lane roads.

(ii) Earth surface roads (Para 10.07) are considered viable up to a vpd of 100. The actual vpd breakpoint between earth and gravel surfaces and gravel and sealed surfaces should be calculated from the benefits derived from the better surfacing (Technical Note 7.3).

(iii) The upper vpd limit considered in this presentation is 200. The Asian single lane paved section noted in Para 10.07 is recommended for vpd below this limit. Paving, e.g., sealing, any road with a vpd less than 100 is unlikely to be justified by vehicle operator savings. Therefore, other justification, such as environmental or material considerations, is required for paving roads with a vpd of less than 100.

II. CHOICE OF ALTERNATIVE TRANSPORT AIDS

10.09 As discussed in Chapter 9, the levels of accessibility are dependent on two principal factors, the types of transport means available and used in the area, and the features of the infrastucture over which they operate. Different infrastructure features affect some transport means more than others. Some examples of the different applications of transport aids, and the effects of infrastructure, illustrate these points.

10.10 In India [62] human porterage is the common transport mode in rugged areas where accessibility is limited by hills, ridges, ravines and valleys. Indian porters carry loads ranging from 20 kg for women to 30 kg for men, over distances up to 25 to 40 km per day. In the same conditions, horses and mules carry from 100 kg to 150 kg. Camels normally carry 200 kg to 300 kg in flat areas

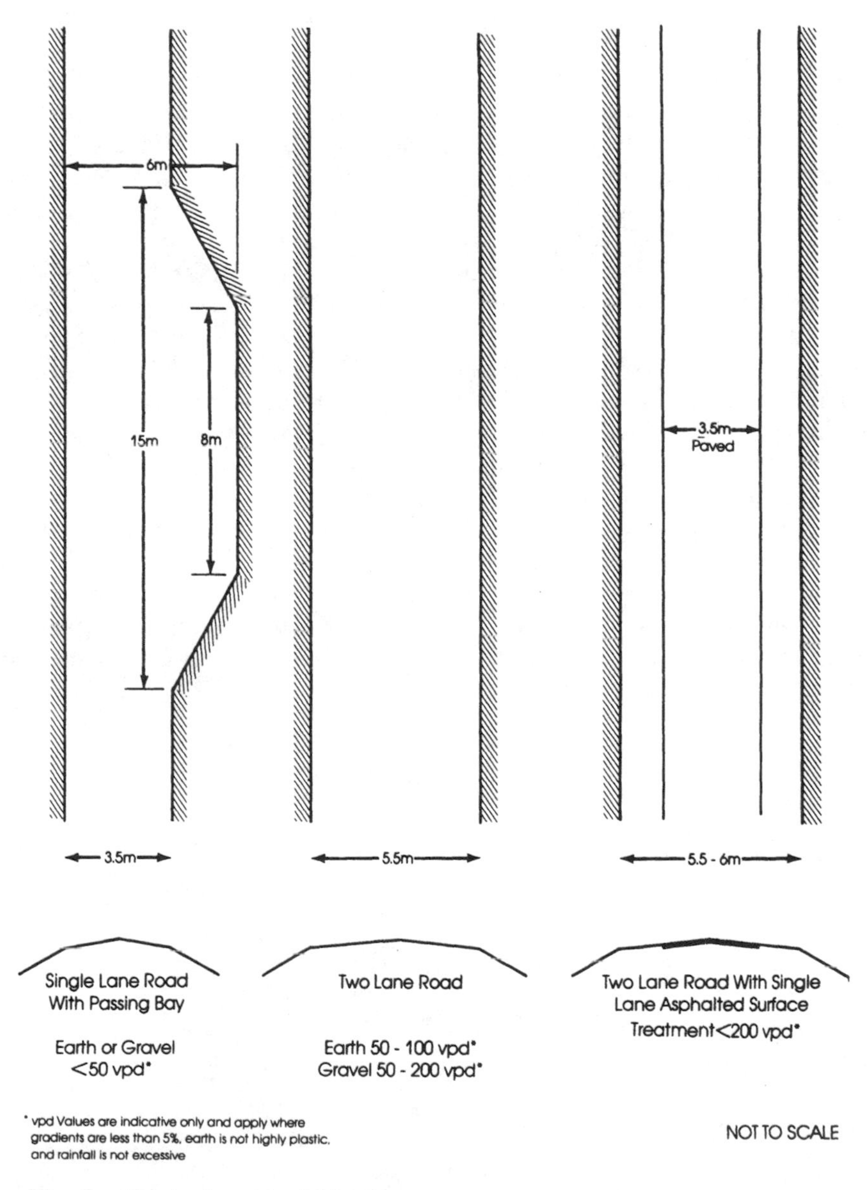

Fig. 2 *Typical road widths for access*

where the soil is light and friable. However, they have trouble negotiating wet sand, so donkeys, which normally carry 50 kg, are more commonly used in these areas during the rainy seasons. Bicycles in India [62] are used to carry up to 80 kg of compact, non-bulky commodities. Indonesians carry up to 150 kg on strengthened bicycles [67]. At times when the loads are heavy or the slope is steep, the bicycle is used successfully like a barrow, with the loaded bicycle being pushed rather than ridden (Figure 3). Three-wheeled cycle use in India (Figure 4) is mainly limited to relatively flat urban areas [62]. Indonesians [47] favour motorcycles in remote areas even when conventional vehicular access is feasible, because of the very limited fuel supply and limited capacity for mechanical repair. Animal carts can be pulled by most types of working farm animals such as bullocks, buffaloes, cows, mules, donkeys and light horses [29]. Asian bullock carts with two large-diameter (1.0 to 1.75 m) wooden wheels have capacities of about one ton when pulled by two animals. However, the steel rims on the wooden wheels tend to damage surfaced roads because of the high contract pressure on the narrow rims. Wooden wheeled cart performance on soft or muddy soils is superior to that of carts with rubber tyred wheels. Technical Note 2.5, Improved Animal-Drawn Vehicles (ADV), outlines the preferred choices and applications of different bullock cart wheel types in India.

10.11 The overwhelming preference in India [62] is for cheaper transport modes with relatively low capacity, to which the farmer already has access. Meagre earnings and lack of surplus cash make it impossible for the average farmer to take advantage of economies of scale and profit from hiring motorized bulk transport, except for cash crops such as onions, cotton, tobacco and sugar cane, which are either perishable or relatively valuable. Indian farmers have shown a decided preference for transport modes which incorporate the multipurpose use of available resources and which can travel on any type of surface, including rough farm tracks. This preference is not only apparent among the poorer farmers who use oxen for ploughing and pulling carts, but even among the more prosperous farmers who tow trailers (Figure 5) with two-

Fig. 3 *Bicycle as a goods carrier*

Fig. 4 *Indian three-wheeled cycle*

three- and four-wheel tractors, which are also used for their other farming activities [62]. Tractors in Nepal tow two axle carts carrying people for quite long distances [41] although tractors are not an efficient over-the-road transport vehicle [67].

10.12 Irrespective of the existence of alternative types of transport that may: (i) be applicable with less effort or cost or time, (ii) relieve the transport burden of farmers and other rural residents, or (iii) enable them to increase their production, these aids will not necessarily be adopted and used. A World Bank report [10] indicates that some current governmental policies, or lack thereof, such as availability of credit for purchasing transport aids, may be a major deterrent to their expanded use. Technical Note 2.6, The Supply and Quality of Rural Transport Services in Developing Countries: A Comparative Review, contains quotes from that report summarizing its authors' views.

Fig. 5 *Two-wheeled tractor with trailer*

III. INTRODUCTION OF ALTERNATIVE TRANSPORT AIDS

10.13 Innovations in a rural population's activities will only be accepted if the local inhabitants are personally motivated to adopt them. This is most likely to happen where the people receive actual or perceived social and/or economic benefits [62]. Introducing alternative transport aids and developing local rural access infrastructure to accommodate those aids, are no exception. Technical Note 2.7, Constraints to Existing or New Transport Aid Usage, lists technical, user, and institutional constraints and their possible solutions. A governmental agency sponsoring a transport aids progamme will need to eliminate such constraints to ensure the local population: (i) accepts the new transport aids, and (ii) has the capability and motivation to continue operating the improved transport system [31]. This is especially true since both the maintenance of the new transport aids, and the maintenance of the tracks, trails and paths essential to the successful operation of these aids, are normally the continuing responsibility of the local population, even when their construction is assisted or financed through non-local efforts. The necessary ingredients of a successful programme for introducing alternative transport aids, including several steps requiring overt governmental support, are suggested in Technical Note 2.8, Requirements for Developing the Use of Alternative Transport Aids [30].

10.14 Before any local industry is proposed for building and maintaining a new alternative transport aid, it also should be tested for local acceptance. When such industries require training artisans in crafts which were once popular but are now outmoded, the historical and social factors leading to their reduced status should be determined. For example, if the village blacksmith is no longer an honorable profession, it may be difficult to find candidates to learn wheel rim repair for animal carts. If the local donkey population is rapidly dwindling, any historical or social factors that may influence their acceptance as a draught animal should also be investigated before introducing a donkey cart assembly plant.

IV. PRACTICAL TRANSPORT AID OPERATIONS

10.15 In practical application, certain types of terrain and some types of infrastructure favour particular transport aids. Frequently, where various aids are in use, they are complementary to each other rather than competitive. In spite of such observations, a common assumption among some donors is that the preferred transport aid is a motorized vehicle where its use is practicable.

10.16 The basis for this assumption is the available evidence of the comparative costs of transport by different aids. Such cost data for the various aids are collected from actual applications of those aids. Motor vehicle costs have been widely evaluated [20, 26], while the operating costs of other transport aids have been the subject of but a few studies that are even less replicative than those of motor vehicles. Obviously, traditional transport aids carry smaller loads for shorter distances at slower speeds. However, when comparisons are made, they assume complete transferability and competitiveness in application. Frequently, motor vehicles cannot operate in the same locations as animal carts. If motor vehicles were used on infrastructure best suited for ox-carts, the motor vehicle performance parameters and costs would change significantly from the averages normally cited.

10.17 On low-volume roads and tracks, vehicle operating costs vary significantly from those incurred on conventional highways [47]. Fuel costs on tracks and poor earth or gravel roads are not a function of speed, since trucks are continuously being operated at upper levels of engine rpm. This high rpm rate is more constant than the actual vehicle road speed. At 10 kph the operational speed is maintained within first or second gear at high rpm, while under better conditions the vehicle will travel at 20 to 30 kph in a higher gear ratio using the same high range of rpm. Therefore, fuel consumption and engine wear are the same at either speed. Tyre life may only be 25% to 50% of normal tyre life under good road conditions, due chiefly to sidewall and carcass injury rather than tread wear [47].

10.18 Given the short distances travelled on most rural feeder roads and the labour-based methods of loading and unloading trucks, the fixed costs, such as depreciation, wages, insurance, etc., which are affected by the degree of utilization, are even greater penalties than the operating costs. Consequently, large trucks are seldom found operating in rural areas populated by poor or marginal farmers with small surplus produce to market, even when the roads are in good condition. Instead, the truck use in such areas tends to be limited to older, medium to small capacity vehicles that have a low capital cost, are easy to maintain even in rural areas, and do not need a capability for high speed travel [47].

10.19 Local passenger transport is often provided by minibuses, small converted goods carriers, and mixed purpose goods/freight vehicles, rather than the conventional buses of the type that travel the major traffic corridors. Because of the frequent repair needs due to local road conditions, the extra repairs caused by overloading the types of minibuses, trucks and converted goods carriers in local use on rural roads is usually not a deterrent to such practices [47]. Multipurpose transport aids, such as animal carts and tractor trailers, are used to carry heavy or bulky material not requiring fast transport, and involving long collection, loading, and unloading times that negate the use of high investment transport vehicles.

10.20 Technical Note 2.9, Illustrative Comparison of Transport Costs Using Optional Aids, presents a chart of hypothetical comparative costs of three types of transport with typical operational parameters. It shows that under circumstances that force reduced loads on trucks or pickups such as a muddy or slippery surface, a narrow speed-restricting infrastructure, etc., and reduced annual utilization, the typical costs per ton-km for animal carts are likely to be lower than those of motorized vehicles. Although no field research was uncovered to either validate or refute the illustrative example in Technical Note 2.9, the implication is that there may be good economic reasons for adapting and using lower technology transport aids in many remote rural areas.

V. CONCLUSIONS

10.21 Transport aids are any motorized, animal-powered or human-powered devices used to carry goods or people. Alternative modes of transport use transport aids other than conventional cars, trucks and buses. Their use in developing countries holds the potential of both providing a much needed transport capability for the rural poor and reducing government investment in future infrastructure requirements needed to improve the standard of living of the rural poor.

10.22 A wide variety of proven traditional transport aids exists in various parts of the world which can and do serve valuable roles in the lives and economies of the people that use them. The technology of these aids is readily transferable to other places and countries where there are wide, uneconomic gaps in the spectrums of available and used means of transprort. However, the introduction of suitable transport aids into new areas will require both governmental support and positive direct action.

10.23 Governments must first take an active role in identifying the real transport and travel needs of the rural small farmer who lacks physical or financial access to motorcar, truck or bus service. The adequacy of the transport means currently available to this target group must next be assessed, along with the extent, and appropriateness, of the existing infrastructure for these transport aids.

10.24 If such an investigation shows the need for the introduction of either improvements to existing transport aids or new transport aids, a systematic evaluation must be undertaken of both the selection of the innovations and the resource commitments required. Not only must the required skills and technology for manufacturing and/or assembling the aids be promoted, but the skills for maintaining and operating the aids must be developed, and the proper infrastructure provided. These resource investments must be evaluated within the framework of existing government regulations and credit policies.

10.25 Once these planning obstacles have been overcome, and a policy commitment made to invest sufficient resources to assure provision of an alternative transport aid, the government must systematically introduce the concept of the new aid to the target population. Although the initial evaluation should have considered the likelihood of local acceptance of the alternative transport aid, final acceptance of any innovation will require a programmed promotional activity to convince the local farmer that the aid will benefit him/her economically while incurring no social, ethnic or religious disadvantages.

CHAPTER 11

Infrastructure Aspects

11.01 A country's transportation infrastructure network consists of all the transport-related civil works within that country. It includes airfields, railroad facilities, ports and harbours, and all roads, tracks, trails and paths. The transportation infrastructure being considered in this handbook is chiefly concerned with that portion which services low volumes of motorized, animal-tractive and human-tractive forms of transport. However, the planning and expansion of this network must be carried out with proper regard to the other highway, air, rail and sea components that make up the overall transportation system. The priorities and procedures for carrying out such general transport sector planning and intermodal coordination are not included in this book as they are adequately dealt with in numerous other sources [4, 5].

I. INFRASTRUCTURE PLANNING

11.02 Rural tracks, trails and paths have historically been omitted from national consideration of transport infrastructure because they are often considered as a separate, unidentifiable network which is outside the national interests in transportation. Even as concern for the rural poor has increased, many governments have attempted to improve rural transport mainly through the increase of conventional roads for motorized traffic. The track/trail/path segment of the infrastructure has very seldom been properly acknowledged or recorded, in part because it is so difficult to gather such data. However, a major reason for this neglect is that roads and vehicles can be provided by periodic injections of capital, often through multilateral or bilateral aid projects. Non-motorized transport does not attract such inputs. The promotion of transport aids and their relevant infrastructure is therefore likely to escape the notice of planners concentrating on national development activities that have attracted outside donors. This neglect is becoming a critical factor in rural development. For the large numbers of farmers and other residents living in the rural areas away from the motorable road system, the vast, unmeasured network of paths, trails and tracks are their lifelines. This part of the system is as crucial to the existence and survival of its users as are motorable roads to their users.

11.03 As with decisions for most types of investments, planning and analysis of rural transport problems needs to be considered at two separate levels. At one level, discussed below, it is necessary to look at the overall network within an area, taking account of existing settlements and service facilities, terrain and topographical features, proposed developments, and the overall levels of economic and social activities. At the other level, discussed in

Chapter 12, it is necessary to look at individual links in the system and their features, problems and usage.

11.04 The analysis, review and planning of the overall transport network in rural areas is better dealt with in some developing countries, mainly in Asia, than in others. The system most used by those countries that carry out such overall physical planning, is one called the 'Plan Book System' described in Technical Note 3.1. This system records, by a simple process of maps and transparent overlays, the physical features of the area such as the various types of infrastructure and some of their more important general features related to levels of accessibility. Such general features include: significant terrain features; locations of villages, settlements, towns and other population centres; general agricultural activities; and locations of service facilities such as schools, health services, markets, cultural and administrative centres, and other relevant services. Overall decisions about the infrastructure network and its hierarchy of different types of infrastructure, new connections, relocations, detour routes and other general physical planning, can be considerably simplified when all of the necessary data are displayed at one time. Various solutions may be drawn on different overlays to compare their suitability for solving specific access problems.

II. PHYSICAL ENVIRONMENT

11.05 Physical environment has an effect on all rural access infrastructure. This effect is acknowledged in the geometric criteria used for conventional highway design. For specific road classifications, highway geometric design criteria such as minimum vertical and horizontal curvature and maximum allowance gradients are defined as a function of specific design speeds which are in turn specified by terrain classifications. Most highway design criteria are divided into three or more terrain classifications: normally flat, rolling and mountainous. The rougher the terrain becomes, the less restrictive are the design criteria; for example, design speed parameters such as the minimum curvature are reduced and the maximum allowable slope is increased, because of the rapid increase in construction costs for a given standard of road in progressively more difficult topography. Design criteria, of course, are governed by the topography within the route alignment, not the terrain in the general area [73]. As indicated in Technical Notes 2.3 and 2.4, paths, trails and tracks also have minimum geometric configurations; however, they are not regulated by the design speed concept. Chapter 13 suggests formal design speed criteria be waived in the early stages of feeder road development.

11.06 Terrain constraints also limit the infrastructure's zone of access. Physical obstructions such as steep slopes, waterways, swamplands and bodies of water, normally delineate the effective boundaries of the zones of access of rural infrastructure. In addition to these physical obstructions, the limits of the zones of influence may also be based on the distance to a parallel road or on considerations of travel time and/or distance travelled. Insofar as the terrain combined with the features of the access infrastructure directly affect travel speed and time, they are important determinants of the limits of the zones of influence. Research for this book indicates that a currently widely used determination of the limits of such zones of influence in remote areas is about 2 to 3 km on each side of a road where physical barriers do not restrict access. Even where all off-road transport is pedestrian, these limits are lower than the previously popular one half-day travel criterion which, according to Technical Note 2.2, would expand the limits to 6 to 12 km. This text accepts this larger zone of influence as a practical goal in future project evaluations.

11.07 Zones of influence of course do not end abruptly at a specified distance from a road if there is no well-defined physical constraint. The terrain type, soil, vegetation, local habits, etc. influence the people's willingness to travel further to the road. The social and economic effects of a road taper off in a non-lineal manner as the distance from the road increases. For ease in estimating benefits, the distance from the road to the limit of the zone of influence is usually determined by modifying the tapering effect and then assuming the benefits will affect 100% of the population within the modified limits.

11.08 The introduction of alternative means of transport, along with its related suitable infrastructure, can enlarge the effective limits of zones of influence both by making transport easier over a given terrain – for example, introducing animal carts increases the transport speed only slightly but permits larger loads to be transported with less effort by the farmer – and by providing access for transport aids across physical barriers such as streams, that otherwise limit travel. Each transport aid has its own terrain restrictions, maximum trip length capability, and associated infrastructure construction costs, as noted in Chapter 10. However, the use of transport aids and their related infrastructure has the effect of extending the limit the local population is willing and able to travel to enjoy the benefits derived from the presence of a road. This additional mobility in turn increases the theoretical zone of influence of the road.

III. CONCLUSIONS

11.09 The ultimate goal of a transport infrastructure is a balanced network of various components that meet, but do not exceed, the needs of the population in the most cost-effective manner. The weakest links in many developing countries' transportation networks are the smallest branches of the network tree, the low-volume feeder roads and the tracks, trails and paths that join the rural population to those roads. It is important that the planning and decision process for rural transport investments takes account of the overall situation of the targetted area. At the level of rural access planning described in this handbook, the assumptions as to the limits of the zones of influence are a critical factor. These assumptions should be based on the terrain and topographical features; the location of settlement and population centres, and existing service facilities; and the network of roads, tracks, trails and paths that will make up an area's transport infrastructure. For this purpose, a planning technique such as the one described in Technical Note 3.1, has been found to be a valuable tool in the countries where it is used.

CHAPTER 12

Infrastructure — Individual Links

12.01 The level of accessibility for particular transport aids making specific trips is influenced by a number of features of the actual infrastructure route selected for the trips. The features that impinge on accessibility are discussed under the headings of geometrics, soils, climate and drainage. Each operational constraint to access is a physical barrier which denies access to specific transport means like walking, pack animals or motor vehicles. These constraints normally consist of one or more of four types. The first type is water-related or environmental constraints, which can completely deny access or limit access only during rainy periods, as described in Paras 9.17 to 9.19. The other three types of physical barriers – steep grades, sharp curves and inadequate width – either allow access all the time, except as modified by environmental constraints, or not at all.

I. GEOMETRICS

12.02 Chapter 11 presented information about the interrelationship between the general features and development of an area, the network of rural access infrastructure and accessibility. The physical dimensions and shape of the infrastructure also affect the ability of a particular transport aid or vehicle to traverse it. These features include tread widths, clearing widths, gradients and curvature. For the types of infrastructure of prime interest in this handbook, superelevation and curvature are not critical because traffic will tend to be low in volume, weight and speed. Therefore the principal consideration is whether the transport aid can negotiate the route safely when operated prudently. Chapter 10, Transport Aids, presented information about treadway and clearing widths and gradient limitations for various transport aids. Technical Note 4.01, Modified Geometric Design Criteria, presents some background discussion and rationale for these paratmeters.

II. SOILS

12.03 As indicated in Para 9.19, soil conditions, rather than soils *per se*, cause access constraints. However, constraints due to a soil's physical properties such as low wet strength, lack of cohesion, slipperiness and impermeability are sometimes neglected when low-volume rural access infrastructure is considered; either because of lack of trained personnel or the high cost of standard soils evaluation testing. Relatively simple tests of the types described in Technical Note 4.02, Simplified Soil Sampling Procedures, have been developed for the purpose of identifying the pertinent properties of soil materials. These tests can be conducted by field personnel and do not require expensive equipment. Their reliability is sufficient to alert technically-oriented people to the physical properties of a soil which, under adverse conditions, may cause transport constrains on rural access infrastructure.

12.04 Soft or expansive soil may require a granular surface to improve its structural strength when the soil is wet (Figure 6). Granular material may also be placed on top of a subgrade, i.e., the surface of a naturally occurring soil, to solve traction problems. When traction is a soil's only accessibility constraint, that is, when the surface gets slippery but not soft during a rainfall, the subgrade's structural strength is of secondary concern and the applied thickness of granular material is not critical. Technical Note 4.03, Size of Surface Material, notes some granular size considerations for surfacing material. Technical Note 4.04, Surfacing Material Thickness, discusses the amount of granular material to be applied when such surfacing is intended to improve the structural strength of infrastructure carrying motor vehicles. However, it is often more economical to locate, or relocate, rural access infrastructure on more favourable soil than to import borrowed granular materials to an existing alignment. Vehicle operating cost savings for the low traffic volumes found on many feeder roads usually cannot offset the increased construction costs incurred by locating a shorter alignment on poorer soils.

12.05 Many tropical regions have deposits of latteritic gravels which can generally provide a surfacing of good mechanical stability. Other regions often have suitable deposits of alluvial gravels or decomposing rocks [51]. Detrital gravels and sands are often found in low rainfall areas, i.e. less than 750 mm annually. These include; (i) granite, gneissic and quartzitic materials; (ii) sandstone gravels and sands; (iii) doleritic and basaltic materials; (iv) micaeous

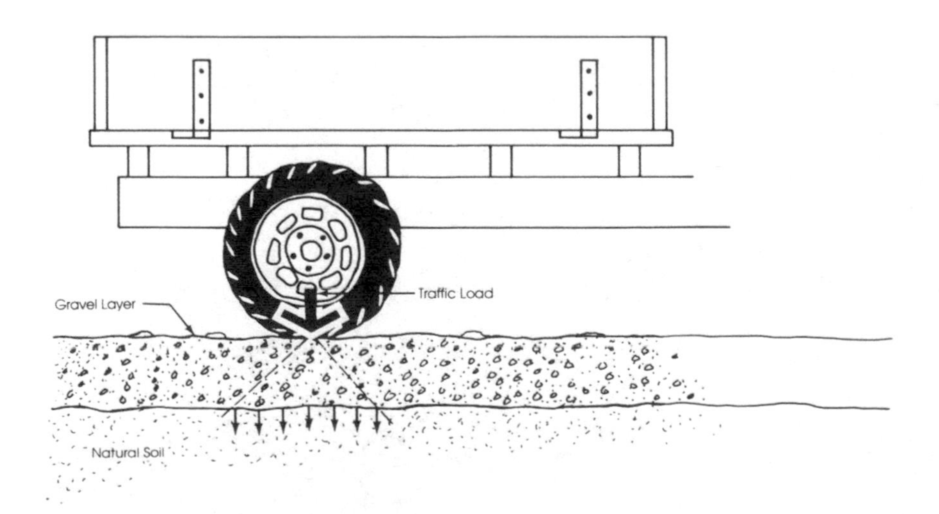

Fig. 6 *Load dispersing granular surface*

and schistose gravels; (v) materials derived from shale; and (vi) limestone gravels [79]. Quartzitic gravels and sand-clays occur in areas of intermediate and high rainfall areas and are most often found where the annual rainfall is between 750 mm (30 in) and 1000 mm (40 in) [79]. Nodular lateritic and calcareous gravels occur most often in sloping areas where drainage is impeded and evaporation predominates as indicated in Technical Note 4.05, Rainfall and Soil Formation.

III. CLIMATE AND DRAINAGE

12.06 Climatic effects influencing transport conditions are mainly rainfall, temperature and wind velocity. In tropical countries, climatic conditions may vary considerably. Ghana, for example, has four climatic regions: a coastal savannah with a rainfall of 600–750 mm/year, a rain forest zone with a rainfall of 1750–3000 mm/year, a forest zone with a rainfall of 1250–1750 mm/year and an interior savannah with a rainfall of 1000–1250 mm/year [79]. Temperature variations, due to altitude and location in relation to water bodies, and wind velocities which affect drying and erosion of fine materials, can similarly vary within a given country. Climatic conditions are quite site-specific; each infrastructure project, and sometimes even each infrastructure link, is influenced by climatic conditions quite differently than similar infrastructure in other nearby locations.

12.07 Rainfall is often the most important climatic factor affecting low-volume rural access infrastructure. Rainfall intensity of 25 mm/hour is generally considered as the threshold intensity for erosive destruction, [51, 70], although lesser intensities can cause erosion when the runoff flow is concentrated in constrictive channels (Figure 7). Unfortunately, very little data for

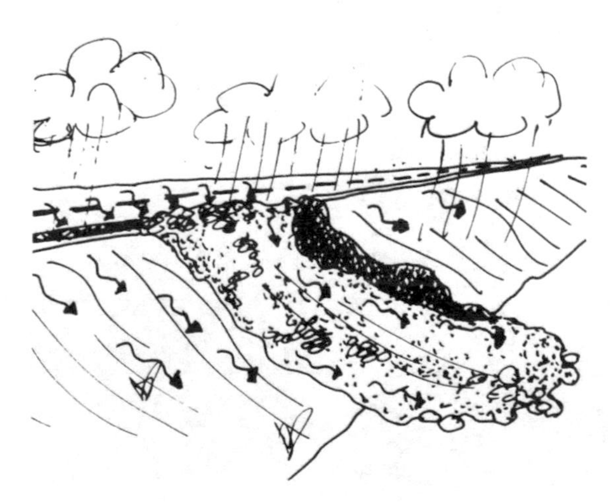

Fig. 7 *Slope erosion from heavy rainfall*

rainfall intensity is available in most developing countries although total daily, monthly and annual rainfall data are frequently recorded. Erosive rainfall intensity in temperate zones is thought to be about 5% of the total rainfall amount, while in tropical zones it is thought to be about 40% of the total rainfall [51, 70] (Figure 8).

12.08 The historical collection of general rainfall data covers a shorter period in many developing countries than in the majority of developed countries where most of the so-called drainage quantity formulae were developed. This lack of data seriously limits the value of the hydrologic analysis of drainage area run-off and stream-flow characteristics. Furthermore, many accepted principles of hydrology were developed for large flood-control or water-supply projects involving extremely large watershed areas. Variations that are insignificant in large watershed areas become very important in small drainage areas. Engineering judgement and approximations must, therefore, be applied to the basic principles of hydrology when smaller drainage areas are being analyzed [81]. Technical Note 4.06, Field Evaluation for Drainage Structures, describes some of the judgements that may be required in the field.

12.09 Hydrological analysis is approximate, and the hydraulic design of drainage systems to carry the maximum flow possible, even if it could be determined, has long been accepted as uneconomical. Consequently, all drainage design inherently contains an accepted risk of occasional failure. The risk factor can be defined as the probability that a given runoff quantity, i.e., a volume of water measured in cubic metres/second, will be equalled or exceeded at least once in some given period of years.

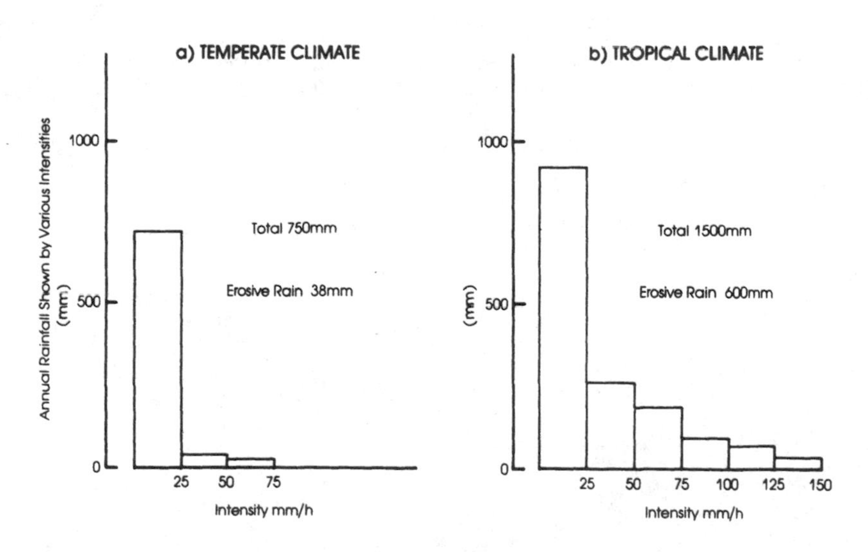

Fig. 8 *Comparison of typical annual rainfall distribution for temperate and tropical climates*

12.10 Because of the acknowledged inaccuracies in hydrological evaluation and the non-life-threatening nature of small drainage area component failure, a one- or two-year storm as described in Technical Note 4.07, Design Storm Definition, for drainage ditches and small culverts seems to be adequate for low-volume rural access infrastructure. More costly drainage facilities warrant larger design storm selection. However, enlarging of underdesigned ditches or replacing small culverts showing early signs of failure is normally a much less costly alternative then the unmeasurable overdesign of small drainage systems due to the selection of large design storms for all drainage features. The choice of design storms must, of course, be weighed against any unusual effects from possible failure such as danger to life or extensive damage to adjacent areas, and against the availability of alternate routes or the capability to make timely repairs.

12.11 The importance of drainage-related features is paramount during early construction phases. At the low traffic volumes considered in this handbook, environmental damage is the greatest threat to the satisfactory operation of any improvement. Environmental damage occurs in three forms: erosion of the infrastructure itself; cut slope slides where excavation interferes with the natural drainage patterns; or as structural failure due to traffic action on surfaces softened by retained water.

12.12 Drainage improvements must include consideration of the interrelationship of flowing water and soil characteristics with each feature of the infrastructure alignment and cross-section [70] (Figure 9). Flowing water includes both water that flows in varying volumes throughout the year in live streams and rivers, and runoff of rain water that falls on or near the route. Since the quantity of water that must be handled is not a function of traffic type or volume, drainage design methods are the same for all types of infrastructure [81].

12.13 The travelled surface must be sloped to remove standing water as quickly as possible (Figure 10) to prevent structural failure, i.e., softening, and loss of traction, i.e., slipperiness However, excessive cross slope on the

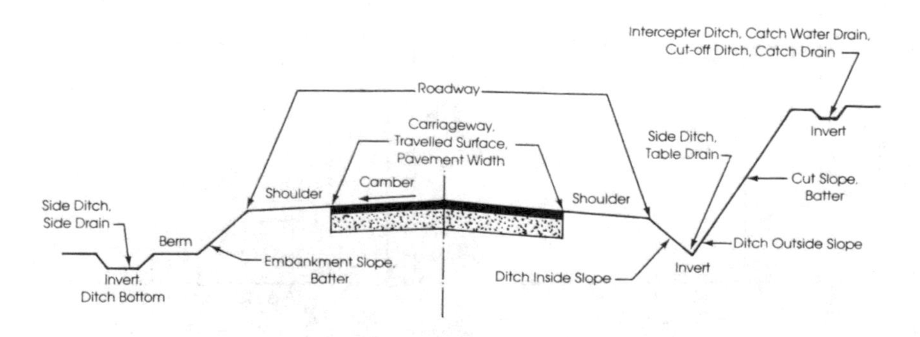

Fig. 9 *Cross-section drainage terminology*

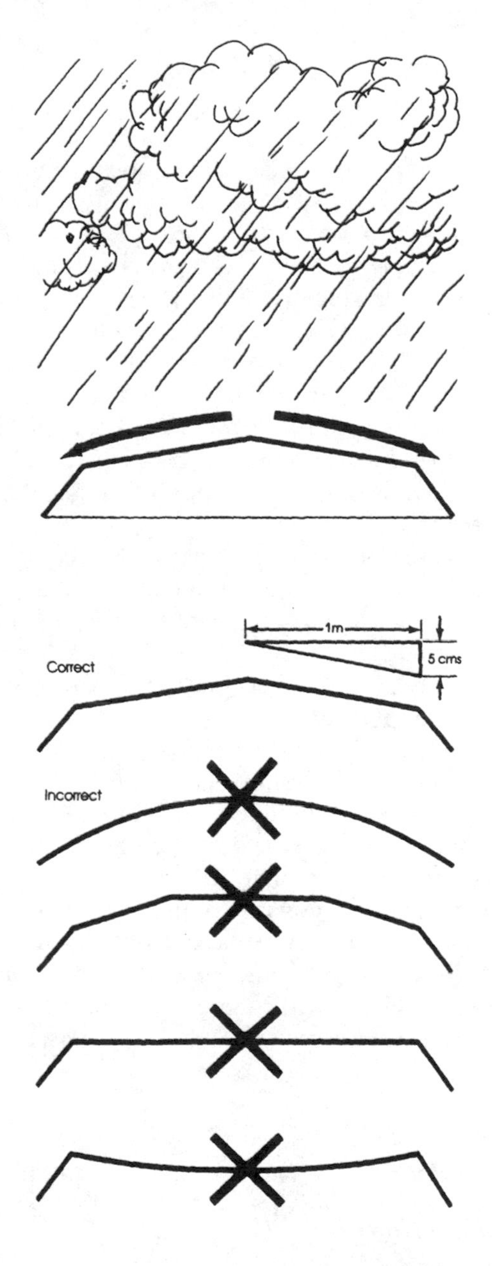

Fig. 10 *Proper camber configuration for drainage*

travelled surface will cause surface erosion. Longitudinal travelway slopes also erode. Technical Note 4.08, Controlling Travelway Erosion addresses the problems of surface erosion on all types of rural infrastructure. Technical Note 4.09, Surface Water Diversion Spacing, suggests suitable spacing for water diversion locations on paths, tracks and trails. The same spacing is suitable for trapezoidal ditches.

12.14 Roadway side ditches, side drains or table drains, often have an invert or bottom slope that is parallel to the gradient of the roadway, although in flat areas it may be necessary to slope the invert down towards its outlet to make the water flow. Drainage ditch invert slopes should normally be 1 in 100 or steeper. Since feeder road side ditches are often one standard size, their capacity is limited. Ditch relief systems are therefore required to divert the water from the ditches before they overflow, erode, or accumulate silt. Such systems allow the water flowing in a ditch to: (i) pass under the travelway from an uphill side drain (Figure 11); (ii) cross the travelway in a controlled location, i.e. a dipout; or (iii) flow from the ditch onto the surrounding terrain for dispersal, i.e. a turn-out or mitre drain. In this presentation, interceptor ditches are also considered as an element in the ditch relief system because one of their functions is to reduce the water a roadside ditch must handle. Technical Note 4.10, Drainage Relief Systems, further details and depicts of the elements making up such relief systems.

12.15 Stage construction of a drainage ditch relief system may consist of putting in a culvert under a road in conjunction with a modified dipout for a period of time before the road is scheduled to receive granular surfacing [70]. This technique (Figure 12) not only reduces costly culvert overdesign, but also reduces the probability of a road washout if the culvert overflows during the 'review and correction' period.

12.16 Timber or log culverts, masonry culverts and pipe culverts can be used for both small live water crossings and in ditch relief systems on motorable roads. Timber, log and masonry culverts, in many cases, can utilize both local resources and technology. Two types of pipe are usually used for culverts on rural roads in remote locations. The first is concrete pipe. Many countries already have a manufacturing industry for precast concrete products which can supply concrete pipes suitable for use as culverts. Since the shipping cost of a unit of precast concrete pipe to a remote location can be substantial, it may well be worth while to establish a local concrete pipe manufacturing site. Both the quality of the concrete and the casting of the pipes must be carefully controlled

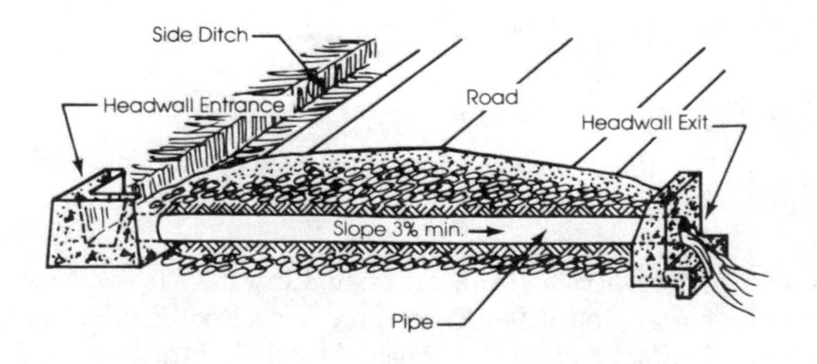

Fig. 11 *Concrete pipe ditch relief system*

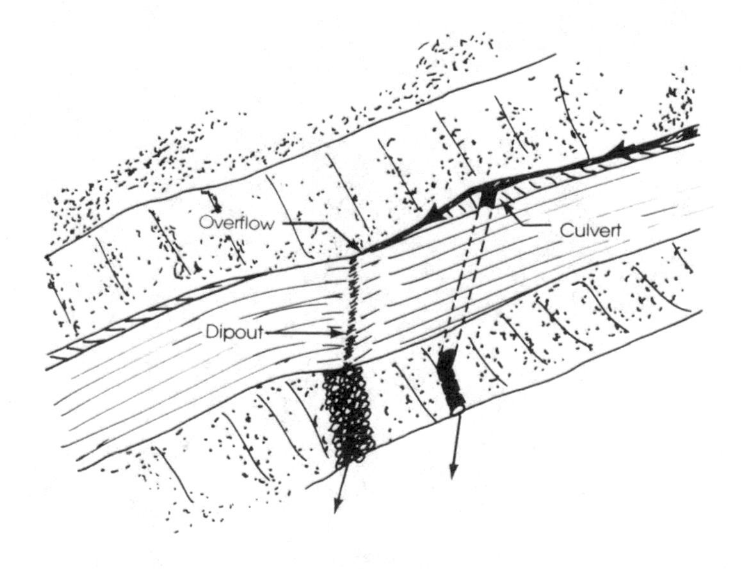

Fig. 12 *Dipout provides safety overflow*

in order to avoid the risk of breakage during installation and under traffic loads. The cost of these controls must be added to the base cost for manufacturing and installing such concrete pipe. Nestable corrugated metal pipe, the second type, often requires foreign exchange and a long lead time for dependable delivery, but can be easily installed by unskilled labourers, is more tolerant of poor installation practices, and is reusable in most cases [81].

12.17 Fords and Irish bridges (vented fords) can be extended should the flows of streams and rivers be underestimated. But it is difficult and expensive to increase the flow capacity of large culverts and bridges. However the initial costs of providing such structures can be reduced by stage construction of different types of water crossings as outlined in Technical Note 4.11, Stage Construction of Water Crossings.

12.18 Compaction is imperative around all drainage structures to prevent destructive erosion as well as to prevent movement or deformation (Figure 13). Compaction is highly desirable on all disturbed soil to increase soil strength and to reduce subsequent settlement and deformation under repeated wheel or hoof loads. Such deformations produce ruts which retain water thus weakening the surface and promoting rapid deterioration under traffic and surface erosion along the travelled way (Figure 14).

12.19 When erosion occurs, the source of the erosive force must be eliminated at the time the eroded area is repaired to prevent repeated failures [70]. High maintenance costs usually reflect poor drainage design or execution on rural access infrastructure improvements. Horizontal and vertical alignment of existing routes should be retained wherever possible to preserve existing environmental balance and to prevent additional erosion problems [94]. When new alignment is required, the route should follow the existing terrain where-

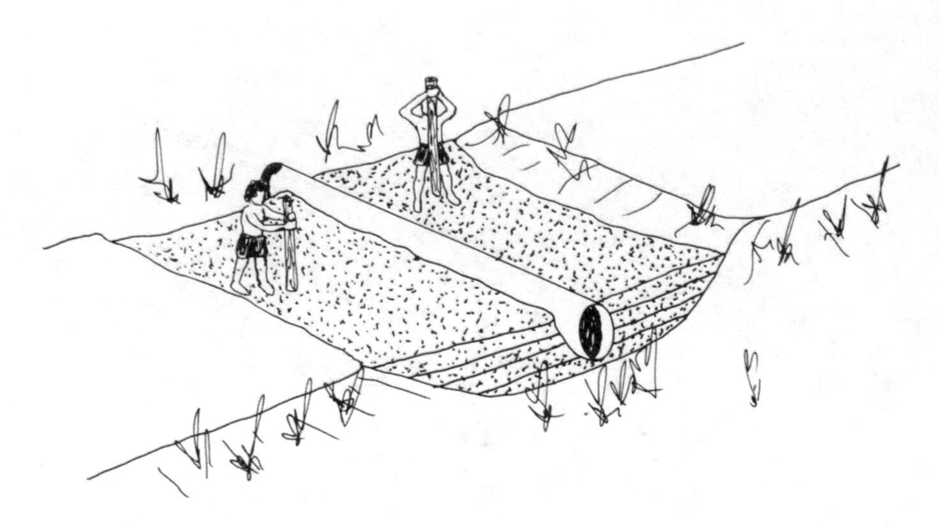

Fig. 13 *Compacting culvert backfill*

Fig. 14 *Longitudinal erosion gullies*

ver possible, and present minimum interference to natural drainage courses. Traditional transport aids often operate over lower quality infrastructure which is not only less costly to construct or replace, but is also less disruptive to the established equilibrium between existing topographical features and climatic conditions.

IV. INVENTORYING SPECIFIC LINKS

12.20 When rural access infrastructure inventories are required, they must include each road, track, trail and path that will be considered for improvement. Consequently, project planners must carefully define the types and locations of the infrastructure links whose improvement will satisfy the project objectives in order to limit inventory activities in heavily populated areas. Any traffic counts must also be expanded to include all types of traffic and transport aids currently using the specific infrastructure links. The inventories may be classified into three levels depending on the type of infrastructure being inventoried:

Level 1. Feeder roads and important tracks that may be substantially upgraded at some time during the project;
Level 2. Tracks and trails that may be substantially improved for the transport aids that they now accommodate, immediately or during some later development phase; and
Level 3. Trails and paths that may be in need of some improvement.

These infrastructure inventory levels determine the extent of the physical, condition, and needs inventories required. However, any inventory should include information about the location and nature of obvious weak points where traffic movement can be interrupted. Technical Note. 3.1 describes physical inventories in Para 5, condition inventories in Para 6, and needs inventories in Para 7.

12.21 A physical inventory for Level 1 infrastructure (Para 12.20) should record locations of all road features, including: surface type; surface and platform width, which are frequently the same for earth and gravel roads; locations of cuts and fills; side ditch locations; alignment changes; intersection locations; all drainage structures; and land use on each side of the road, including the location of borrow pits, streams, rivers, villages and other significant landmarks. A condition inventory may be made at the same time or may be undertaken as a later separate activity. The key to making any inventory is to amass data that is replicable, that is to record information of specific locations which can be found by later inventory crews, evaluators, and construction personnel. Therefore odometers used in inventory work should have a calibration factor which is applied to all results to give true distances. Technical Note 4.12, Recording Road Physical Inventory Notes, includes a method for determining the proper location of inventoried features.

12.22 Condition and needs inventories for Level 1 infrastructure (Para 12.20) may be recorded in a survey notebook or on prepared forms similar to the one shown in Technical Note 4.13, Recording Condition and Needs Road Inventories. This inventory requires more judgement than a physical inventory. It should be carried out by an experienced engineer, while a physical inventory can be undertaken by trained inventory technicians. When both inventories are combined into one operation, it must therefore be done by an experienced engineer or road supervisor. The normal sequence of inventories for Level 1 infrastructure would begin with a physical inventory of existing tracks and roads in the project area. When a development concept has been decided, the condition and needs inventory would be conducted on the candidate Level 1 road and tracks. This sequential approach has several advantages:

(i) the inventorying engineer can check the accuracy of the physical inventory;
(ii) the inventorying engineer is aware of the types of transport anticipated on each infrastructure link so that the conditions and needs can be evaluated in light of the proposed levels of access;
(iii) the inventorying engineer can concentrate on evaluating access constraints on the candidate roads; and
(iv) the inventory is carried out nearer the time when the improvements will be undertaken.

12.23 Inventories for Level 2 infrastructure need not be as extensive as the inventories described above. They can be conducted by vehicle if the route is motorable, otherwise a motorcycle or bicycle with an odometer may be used. If a link must be inventoried on foot, a pedometer should be carried. Longer links may be inventoried by horseback if mapping or air photographs are available that permit distances to be scaled. Since the improvements are likely to be small and dispersed, the physical inventory should be minimized and the condition and needs inventory limited to locations where access is a problem; therefore the physical, condition and needs inventories should be combined after a preliminary determination of the link's future status has been made. Technical Note 4.14, Level 2 RAI Inventory Form, is an abbreviated version of the Level 1 Form shown in Technical Note 4.13. The only added feature is a

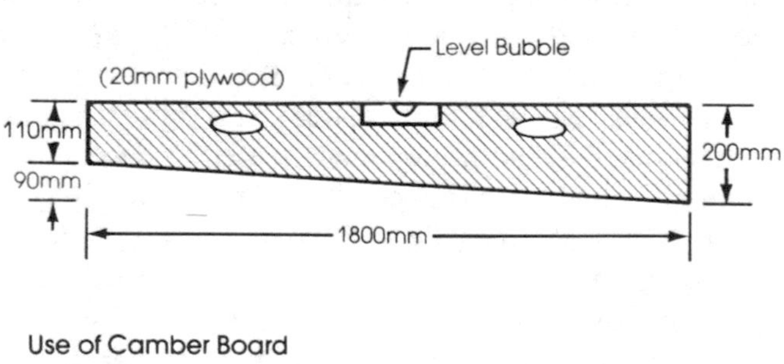

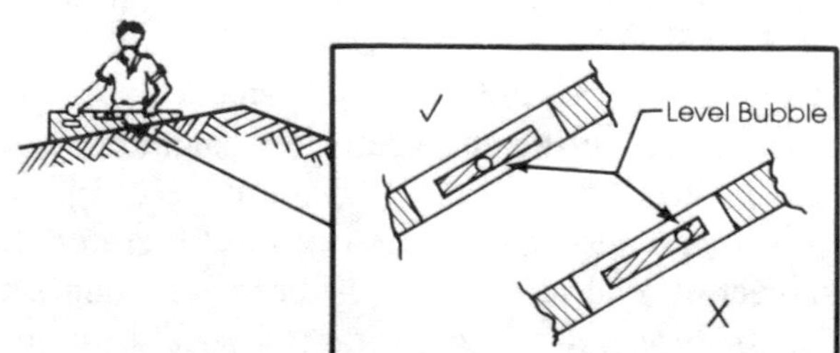

Fig. 15 *Using a camber board*

subdivision between features preventing reliable and minimum access (Para 9.26). Each transport aid considered may be affected differently by specific features. Consequently any aid-specific access constraints should be so noted on the form.

12.24 Inventories for Level 3 infrastructure may be presented in even less detail, although they still require someone to travel the route. The inventory may be presented in narrative form but should include the following topics: (i) route description including terrain, length, average treadway and cleared widths, and soil comments; (ii) types of transport-aids being used and being considered for use; (iii) location, length and description of any access constraints listed by individual transport aids, subdivided into reliable and minimum accessibility; and (iv) suggested solutions for each access constraint listed in item (iii). Locations may be listed by using references to existing features, for example: additional one metre of clearing for pack animals required from 500 m south of third stream crossing to 250 m north of fourth stream crossing, a total distance of about 840 m.

12.25 All condition and needs inventories described in this section are concerned with identifying the geometric, soils, climatic and drainage factors described earlier in this chapter which hamper or prevent accessibility. The use of (i) a camber board, which is a carpenter-type level mounted on a template of the proper slope (Figure 15), (ii) a hand auger to determine subgrade material and groundwater depth, (iii) a hand level, (iv) an engineer's tape, and (v) two or three soil screens, will greatly assist an engineer in his or her efforts. However, such aids are required less and less as an engineer gains experience and practice. Inventory time constraints must be flexible because if surface flooding is a suspected cause of access problems, the infrastructure must be viewed during a heavy rain. The best time to undertake an investigation of needs for spot improvements is during the rainy season, but poor access problems may dictate a first review during a drier period with a return visit only to any questionable areas.

12.26 The existing road network beyond the project area must also be evaluated, even if it is not formally inventoried, to make sure that overall network accessibility is at least as good as the improved accessibility within the project area. Common sense indicates that investment in reliable access, rather than in minimum access and improved storage facilities, is wasted if transport to and from the project area is impossible throughout part of the year.

V. RELOCATION OF EXISTING EARTH ROADS

12.27 In spite of all attempts to improve accessibility by spot improvements to existing infrastructure, some circumstances will require the relocation of existing earth roads to improve access reliability by eliminating environment contraints. Such relocations should be limited to locations where they would provide a lower cost solution than would improving the access infrastructure in its existing alignment. Thus, relocations may be appropriate where the existing alignment is [58]:

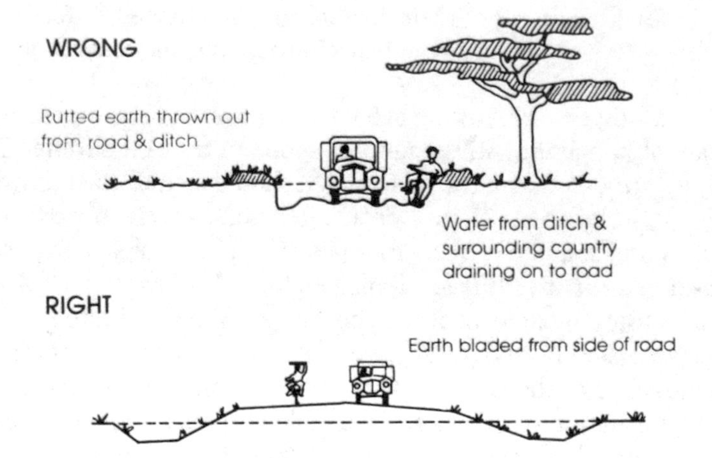

Fig. 16 *Typical sunken road*

(i) in an area of frequent flooding;
(ii) in flat terrain and below ground level, causing water from the ditches and surrounding land to drain onto the road (Figure 16);
(iii) crossing a curved water channel where erosion may attack the outer channel bank;
(iv) located at a wide stream crossing, unless the additional width permits installation of a suitable ford;
(v) on an earth surface overlying very expansive clay;
(vi) on a side hill location where the hillside shows signs of instability;
(vii) located where widening will require excavating into a hill-side where artesian springs are present; or
(viii) located where widening will require cuts or embankments in sand.

12.28 Relocation of existing earth roads to eliminate major operational hazards should be limited to locations where the existing alignment [58]:

(i) has very steep grades;
(ii) has very sharp curves; or
(iii) is on an existing embankment or in an existing cut which has insufficient width and widening would be more expensive than realignment.

12.29 Such earth road realignments should be staked out in the field to permit suitable right-of-way taking for any staged improvements. Drainage ditches should be staked and their profiles set with a hand level to assure proper flow. If the alignment is in flat terrain, the road surface should be above the existing ground, using side borrow if possible (Figure 17). If an engineer is reluctant to eyeball curves on earth roads, curves can be laid out for a 30 to 40 kph design when such curvature does not add to the construction cost. Speeds

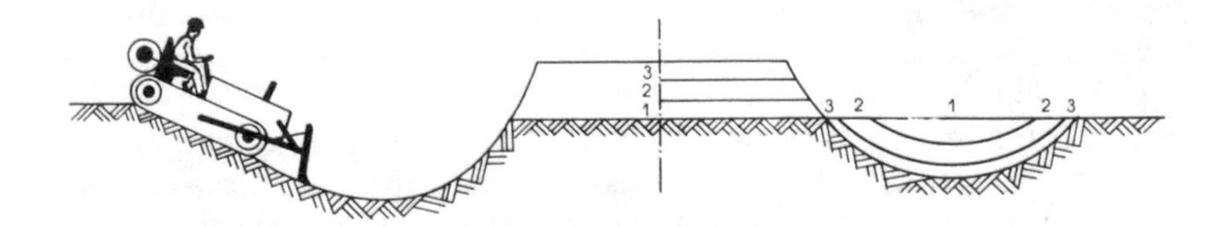

Fig. 17 *Constructing a roadway on level ground*

greater than 30–40 kph [34] are neither essential, given the low number of truck trips anticipated, nor desirable, given the nonmotorized traffic demand, on most rural earth roads. Technical Note 4.15, Field Engineering for Realignment, describes a suitable level of engineer effort to realign or relocate sections of low-volume rural feeder roads.

VI. CONCLUSIONS

12.30 To apply standard highway design concepts and criteria to rural access infrastructure is to seriously overcommit, and possibly waste, scarce resources, both technical and financial. These same resources can improve access for a much larger segment of the rural population if they are applied judiciously to eliminate bottlenecks in local infrastructure only to the level required to satisfy foreseeable local access needs. However, such improvements should not close out options for further upgrading (stage construction).

12.31 Physical environment impacts rural transport infrastructure in several diverse ways:

- (i) topography affects the cost of constructing infrastructure to specific standards;
- (ii) topography constrains the practical zone of influence by hampering travel, and the absolute zone of access by imposing physical barriers;
- (iii) climatic seasons determine the total time dry-weather roads are accessible;
- (iv) rainfall patterns govern the frequency and length of intermittent infrastructure closures;
- (v) rainfall intensity controls the size and design of small drainage facilities, and the requirements for erosion control measures; and
- (vi) yearly rainfall amounts affect the soil quality at specific infrastructure locations.

12.32 Technical considerations should be directed towards improving drainage-related constraints since accessibility levels are usually intimately related to environmental problems, either in the condition of the infrastructure surface or as physical closures and failures due to flowing water. However, other operational constraints may be present in the form of physical barriers, specifically inadequate widths, steep grades and sharp curves.

12.33 The technical solutions should be directed at improving accessibility for the particular transport aids that will make use of the infrastructure. If the need for a specific area can be met wtih four-wheel drive vehicles that are currently available, truck passage may not presently be a consideration, especially if large trucks cannot easily reach the area over the existing highway network. Improvements can also be directed at one or more of the transport aids usable in the specific area, for example, by improving access for pack animals during seasons when trucks cannot travel the route. Therefore, each technical improvement must be evaluated not only for a specific site but also for specific transport aids and specific seasons, to properly meet an area's immediate access needs with a minimum resource commitment. Alternative or traditional transport aids often operate over lower quality infrastructure which not only is less costly to construct or replace, but also is less disruptive to the established equilibrium between the existing topographical features and climatic conditions.

12.34 Rural access infrastructure projects should begin with an expanded inventory of existing conditions to provide a basis both for the selection of alternatives and the allocation of resources. Such inventories should consider all modes of transport using the corridors under consideration, and the location of specific access deterrents whose removal will most economically improve accessibility within the project area. When relocating sections of individual rural access infrastructure links is the most economic solution to improved access, such relocations should apply the same principles as those used to determine spot improvements, i.e. the relocation should provide only the accessibility necessary to satisfy foreseeable local access needs while retaining the options for future upgrading.

CHAPTER 13

Stage Development and Spot Improvements

13.01 The objectives of rural access project components are realized not when the infrastructure is constructed but when production volumes are high enough to warrant marketing and when transport aids are available to haul the produce. Improved personal transport occurs only when people have enough money to be able to afford the fare, or to purchase their own transport aid, such as a bicycle [19].

I. UNCERTAINTIES

13.02 The assessment of the probabilities of a rural access project component producing the desired impacts has a greater degree of uncertainty than similar assessments for more conventional highway projects. Estimates of future economic and social development include substantial uncertainties in many elements of a development project, such as [15]:

(i) implementation of the necessary complementary investments;
(ii) actions of other cooperating governmental agencies;
(iii) individual's response to risk and innovation;
(iv) variations in weather;
(v) variations in gestation periods of different types of development project goals; and
(vi) fluctuations in international market prices.

In view of these uncertainties, a minimum RAI investment should be made to encourage development, and further investments should respond to the increases in transport demand brought about by such development.

13.03 Minimum initial investment often means foregoing conventional infrastructure design standards in order to improve existing rural access infrastructure (RAI) to the level that most economically fulfills the immediate development objectives. Stage construction, as defined in this presentation, involves a progressive series of improvements undertaken under an investment programme that concentrates resource expenditures where they are immediately and obviously more useful. As development continues, increasing transport demand warrants further stage construction. However, if the impact of a specific infrastructure link does not materialize as rapidly as expected, its staged improvement can be delayed. Stage construction therefore eliminates the cost of initial overdesign if agricultural production forecasts prove incorrect.

13.04 In an economy with capital scarcity and uncertainty, even the supposedly higher cost of stage construction of feeder roads which must eventually be upgraded to design standards that require extensive reworking may be less than the initial high cost of using standard road design criteria when the appropriate discount rate of future costs and benefits is applied [65]. Furthermore, initially building several rural feeder roads to specific design standards in a development project area represents a considerable investment in anticipation that the development process will generate traffic volumes in every location sufficient to warrant the chosen design standards [12, 87].

13.05 The betterment of existing infrastructure is the most cost-effective minimum investment in access improvement since existing infrastructure represents a significant potential resource. Spot improvements – minimum localized investments in access betterment – are upgrading activities undertaken to improve accessibility by correcting or improving drainage-related access constrictions, traction problems, and other operational constraints. They do not require formal design using a standard width travelway with specific geometric criteria to accommodate an arbitrary design speed. Spot improvements are the first step in the stage construction concept of incremental route improvement which should continue only as the traffic demand increases. They are in fact the method whereby rural road networks have been evolved in almost every country in the world. Technical Note 5.01, Cost Evaluation of Spot Improvements, offers guidance in determining when feeder road spot improvements are preferred to complete construction. It ignores user costs since low traffic volumes do not offer significant savings, but includes a provision for maintenance costs which may be higher for feeder roads with spot improvements than for roads built to higher standards.

13.06 As a consequence of providing minimum cost solutions, or spot improvements, often with a low safety margin, the risk of some damage to the road after completion of the improvement must be accepted by politicians and planners as well as technicians. Spot improvements, because of their low cost and consequent lack of formal design, can be considered a trial and error design process; however, proper construction practices must still be followed. Improvements are continued until satisfactory access has been achieved. Therefore, the subsequent repair or minor additional work needed within the first or second year after the improvements are initiated, i.e., the 'review and correction' period, should be considered as part of the improvement investment rather than a road maintenance activity.

13.07 Such an approach often blurs the distinction between construction and maintenance. Maintenance policies, in fact, influence the choice of construction possibilities. Single stage construction normally indicates that nothing except routine and periodic maintenance will be required over the selected design life of a roadway. Multi-staged construction anticipates routine maintenance coupled with a review of the road's current access suitability before undertaking periodic maintenance. Periodic maintenance and upgrading for improved access are demand-oriented where funding is severely limited, so spot improvements represent the beginning of a multi-staged construction programme in which the allocation of funds between upgrading and periodic maintenance can be selectively drawn over an unfixed road design life.

13.08 The cost of a spot improvement programme is project-specific. Since each spot improvement is site-specific, intiial spot improvement programmes need a field analysis before any funding is appropriated. Otherwise, funding will have no relationship whatever to the actual access needs of an area or to the types of spot improvement required. There is little historical data concerning the magnitude of costs involved in spot improvement programmes, because major donors have not yet concentrated on the economic advantages of such programmes. However, spot improvements have received substantial funding in the guise of rehabilitation of previously financed highway construction that has deteriorated due to insufficient maintenance.

II. STAGE CONSTRUCTION FOR TRANSPORT MODES

13.09 Infrastructure stage development can consist of several steps involving a hierarchy of rural access infrastructure, i.e., paths, trails, tracks and feeder roads. It does not have to progress through each step however; it can begin with existing infrastructure in any one of the categories, improve that infrastructure's accessibility within the category, or raise the category upward through projected selected increments. This is often the case when more than one type of transport aid will be used in a project area. For example, a trail could be improved to accommodate pack animals at the beginning of a project and later be developed into a track when increasing production requires the use of animal carts, providing the initial infrastructure gradients are suitable for cart traffic. Each improvement must result in a usable facility that is protected from excessive deterioration caused by weather and usage.

13.10 The complete hirerachy of stage development would include the following steps. However, it is impractical to assume, either technically or economically, that any single infrastructure link would pass through more than two or three improvement stages:

(i) creation of a footpath;
(ii) upgrading to an all-weather footpath, and/or upgrading to a serviceable trail;
(iii) upgrading to an all-weather trail, and/or upgrading to a serviceable track;
(iv) upgrading to an all-weather track, and/or upgrading to a serviceable single-lane earth road;
(v) upgrading to single-lane granular surfaced road, or to a two-lane earth road;
(vi) upgrading to an all-weather single-lane road, or upgrading to two-lane granular surfaced road;
(vii) upgrading to an all-weather paved single-lane road, or upgrading to an all-weather single paved lane on a two-lane granular platform; and
(viii) upgrading to an all-weather two-lane paved road. Two-lane paved roads are primarily for traffic volumes which exceed the access demand considered in this handbook.

III. STAGE CONSTRUCTION FUNDAMENTALS

13.11 While low-volume RAI cannot justify extensive costs and manpower during the formal planning stage, it is essential that certain basic information be available before stage construction is undertaken. The site should be inspected to determine site-specific field conditions such as topography and climate; and information about unusual problems, local flooding, location of useful construction material, and labour availability at likely wage rates should be gathered from local residents and those responsible for the maintenance of existing infrastructure. Technical Note 5.02, Stage Construction Features, contains more information on specific features of stage construction that should be considered when gathering and evaluating site-specific information.

13.12 The determination of the construction method, either labour- or equipment-based, for stage construction, should be made early in the planning process, but not before the site-specific conditions have been determined to be suitable for that specific construction technology [16]. Since later stages usually incorporate work completed in the early stages to the greatest extent possible, the construction technology chosen for the first stage of motorable roads will usually be the method used in latter stages. The construction method will also influence the maintenance method to some degree. While motorable roads built by equipment-based methods can be maintained by labour-based methods, roads constructed by labour are sometimes too uneven for mechanized grading [34] and often have a side-ditch configuration that prevents mechanized ditch cleaning. Figure 18 shows a typical hand dug and grader dug ditch shape to illustrate the latter point. These are common configurations but one should not infer that the ditches as shown are interchangeable since they have different capacities for the same invert gradient. Ditch configurations vary not only by technology employed, but also because of capacity requirements, soil constraints, velocity considerations and personal preference. This handbook includes a variety of ditch configurations used throughout the world.

13.13 There are exceptions to the above principle of using the same technology as stage construction continues. They include the improvement of paths and trails, which when built to their minimum widths are too narrow for common construction equipment; the upgrading of paths and trails to tracks or roads when most of the existing construction will be unsalvageable; and the construction of drainage structures, which can be labour-based even when built in conjunction with equipment-based road construction.

IV. ENGINEERING SPOT IMPROVEMENTS

13.14 The magnitude of spot improvement activities required for a specific project are site- and transport aid-specific. The most economical method of determining the investment required to improve access by spot improvements is through use or development of an infrastructure needs inventory (Technical Note 4.13). Such an inventory sometimes exists for motorable infrastructure, but unless it has been carefully designed, it may not contain sufficient information for intelligent spot improvement delineation. Inventories of other trans-

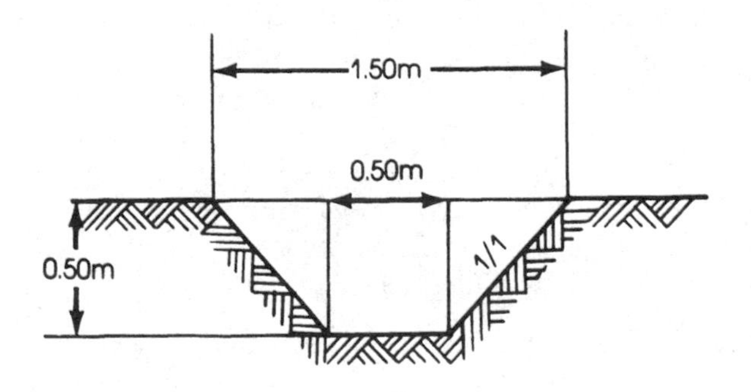

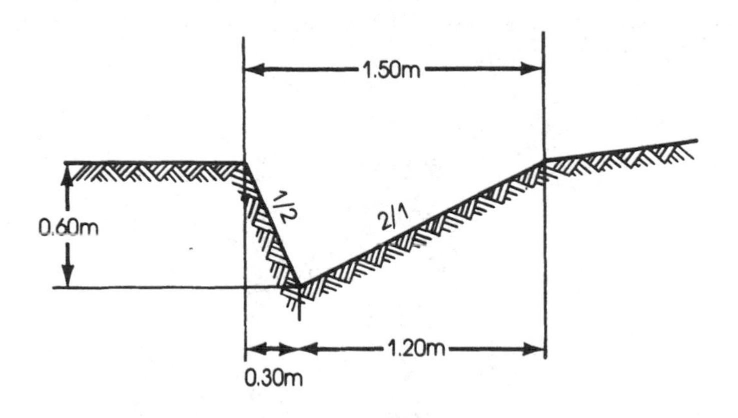

Fig. 18 *Side ditch configurations*

port infrastructures are usually non-existent and must therefore be included as a project implementation cost.

13.15 Chapter 12 contains suggested inventory procedures for rural access infrastructure. Two methods are outlined: in the first a physical inventory is carried out by technicians who specialize in inventory work, while the condition inventory and needs inventory are undertaken at a later time by an engineer; in the second, the physical, condition, and needs inventories are carried out simultaneously by an experienced engineer.

13.16 Since a large portion of spot improvements are drainage-related, several Technical Notes are provided in Annexe 5 to assist the inventory crew and/or engineers in identifying the possible causes and remedial actions for the more common drainage-related problems that may be ameliorated by spot improvements (Figure 19). They include:

Technical Note 5.03 – Road Surface Potholes
Technical Note 5.04 – Travel Surface Longitudinal Erosion
Technical Note 5.05 – Muddy Surface, Rutting under Traffic
Technical Note 5.06 – Slope Erosion
Technical Note 5.07 – Ditch Erosion
Technical Note 5.08 – Waterway Crosses Roadway
Technical Note 5.09 – Bridge Deficiencies
Technical Note 5.10 – Pipe Culvert Deficiencies

Technical Notes 5.03–5.10 can assist in both the infrastructure condition inventory and the needs inventory, especially when these inventory activities are undertaken by different personnel. A careful reading of the suggested remedial actions will show that they are often similar for more than one drainage problem. For example, proper shaping and compacting a surface may correct both potholing and muddy surface problems. However, all actions cannot be used to correct the same problem in different infrastructure locations. For instance, solutions that are valid for a hillside location may be unsuitable for a depressed section in flat terrain.

Fig. 19 *Road failures are often due to the combination of traffic and water*

V. EXECUTING SPOT IMPROVEMENTS

13.17 The final determination of the proper sequence of spot improvements must be approached logically. The accompanying Technical Notes 5.03 through 5.10 indicate some possible corrective actions, but their order of execution and applicability are problem-specific. All necessary drainage activities should be completed first, and obviously any work underneath the surfacing, such as raising the road or reshaping and compacting the subgrade, must take place before any granular surfacing is applied. Alternatives that are not available for a depressed road section in flat terrain become possible if the road surface is elevated above the surrounding terrain. Depressed roads in flat

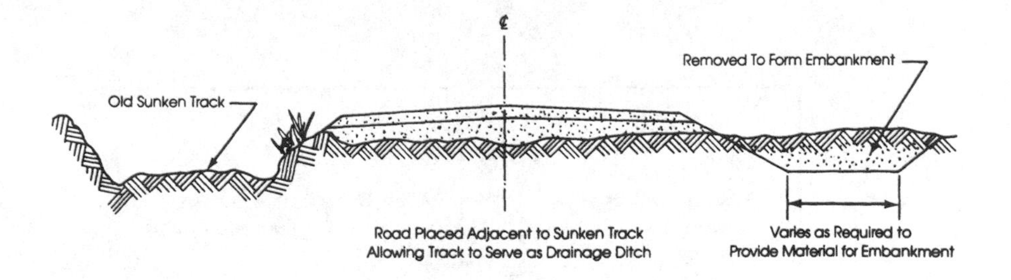

Fig. 20 Relocation of a sunken road section

terrain are a major problem in many countries. A common solution is to use the existing road as a drainage area adjacent to a new elevated alignment built with side borrow (Figure 20), since such depressed sections are usually not maintainable. Providing seepage pits along the original depressed alignment is a less costly solution when the terrain is not flat; but in level areas it is a less suitable solution because the side ditches are usually ineffective. The side ditches of an elevated alignment sometimes act as a continuous seepage pit and evaporation bed in flat terrain.

13.18 Any actions involving drainage spot improvements should take place upon initiation of work in a given area. The addition of granular surfacing may be delayed if there is a reasonable chance that drainage improvements will eliminate the need for such surfacing, or if surfacing can be done more cost-effectively for several locations at one time. The division of spot improvement activities into discrete tasks, e.g., cutting ditches, reshaping and compacting subgrade, applying and compacting granular surfacing material, permits work division either laterally by having different crews perform different activities, or horizontally by having the same crew do different activities at different times. In any case, careful evaluation of the total workload is critical if minimum costs are to be achieved.

13.19 Spot improvements also include removal of physical barriers that constitute operational constraints. Para 12.02 describes such physical barriers. These barriers are both site- and transport aid-specific and cannot be treated in generalities in the same manner that environmental constraints are addressed in Technical Notes 5.03 to 5.10. Any improvements made to remove physical barriers for motor vehicles should include provision of minimum sight distances for safety at low speed by eliminating blind spots, or provision of sufficient width for two-way traffic when minimum sight distance cannot be achieved. Technical Note 4.01 contains more detailed information about physical barriers.

13.20 Spot improvements can be undertaken by existing maintenance or construction crews, by special brigades created to do the work, or by contractor. They can be carried out by either labour- or equipment-based construction technology. However, existing maintenance crews cannot usually undertake

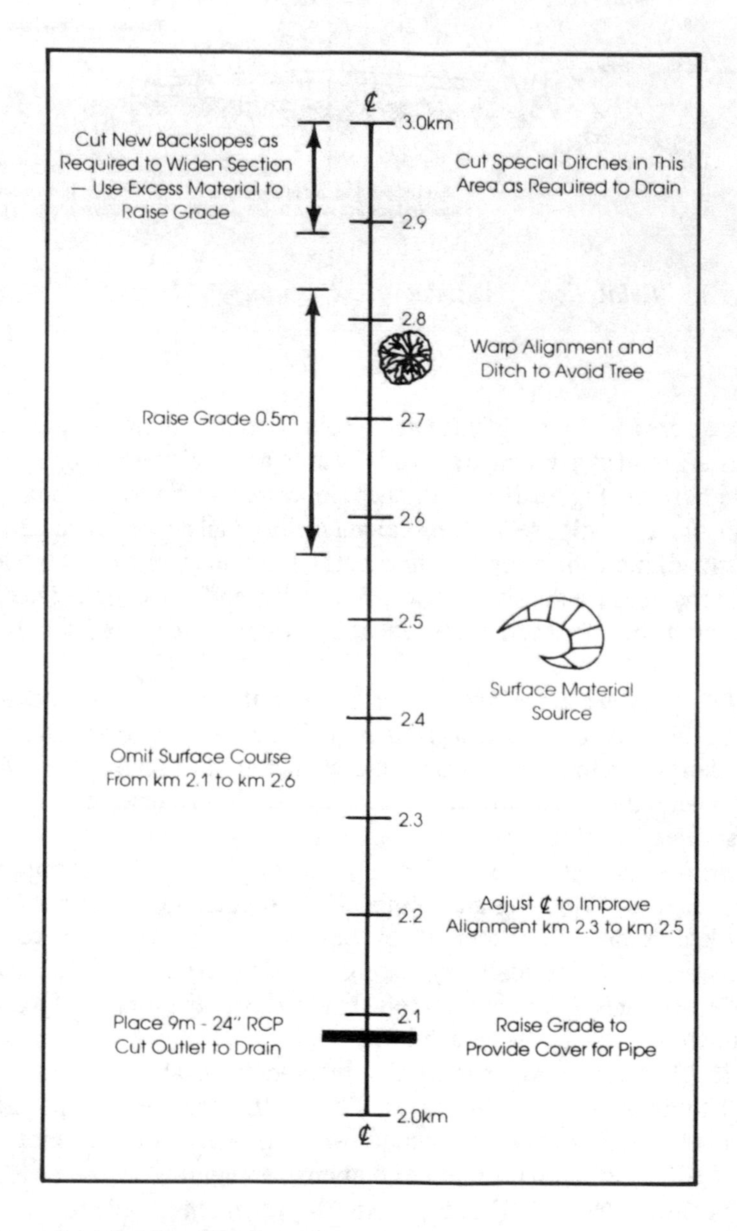

Fig. 21 *Straight line plan*

an ambitious spot improvement programme without unacceptably neglecting their assigned maintenance activities. Contractors need to be highly competent or very closely supervised, since spot improvements are often designed and estimated in the field, are not standardized, and lack formal design drawings (Figure 21).

13.21 Logistical planning is a key to minimum costs for spot improvement activities. Because of the high mobilization costs, compacting reshaped cross-slopes in order to spot improve drainage-related constraints should not be neglected at the time the reshaping is undertaken. The cost of returning all the equipment plus a compactor to rework previously spot improved sections is far higher than transporting the compactor with the original work crew. This is particularly true when working on problems that were originally caused by improper compaction.

13.22 Feeder road sections that are being spot improved for better access are not expected to have the same traffic safety features as conventional highways. When it is not cost-effective to eliminate existing safety hazards for low traffic volumes, serious considerations should be given to providing warning signs for steep grades, sharp curves and unusually narrow bridges and culverts. Bumps built into the road surface effectively slow vehicular traffic approaching dangerous areas. However, they are difficult to maintain on earth and gravel surfaces. A more positive indication of narrow bridges and culverts than signs can be given by tapering the road surface and cleared width of their approaches. Both warning signs and depth gauges are appropriate for fords where water depths and/or water velocities may present problems.

13.23 Spot improvement to upgrade access for specific transport aids, such as providing access for only four-wheel drive vehicles during the rainy season, must have some sort of protection against unauthorized usage during critical periods. Unfortunately, signs do not usually provide such protection. Road surfaces that cannot support heavy vehicles during certain seasons can only be protected by physically barring such vehicles. The use of gates as rain barriers require the presence of someone to open and shut them at the appropriate time, which can be quite costly or ineffective, depending on the calibre of the attendants. Truck bars, which are templates that prohibit entrance of vehicles over a certain width and height, have proven to be a satisfactory control for both road surface usage and bridge restrictions in some developing countries.

VI. CONCLUSIONS

13.24 Stage construction can be undertaken either to improve accessibility of existing transport aids such as animals, carts or vehicles, by enhancing the infrastructure's ability to provide access in inclement weather; or to introduce access for a higher transport mode, for example, improving a trail to a track. Often it can achieve both objectives at the same time.

13.25 Stage development of rural access infrastructure is an economical approach to: (i) reducing the risk of excessive construction costs inherent to the uncertainties of future development; (ii) keeping transport accessibility in step with the uncertain timing of future development; and (iii) preserving future options for the expenditure of transport-related resources in developing areas. Stage construction often consists of spot improvements to the infrastructure throughout an entire development project area to provide better access during the early development process.

13.26 While the concept of spot improvements as the initial step in stage construction adopted in this handbook waives the application of formal road design standards based on a uniform design speed throughout an entire alignment, each staged improvement must be properly constructed, completely usable, maintainable and economically viable. If possible, spot improvements should not eliminate options for any anticipated further staged construction; but their selection need not be influenced by options that are unlikely or not intended to occur. For instance, a feeder road as defined in Table 1, is intended only to carry local traffic to and from the highway network, so the geometric considerations necessary for through traffic need not complicate the planning of such feeder roads.

CHAPTER 14

Construction of Rural Access Infrastructure

14.01 Appropriate technology selection is a critical step in the evaluation of rural access infrastructure (RAI) construction projects. Appropriate technology is defined as one that is both efficient and least costly [16]. Construction technology covers a spectrum of construction methods using various mixes of labour, animals, and equipment, with equipment-based construction being considered as high technology. The characterization of a construction methods as labour-based does not mean that equipment can not be used. Nor does equipment-based construction preclude using labour. The terms used simply identify the principal source of motive power used. The appropriate mix of labourers, animals and machines can be determined by the technical nature of the project, available resources, prevailing factor prices, and the socioeconomic environment in which the project is executed [16].

I. APPROPRIATE TECHNOLOGY ISSUES

14.02 Small dispersed projects such as paths, trails, tracks and short earth, gravel and sealed roads, make the costs of regular highway construction-equipment delivery, mobilization, and operation high per unit of work performed. There is seldom enough work to use equipment at full capacity. Equipment maintenance costs are very high in areas far removed from established workshops [16]. Replacing standard highway construction equipment with smaller units provides a higher utilization rate which may outweigh the higher unit costs of the smaller equipment. For example, one or more tractor-towed rollers with more suitable production outputs may result in a high utilization rate than one large self-propelled roller. Paths and trails may require so little work or be so narrow and steep that any normal mechanical construction method is impractical. Appropriate use of locally available material may further reduce costs; for instance, building timber bridges instead of reinforced concrete or corrugated metal culverts or arches may be more cost-effective despite the shorter design life of wooden structures.

14.03 Technology implementation requires further evaluation of rural infrastructure construction projects. Contracting, force account and self-help are the three most common organizational arrangements used for local projects. However, as with technology, there is a whole spectrum of alternatives and combinations. Construction implementation can include construction activities by:

(i) international contractors;
(ii) domestic or private sector contractors;

(iii) parastatal contractors (government-owned, autonomous organizations);
(iv) force account, defined here as work performed by permanent government employees;
(v) day labour, defined here as temporary employment of local people to supplement force account personnel constructing local projects;
(vi) self-help, often defined as a work force made up from the local population, organized to build infrastructure with technical and/or supply assistance from government, but with no wages; and
(vii) subcontractors under items (i) to (vi) above to:
   (a) provide labour or materials only;
   (b) construct certain discrete items of work such as culverts, i.e., horizontal slicing; or
   (c) construct certain sections of work such as a complete link between points A and B, i.e., vertical slicing.

14.04 The various implementation alternatives are not universally appropriate to all developing countries. However, some of the alternatives, such as domestic contracting resources, when available, may be more suitable for RAI construction than for major high-volume highway construction. Conversely, large international or larger local contractors may show little interest in bidding on RAI construction except as it is required as part of a larger development package. In such cases large contractors tend to be biased towards using the same level of technology they will use for the rest of the construction programme. The availability of resources such as labour, equipment, materials, and management and supervisor expertise must play a major role in the final selection of both appropriate technology and implementation methods [16]. Furthermore, no rural infrastructure construction activity should begin before the method for maintaining the infrastructure has been determined (Chapter 15). Maintenance considerations therefore become a further qualifying factor in the choice of appropriate technology and implementation.

II. LABOUR-BASED CONSTRUCTION

14.05 The most immediate application of labour-based construction is in rural environments where one or more of the following circumstances prevail [16]:

(i) the population includes many poor people, most of whom are underemployed or unemployed;
(ii) large numbers of infrastructure improvements are needed, some of which are small, technically simple, and geographically dispersed;
(iii) there is a shortage of foreign exchange to buy and operate machinery for the construction and maintenance of all the needed infrastructure improvements; and
(iv) there are no workshops, mechanics, operators and supervisors used to working with machinery.

These same circumstances encourage the introduction and/or use of traditional transport aids and construction of their associated infrastructure needs. The advantages and disadvantages of labour-based construction are outlined in Technical Note 6.1, Pros and Cons of Labour-Based Construction.

14.06 Design criteria appropriate to the type of infrastructure required for rural access are conducive to labour-based construction techniques [65]. However, when governmental transportation organizations are receiving funding from outside donors for equipment-based construction programmes, the local contributions to these programmes absorb the available local funds. This creates a shortage of local funds to pay the large and frequent wage expenditures necessary for labour-based construction programmes. This becomes a deterrent to the introduction of labour-based technology. It has been suggested [65] that labour-based rural infrastructure projects should be placed in government agencies other than the highway department, such as regional development authorities or integrated development programmes, whose principal objectives are to employ the rural poor and to develop local organizations for community construction projects. A counter-suggestion [75] contends that the creation of a new operational division within the road authority devoted exclusively to labour-based rural infrastructure construction will demonstrate the government's commitment to labour-based methods and provide new career-track opportunities for the agency's personnel. For countries with no traditional experience, a minimum of three years of preparation is required for mobilization, staff training and institution building before a large-scale use of unskilled labour can be successfully attempted [16]. However, major technical assistance inputs may reduce this preparation time somewhat.

14.07 Proper motivation is essential for successful labour-based construction. Workers paid under a task-rate or piece-rate system (called incentive payment systems) produce a much higher daily output than daily-paid labour [35]. Daily payment is commonly used when no productivity data for major operations in infrastructure construction are available, or where forbidden by law or labour agreement. Task-rate payments consist of a fixed daily wage given for a fixed quantity of work. A task is correctly defined if a good worker finishes in less than a full day; task work provides a high incentive for local labourers who can return home when they complete their task. Furthermore, the high cost of camps makes it advisable to recruit labourers who live within walking distance of the site, sometimes assumed to be a maximum of 6 km [16].

14.08 Payment by task-rate may be unsatisfactory for sites where labour and machines have to work together, since the labourers' goal is to finish early. If they succeed, the equipment will be underutilized unless it can be worked independently during the last few hours of the work day. It is also unwise to have labour and heavy equipment on the same site performing identical or interrelated activities, because it demoralizes labourers to observe the machines' much greater output [16].

14.09 The piece-rate system pays the worker a fixed sum per unit of work. Unit costs are normally lower than for task work and productivity is higher [35]. This system works well with migrant labourers who lack incentive to finish early because they live in a labour camp. In some countries it is not possible to introduce this system because of administrative problems with a non-fixed monthly wage. The piece-rate system is more suitable for projects executed by contracted labour because the contractor bears the overhead costs of the higher supervision needs generated by piece work [35]. But, if the government intends to employ additional supervisors for a continuing force account piece work programme, the government's additional supervision overhead costs will be outweighed by lower production costs. Direct unit costs under a piece work system can be accurately predicted although the rate of progress is less predictable.

14.10 Whatever system is used, it is imperative to pay wages on time. Despite the administrative difficulties for the site staff, workers should be paid at least once every two weeks, once a week is better. Under no circumstances should the payment period extend beyond one month. Payment should be in cash rather than by cheque [16]. Some labour-based construction programmes also provide food as part of each payment.

14.11 The design methodology for labour-based construction projects [16] is the same as the design methodology recommended in this presentation for RAI spot improvements. The major design issues should be laid down in instructions from headquarters. The instructions should also specify those details to be left to the discretion of the field staff. The principal objective of the site engineer is to design the details of the project in a manner that minimizes construction costs.

14.12 Earthworks often account for half of a project's costs [16]. Longitudinal balancing of cut and fill (Figure 22) is therefore critical. However, for labour-based construction using wheelbarrows, earthwork balancing is impractical for distances over 100 m for soft soils and over 200 m for hard soils. Earth moving using work animals is impractical for distances over 500 m. For labour-based construction, the site engineer should therefore select an alignment that:

(i) requires the least amount of material movement;

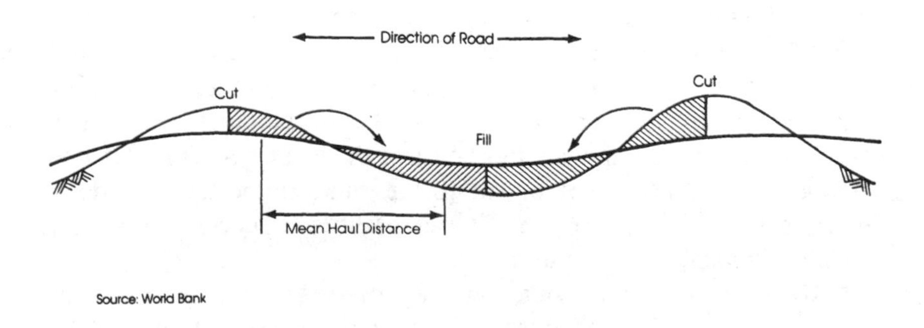

Fig. 22 *Longitudinally balanced cut-and-fill*

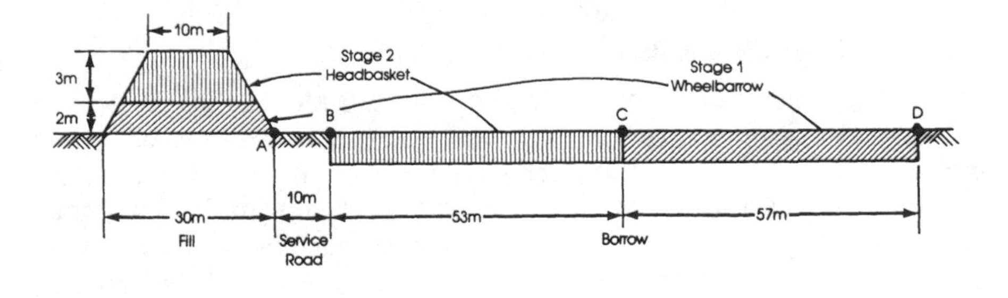

Fig. 23 *Lateral material haulage*

(ii) permits lateral material haulage (at right angles to the route) (Figure 23);
(iii) favours more easily worked soils in place of harder materials;
(iv) reduces the grades over which the material must be moved;
(v) avoids expensive river crossings;
(vi) avoids rock excavation; and
(vii) is close to water sources for compaction when feeder roads are being built.

This alignment selection procedure should be followed even if these requirements result in somewhat greater route lengths. Technical Note 6.2 contains more information about earthwork considerations for labour-based feeder road construction.

14.13 A 'nominal' [16] labour force must be recruited for labour-based construction projects. The size of the nominal labour force includes allowances for absenteeism, holidays, weather, etc. The 'effective' work force, after these allowances have been deducted, must be able to meet the construction requirements developed from the productivity rates in use without unduly interfering with agricultural activity in the area. The nominal force size, however, dictates the logistical and supervision requirements. The willingness of labourers to work on rural access infrastructure construction depends on a variety of factors:

(i) availability of alternative work opportunities and the wage differential between other forms of employment and RAI construction work;
(ii) personal wealth of the labourers;
(iii) work-related costs for food and transportation to the site; and
(iv) social attitudes that may allow one form of work but not another.

In general, labour supply presents few problems provided sufficient care is taken to ensure the project is not too big for an available labour pool based on the above considerations.

III. EQUIPMENT-BASED CONSTRUCTION

14.14 Equipment-based construction does not exclude the use of labour; it takes advantage of mechanizing the high energy requirements of certain infrastructure construction tasks. The output of a single piece of mechanized equipment may be equivalent to several hundred labourers [16, 75]. Infrastructure construction involving the movement of large quantities of materials over long distances, or substantial production within a short period of time, are best suited to equipment-based construction. Equipment-based construction is often substituted for proposed labour-based construction if the construction schedule does not have the flexibility to accommodate the time and manpower requirements for organizing the labour-based components of the construction work [2].

14.15 Material transport is perhaps the most common reason for using construction equipment; manufacturing aggregate smaller than 25 mm and earthwork and surface compaction are other tasks that have been identified [16] as too energy-intensive to be economically carried out by labour. The accepted material movement breakpoints, i.e., the most effective earth-moving ranges, for various types of mechanized construction equipment, are [75]:

(i) bulldozers, up to 100 metres;
(ii) towed scrapers, up to 300 metres;
(iii) motorized scrapers, up to 3 km;
(iv) tractor/trailer combinations, up to 5 km; and
(v) dump (tipper) trucks over 3 km.

All except the trailer and the truck are self-loading, perhaps with some assistance, and all are self-unloading. Small jaw crushers may also be useful in areas where rock is the only available roadmaking material. Rock includes any material that must be crushed, gravels are coarse grained materials which can be used 'as dug'.

IV. INTERMEDIATE TECHNOLOGY

14.16 Intermediate technology is a mix of equipment, labour and, in some cases, animals. In many countries which use equipment-based methods for other infrastructure construction projects, RAI construction may be best carried out by intermediate technology if this increases utilization of their existing equipment fleets. This can be accomplished by allocating specific tasks to the existing equipment best suited to carry out those tasks, and/or by using smaller equipment that will be more fully utilized. However, the purchase of smaller equipment should be evaluated against the in-country capability for spare parts and repairs to assure that any new type of light equipment will be productive. Productivity is further discussed in Chapter 17, beginning with Para. 17.16.

14.17 The allocation of existing equipment should be done in packages (Figure 24). Loading dump (tipper) trucks by hand is inefficient because of the

Fig. 24 *Equipment package in a borrow pit*

standing charges for the truck while it is being loaded [16, 44]. Therefore, if trucks are to be used, the proper allocation of resources frequently indicates that loading must also be mechanized by using a front-end loader or a bulldozer with a loading ramp. High utilization of such equipment packages requires considerable planning when required quantities are small.

14.18 Substituting smaller mechanized transport equipment in the above example both raises the equipment utilization factor considerably and accommodates greater labour participation. A common substitution for haul distances of over 500 m is tractor-drawn trailers. Since the trailer can be unhitched, the more costly tractor does not have to stand idle while the trailer is being loaded (Figure 25). Using two or more trailers permits both hauling and loading operations to be balanced and continuous. The following tractor size/trailer load capacity combinations are recommended for average off-road hauling [16].

35 hp tractor with 3 ton trailer;
45 hp tractor with 4–5 ton trailer;
60 hp tractor with 5–6 ton trailer; and
75 hp tractor with 6–7 ton trailer.

14.19 The evaluation of proper equipment selection should also include secondary tasks that may be undertaken by the same piece of machinery. Dump (tipper) trucks are specialized equipment while tractors are general purpose equipment. A truck may be better suited to the primary task of hauling material than a tractor and trailer, but the tractor is better adapted for such secondary tasks as towing water tankers, small rollers, or small graders [16].

14.20 Intermediate technology is not usually classified as such since it covers such a wide array of labour/equipment mixes. A recent publication [16] states that in average labour-based rural road construction projects, labour plus handtools costs account for 40–45% of the total cost, while equipment costs account for 25–35%. These are order of magnitude percentages, since they are very sensitive to the quantity of gravelling required. For average

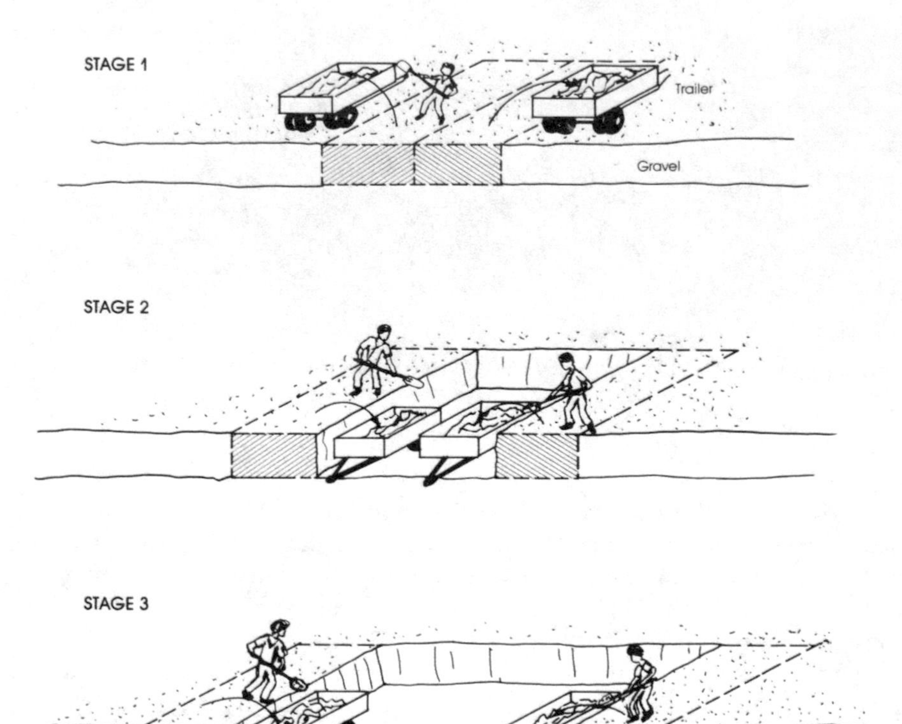

Fig. 25 *Loading trailers in a gravel pit*

equipment-based rural road construction projects, labour plus handtools costs generally account for 10–25% of the total cost, while equipment costs account for approximately 60–70%. Equipment-based operations for spot improvement work must be highly mobile, versatile and self-contained, indicating their labour-equipment ratio will be in the higher equipment percentage range. Many spot-improvement operations are small, therefore the equipment should be small enough so it will not damage the sections of RAI that are not in need of improvement.

V. CONTRACT CONSTRUCTION

14.21 Contract construction is normally undertaken within a legal agreement between the owner, usually a governmental agency, and a construction contractor, who frequently promises to [74]:

(i) furnish all labour, materials, and equipment;

(ii) complete the work according to plans and specifications;
(iii) complete the work within a specified time and price;
(iv) protect the owner from all losses due to damage suits, liens, or other causes; and
(v) assume financial responsibility for completion of the work, usually by means of a contract performance bond. These last two requirements may have to be waived, however, to encourage the use of small contractors in rural areas.

The owner in return is normally obligated to provide detailed plans, specifications, and bills of quantities; and to pay within an agreed-upon time. The requirements for detailed plans, specifications and estimates are somewhat at odds with the *ad hoc* design procedures recommended for spot improvements, but can be overcome by simple standard sketches for items like headwalls, method rather than performance standards (Para 2.31), and a certain degree of flexibility on the part of both parties concerning the measurement and payment for specific contract items.

14.22 Contracts vary chiefly in the basis upon which the contractor will be paid for his work [74]. The most common forms of contracts include payment of:

(i) contractor's cost plus a percentage;
(ii) contractor's cost plus a fixed fee;
(iii) a lump sum; or
(iv) unit prices.

Contract amounts can be either negotiated or bid. International lending agencies usually insist that contracts involving their funds be awarded competitively using unit prices where possible and lump sums where unit prices are not feasible. It may be feasible for a small contractor to undertake RAI spot improvements for lump sum payments, set by the contractor or offered by the owner.

14.23 Failures of small local contractors of the type most likely to participate in RAI construction as either prime contractors (vertical slicing) or subcontractors (horizontal slicing), are often due to inexperience and/or inability to cope financially with schedule delays and slow payments from the government. Unless careful attention is paid to the needs and capabilities of small contractors, encouraging their participation in rural infrastructure construction can lead to serious damage to a country's small contracting industry [2].

14.24 Small contractors are likely to be equipment-poor and therefore more willing to undertake labour-based construction if equipment is unavailable, labour is cheap, and if the project design, bid advertisment, evaluation and award are neutralized or impartial to construction technology. Contractor prequalification for rural access infrastructure should be based on the contractor's proven ability in the use of resources, either labour or equipment. If a small contractor is required to use certain units of specialized equipment, some governmental agencies rent the contractor the equipment and operator, providing the contractor is otherwise qualified.

14.25 It may be desirable for government agencies to reduce the financial burden of small contractors by providing an operating advance at the start of the contract. The amount of the advance can be reduced, thereby minimizing the risk to the agency [99] by:

(i) reducing the time period between certification and payments;
(ii) including in each certification an advance to cover a portion of the estimated value of the work that will be completed in the period between certification and payment; and
(iii) structuring the bill of quantities on a 'method-related' basis to correspond more closely to the way the contractor incurs costs; for example, providing separate pay items for labour mobilization, labour camps, temporary works and supply of local materials, rather than including these items in the rates paid for completed permanent works.

VI. FORCE ACCOUNT CONSTRUCTION

14.26 Force account work is construction undertaken by government workers using government-owned equipment. Force account operations can also include subcontract elements such as culvert construction, local material or labour supply, or the hourly or daily hire of equipment and operators to supplement government resources. Force account activities are predominant in countries that have no substantial construction contractor base. Technical Note 6.3. Tradeoffs between Using Force Account or Contracting, discusses the relative advantages of using force account or contractors when both are available.

14.27 Day labour force account includes the temporary hire of local people for local projects. These employees do not become a part of the permanent government labour force, thus giving the government more flexibility in payroll expenses. Equipment, operators and supervisors are usually not obtained locally except as sub-contracted items under direct government supervision,

14.28 Many countries maintain a small force account staff to assure a construction capability and contract temporary excess construction requirements to stabilize the size of the departmental forces. Construction of dispersed, small-scale RAI, with its minimal equipment requirements, is well suited to employ day labour or subcontractors in support of a small cadre of permanent employees.

14.29 Self-help projects can be defined as the use of local labour and other available facilities at no cost to the government. However, in most cases, government assistance is required, in which case the text classifies the project activities as a form of the community contract system (Para 15.15). Such projects are usually motivated by the local population's desire for capital improvements from which they feel they will be the ultimate beneficiaries [16]. The required motivation can come from within the community or can be encouraged by proper solicitation from outside the community [59]. The government's share of community contract work can vary from simply supplying tools to partial financing and complete supervision. The government's degree of participation influences the amount of control it can exercise over the quality of the end product. Today, few, if any, self-help projects as defined in the first sentence of this paragraph, i.e., with no governmental assistance, exist in developing countries.

VII. CONSTRUCTION SUPERVISION

14.30 RAI construction is clearly not just scaled-down highway construction. The minimal investment warranted for such construction often does not justify exhaustive engineering investigations and formal design activity at the headquarters level. Consequently these activities must be undertaken at the field level by the site construction engineer or well trained para-professional. This expanded decision responsibility requires experience, maturity, and ingenuity. Technical Note 6.4, Construction Site Supervision, outlines these responsibilities. It is easier and often more attractive for the young engineer to cope with large-scale infrastructure construction with its supporting plans, specifications, bills of materials and back-up staff capabilities, than to work in the isolation of a sole decision-maker. However, since rural access technology is basic, young engineers and technical school graduates quickly adapt to practical supervision techniques with proper tutoring.

14.31 Such experienced tutors are in short supply in many developing countries. Local consulting engineers often lack the depth of construction experience to be successful in the innovation of unique construction techniques and in the necessary interaction with construction workers, although the consultant's engineering capabilities may be well established. The lowest technology construction usually requires the most competent supervisor to achieve lasting results when such technology is first introduced. Unless the proper priority is set for the selection of senior construction supervisors, and adequate motivation and remuneration are offered, most experienced engineers are reluctant to leave the mainstream career track to work in the rural areas most in need of their capabilities to tutor inexperienced future project supervisors and to guide inexperienced local contractors. When such personnel cannot be recruited from within the government they must be hired through an international technical assistance contract.

VIII. OTHER CONSTRUCTION CONSIDERATIONS

14.32 Compaction for rural roads is a concern in much of the literature reviewed for this handbook. Compaction is necessary to mobilize the strength and stability of the soil and the pavement layers to minimize subsequent settlement and erosion of the road. The material must not change volume excessively during wet periods, and it must resist densification and deformation under repeated wheel loads [69]. Ethiopia [35] and the Philippines [3] suggest that proper compaction is achieved when the tyres of loaded trucks do not make grooves on the surface.

14.33 Labour-based road construction often encounters compaction difficulties. Compaction requirements are usually determined as a function of the compactive effort applied. This compactive effort can be readily determined for compaction equipment but is not consistent when manual compaction is utilized. Large mechanical compaction equipment is difficult to move between the many sites being worked by labour crews on large construction projects, and is usually underutilized on smaller projects. Proper compaction not only reduces routine maintenance requirements but also delays the need for periodic maintenance, a major investment in low-volume roads. Uncompacted earth roads frequently must be repaired every year for the first few years after their construction because the environment and traffic cause them to deteriorate before the embankments and surface have stabilized.

14.34 Many countries with major labour-based construction programmes resort to indirect or natural compaction, i.e., the compaction resulting from the action of the environment and traffic on an embankment over time. This is based on the notion [35] that uncompacted roads eventually achieve the same density as compacted roads. However, such countries acknowledge the benefits of direct compaction. For instance, Mexico [58] recommends compaction equipment but suggests the use of loaded trucks or tractors to achieve compaction if specialized equipment is not available. Kenya [97] recommends compacted embankments for their rural roads, including 'free' compaction by loaded trucks followed by compaction with pedestrian-controlled vibratory compactors [32]. Technical Note 6.5 presents a rationale for using indirect compaction. However, given the risk of losing the resource investment due to erosion and traffic that is inherent to the use of natural compaction, it should be supplemented with a direct compactive effort sufficient to ensure stability. This compactive effort need not be performance orientated, i.e., sufficient to provide a specified density, but can be method orientated, e.g., a given number of roller passes, or tamps, per location, once the stability requirements are determined through experience.

14.35 Equipment-based and intermediate technologies have a wide range of mechanical compactors from which to select; however. all rollers are not suitable for all soils. Technical Note 6.6, Productivity Data for Compaction by Equipment [16], lists several units of compaction equipment, their average output, and suitability for various soils.

14.36 Gravel loss occurs on all gravelled roads. The rate of loss is a function of traffic volume, rainfall, material type, road gradient and curvature [77, 83]. If the gravel surface is not replenished, the gravel road will revert to the characteristics of an earth road [71]. Technical Note 6.7, Gravel Surfacing, contains further road gravelling information.

14.37 The search for granular borrow sources such as quarries, pits, and stream beds for construction purposes, must include location of sufficient material to accommodate future regravelling operations. Otherwise any economic analysis to determine the cost of construction and maintenance of a proposed rural access infrastructure project will not be valid. The removal of river or stream bed material causes progressive channel profile degradation both upstream and downstream. This in turn will affect the scour of bridge piers and abutments built in the bed nearby [70]. This factor must be considered in the search for construction material.

14.38 In some areas, notably in some parts of Africa, supplies of good natural gravel for road making are becoming seriously depleted. In such areas the increasing cost of hauling gravels from more remote sites may well become an important factor in benefit/cost calculations to determine the level of traffic at which roads should be provided with an asphalt surfacing. The advantages of stabilizing natural soils rather than further depleting natural gravel deposits are being reevaluated in some of the countries currently experiencing gravel shortages.

IX. CONCLUSIONS

14.39 RAI construction is not scaled-down standard highway construction. Its construction technology and implementation must be evaluated in light of its lower cost, lack of formal engineering input, and variety of small diversified activities. These parameters favour the use of labour, animals and small pieces of equipment by local contractors or cadres of force account personnel supplemented by local inhabitants, either hired as daily paid labour or participating in a community contract programme. They do not favour major contracting organizations specializing in the use of heavy construction equipment. The need to build infrastructure that is amenable to routine maintenance by the local population further restricts technology selection and forces appropriate infrastructure type selection. A separate organizational unit, specializing in rural access infrastructure, either within the public works ministry or in another governmental agency, will help to keep the distinction between construction philosophies clear. There is a need for specialized consultants who can plan and organize rural transport facilities and train local staff. Some international consultants, recognizing this need, are beginning to offer such services.

14.40 Construction technology falls under one of two general classifications, labour-based or equipment-based, which are not mutually exclusive, but which connote the major tractive force employed. The lower levels of rural access infrastructure – paths and trails – are by nature conducive to labour-based construction, either through the use of small contractors or force account. Large road spot improvements projects, due to widely dispersed locations and diversified activities, may be accomplished much more rapidly with an equipment-based cadre of force account or contractor employees specializing in that type of work, although the individual spot improvements favour labour-based activities. In general, rural access infrastructure construction is more conducive to small contractor or subcontractor development than is standard highway construction; but the type of contract used must be tailored to accommodate small contractors both in the technology required and in the method and promptness of payments.

14.41 Construction supervision of rural transport infrastructure is far more demanding than highway construction supervision because of the scattered locations, the varied tasks, the mixes of technology involved, and the frequent lack of formal plans, specifications and bills of material. The choice of a supervisor is critical to its success. Above all, the person must be an organizer, but familiarity with infrastructure construction procedures, a maturity that the local population will respect, and the ability to communicate with and persuade local population groups, are all important attributes.

14.42 RAI construction is a 'no-frills' operation. In most cases, string lines and hand levels are more appropriate than theodolites and electronic distance measuring devices. Experience in working with soils and a box full of sieves and other simple testing devices (Technical Note 4.02), replaces the fully equipped materials laboratory, and common sense more than substitutes for a library full of theoretical textbooks. But underlying all else must be a solid knowledge of the proper use of equipment and resources and of the principles of proper construction methods upon which to improvise.

CHAPTER 15

Maintenance

15.01 Highway maintenance is a problem affecting both rich and poor nations. Low-volume rural road maintenance typically suffers most from the institutional, financial and human resource constraints hampering the maintenance of national highway systems [102]. The spot-improvement, stage-construction approach to improving rural transport infrastructure recommended in this handbook, whereby upgrading takes place only as needs are clearly demonstrated, further increases the need for relatively high levels of maintenance once a proper level of accessibility is achieved. Often the distinction between construction activities, such as spot improvements and stage construction, and the usual concept of maintenance activities is completely blurred.

I. DEFINITIONS

***15.02* Routine maintenance**

consists of those work items regularly performed by maintenance personnel throughout the year [77]. The activities carried out under routine maintenance are [34]:

- (i) filling potholes and ruts with material similar to the material used for the surface layer, and compacting this material;
- (ii) maintaining the correct surface camber by retrieving loose material from the edges and respreading and compacting this material;
- (iii) removing corrugations;
- (iv) cutting vegetation on the shoulders, between the shoulders and the side ditches, and in areas where visibility is hampered;
- (v) repairing erosion channels formed on the travelled way (running surface), the shoulders or the ditch slopes;
- (vi) cleaning waste material such as debris, vegetation and silt from side ditches, interceptor ditches and diversion ditches;
- (vii) maintaining the original grade and cross-sections of all ditches; and
- (viii) cleaning silt and debris from culverts, fords (drifts) and other structures to allow a free flow of water.

***15.03* Periodic maintenance, recurrent maintenance** [75] **or curative maintenance** [37]

consists of more extensive maintenance operations that are required only every several years. Examples include:

- (i) reshaping and, where necessary, raising the level of the crown of earth roads above the surrounding environment [34];

(ii) regravelling gravel roads [17, 34, 77];
(iii) providing a new surface dressing for bituminous surface-treated roads [34];
(iv) reshaping drainage ditches [17]; and
(v) redecking timber bridges [43].

The objective of these maintenance activities is to restore a facility more or less to its original condition but appropriate to the current level of traffic which may well be much larger than its original design level; these activities can therefore be considered as capital improvements although they are usually termed maintenance because they are often undertaken by maintenance personnel. Routine maintenance of tracks or roads is undertaken by regular maintenance crews under any of the appropriate alternatives discussed below (Paras 15.12–15.17). Periodic maintenance is usually carried out by specialized force account or contract work crews.

15.04 **Emergency maintenance** [17], **special maintenance** [34], **emergency repair** [75] **or unplanned periodic maintenance** [77],

are those activities that are required after the partial or total collapse of a facility. Repair of bridge, culvert, or road washouts, and correction of rockfalls and landslides fall within this cateogry. Such special problems [34] can be created by natural disasters or by design or construction mistakes. The repairs should be undertaken only after the cause has been determined, and should include any additional work required to prevent a reoccurrence of the failure.

15.05 Part I of this handbook (Para 2.18), recommends a two-year post-construction *review and correction period* during which the maintenance activities described above (Paras 15.03 and 15.04) would be considered part of the original construction activity. During this period, minor corrections are anticipated to improve the results of minimum cost design. The corrections could include placing additional thickness of surface material in specific locations and the possible addition or relocation of small culverts and drainage channels. Such problems would not be life-threatening and would be easily, quickly and inexpensively repaired. The concept of acceptable risk drainage design was introduced in Para 12.10 as an acceptable alternative to the additional costs incurred through needless overdesign of minor drainage structures to compensate for insufficient design data.

15.06 **Upgrading** [17], **rehabilitation** [2] or **betterment** [77]

involve major restoration of infrastructure or minor construction projects. These activities are capital improvements that, in well-maintained highway infrastructure systems, are often undertaken by maintenance personnel to improve minor bottlenecks. In poorly maintained systems, the terms are often used to described major reinvestments to reconstruct infrastructure that no longer can function as originally intended due to lack of proper maintenance. In this handbook, these terms describe the activities undertaken as spot improvements and stage construction.

15.07 Non-maintainable [77] roads

are existing sections of rural motorable access infrastructure that have deteriorated to the point where routine maintenance activities can no longer ensure accessibility (Figure 26); or are unengineered tracks which become impassable during the rainy season due to inadequate drainage. These types of infrastructure require improvement before routine maintenance can have any impact on continued accessibility. This handbook suggests that the proper approach to improving accessibility is to make any existing infrastructure

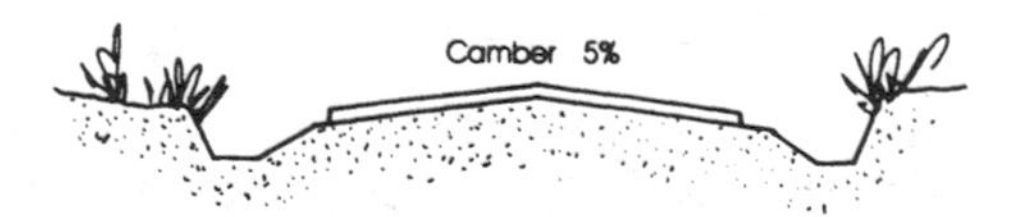

Properly constructed roads (above) deteriorate without adequate maintenance.
As the camber is lost, surface drainage fails and the ditches begin to silt up.
Traffic, combined with surface water retained after every storm, soon begins to form potholes.

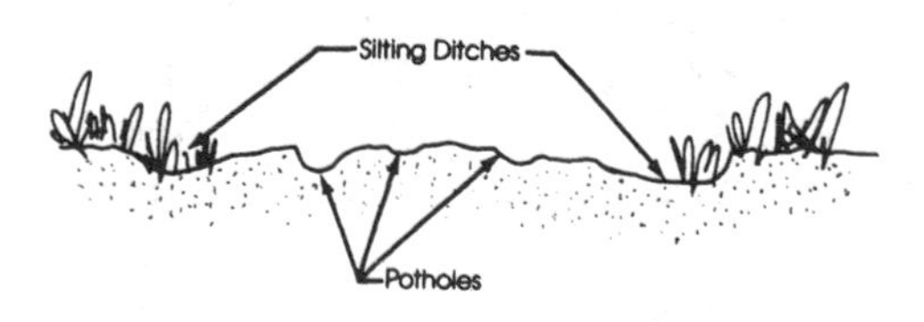

Eventually traffic forms ruts in the rain softened soil which accelerates failure
since the water remains in the ruts in flat sections until it evaporates, or causes
longitudinal erosion in sloping sections, in either case creating a non-maintainable road (below).

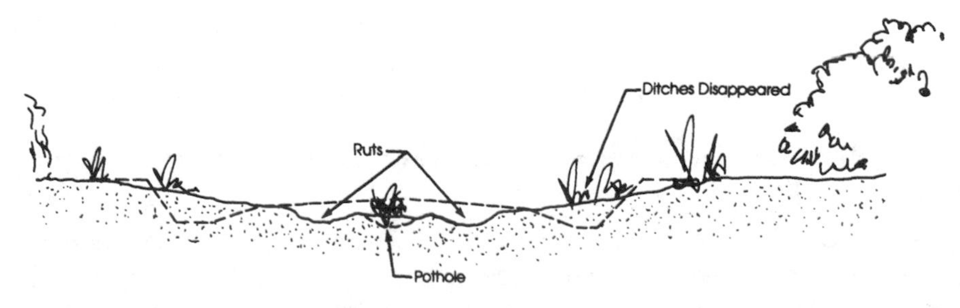

Fig.26 *How roads become non-maintainable*

maintainable by spot improvements as the first step in stage construction. It further contends that once access is available, further improvements to the infrastructure should be limited to those improvements required by increased transport needs introduced by the improved access. However, for those needs to develop, the infrastructure must receive sufficient maintenance to retain a catalytic accessibility.

15.08 The maintenance concepts discussed below are primarily concerned with the routine maintenace (Para 15.02) necessary to keep the access infrastructure operating until the need for periodic maintenance (Para 15.03) or further stage construction is established. Continued maintenance of any existing facility must take into account the use being made of it; therefore, if an appropriate level of maintenance cannot be justified, the facility should be abandoned [17].

II. MAINTENANCE TECHNOLOGY

15.09 Appropriate technology, as defined in Para 14.01, is as important to maintenance as it is to construction. It is possible to construct and maintain by different technologies, but the construction technology and maintenance technology should be selected at the same time [34]. Some of the constraints to varying technologies between construction and maintenance are described in Technical Note 7.1, Maintenance Technology Considerations.

15.10 Many maintenance activities are best carried out by labour under any technology choice. Culvert cleaning, cleaning ditches with obstructions such as ditch checks, removing stones from the travelled surface, trimming and removing tree limbs, removing fallen trees, cleaning drainage channels, repairing masonry, maintaining bridge decks and filling potholes (Figure 27) are common manual tasks [34, 43]. Therefore, the technology choice is actually limited to the manner in which the travelled surface and the areas immediately adjacent to it, including standard ditches on feeder roads, are to be maintained. A major factor to be considered during the evaluation of equipment-based surface maintenance activities is the problem of maintaining, i.e., supplying, servicing and repairing, the maintenance equipment itself [77].

III. MAINTENANCE IMPLEMENTATION

15.11 Technology implementation can include maintenance activities undertaken by contracting, force account, or self-help groups similar to those described in Para 14.03. Due to the dispersed nature of most maintenance activities on rural access infrastructure, these activities are best carried out by small local contractors, small force account groups or self-help efforts.

***15.12* High technology maintenance activities generally use the cyclic system of maintenance** [75]

whereby routine road maintenance is scheduled and maintenance units are organized to cover each infrastructure section at the required time interval. The units may be assigned specific tasks over a large area or all tasks over a

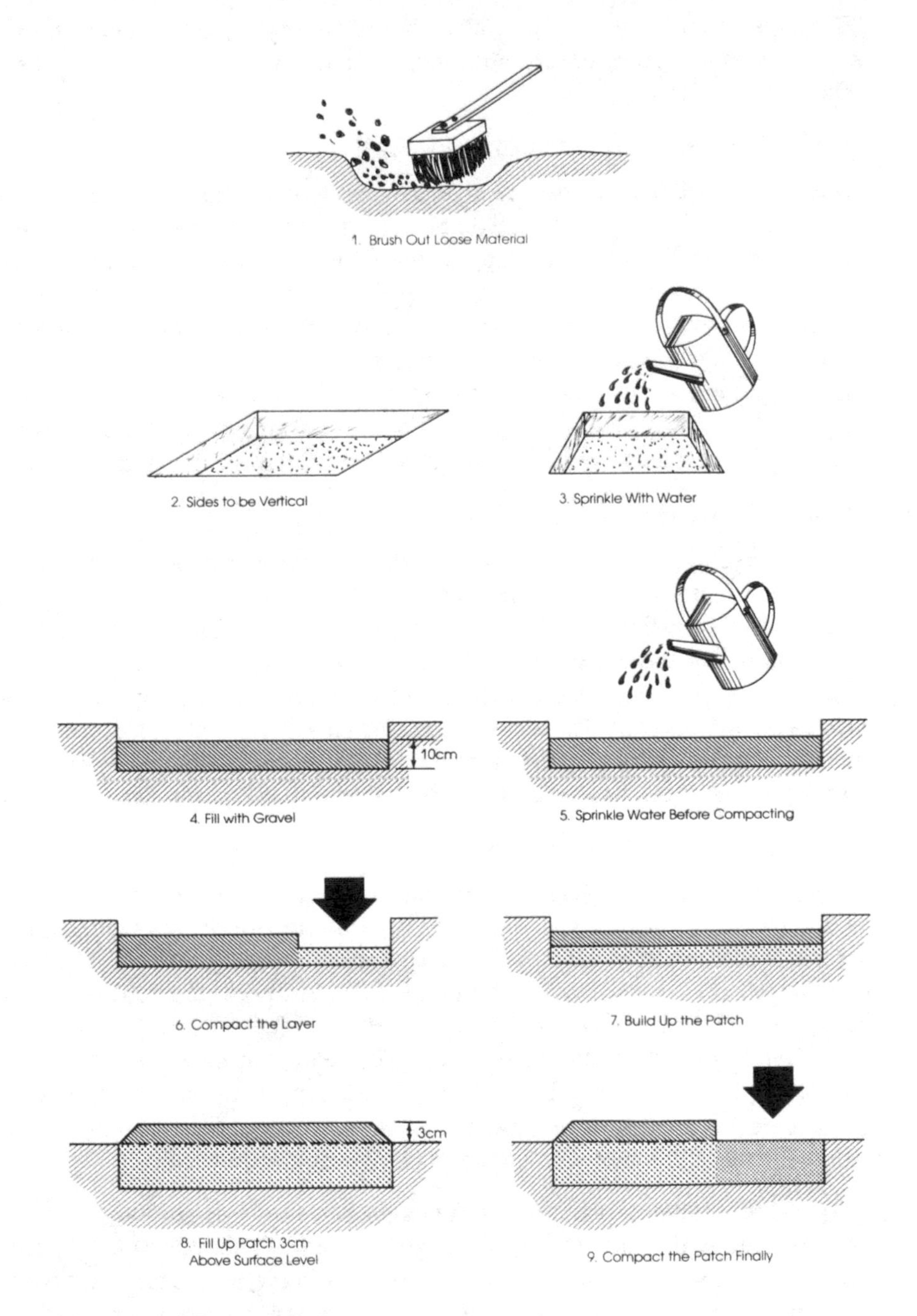

Fig. 27 *Filling potholes*

smaller section of infrastructure. A high degree of mechanization and specialization are possible but the cyclic system is most suited to high-standard roads with known traffic volumes where maintenance needs are predictable.

This type of maintenance is usually undertaken by force account with permanently employed labourers, although maintaining sizable road networks by contract is increasing.

15.13 The area system of maintenance [75]

concentrates all the maintenance resources at one location. Areas usually include 100 to 300 km of road and employ a maintenance force of approximately 10 men. The area is inspected regularly and crews are made up daily to handle the various maintenance tasks as determined by the periodic inspections. This system is most suitable for a close network of roads within a relatively limited area and is usually undertaken by force account with permanently employed labour. It is easy to administer because the crew is based at one site. Like the cyclic system of maintenance, the area system is normally not compatible with the type of infrastructure considered in this presentation.

15.14 The patrol gang system [75]

includes several units, each consisting of three or more men and a motor truck. One field supervisor may be able to direct as many as 10 patrols and allocate tasks to combined mobile gangs when one patrol unit has localized difficulties. Each patrol can inspect and maintain up to 150 km of roadway. This type of maintenance is usually undertaken by force account with permanently employed labour, since contractors are not normally interested in it. If gangs are located in camps 150 km apart, as much as four working hours may be lost to transport when work has to be carried out at a point 75 km from the camp [34].

15.15 The community contract system

is defined in this text as any maintenance activity for which a national or local government provides support to an organization for the maintenance of a section or sections of rural access infrastructure. The organization may take the form of a legally constituted association for community development as is found in Mexico [58], a village, tribe, cooperative, religious group, non-profit organization or merchants' association. The maintenance activities may be equipment-based or labour-based, but are most likely to be labour-based when considered for the types of rural access infrastructure described here. The labourers actually maintaining the infrastructure in turn receive:

(i) all, or a large portion, of the funds paid to the organization;
(ii) more than the amount paid by the government when a government pays a flat fee to a merchant or business organization in return for that organization assuming complete responsibility for infrastructure maintenance. This only occurs in situations where the government payments are insufficient to cover real expenditures and it is in the interests of the organization to maintain the infrastructure for its own purposes;
(iii) payment, or partial payment, in foodstuffs and/or other commodities when the contracting group is socially oriented, such as many communal, cooperative, religious and non-profit organizations; or

(iv) no payment, if the funds are to be allocated for organizational purposes, purchase of materials, etc., and the labourers are instructed by their leaders to perform the work for the good of the organization. In such cases it may be necessary for the government to employ 'Promoters' who act as *de facto* maintenance supervisors; their job would include motivating, mobilizing and training the local maintenance workers.

15.16 **The lengthman system** [75]

of maintenance involves a contract between a local individual (the lengthman) and a government agency, whereby the individual receives payment to carry out all routine maintenance over a fixed length of infrastructure using government-furnished hand tools. Payment is contingent on satisfactory performance. This system operates satisfactorily in populated areas where the lengthman lives close to his assigned section of infrastructure. Lengthman maintenance is suitable for all types of rural transport infrastructure, including feeder roads with low traffic volumes. The lengthman is motivated to do good work because he can earn local status and recognition as a valued community member. A large group of lengthmen are assigned to one supervisor who can usually provide a supplementary mobile gang to handle larger tasks. One supervisor and vehicle may handle inspection and wage payment for a 200 to 300 km system (150+ men) of low-volume roads if all the roads lie within a radius of about 100 km from his base station [16].

15.17 **Self-help maintenance**

is rural access infrastructure maintenance undertaken voluntarily, without payment, by the local population. In the context of this text, self-help maintenance activities are defined as occurring only when no payment what-so-ever is received by the participants. When tools are supplied or any payments are involved, this handbook considers such maintenance activities as community contract maintenance undertaken by a definable organizational entity even though self-help may be a motivating factor to the individuals involved. Today, few, if any, self-help road maintenance operations exist in developing countries. In the past, it was most commonly found in areas where the residents have a strong sense and history of community betterment activities, and where the population perceived the benefits from maintenance as accruing specifically to themselves. Self-help maintenance is more likely to succeed when the original infrastructure is built by villagers using self-help construction since (i) at village level the community has no other roadbuilding activity to compete for attention, (ii) the village will suffer directly the costs of inadequate infrastructure maintenance, and (iii) the village residents have no certainty that a rehabilitation project will be undertaken by the government if the road is allowed to deteriorate [65]. The probability of self-help maintenance programmes succeeding decreases considerably if the village population has no self-help history, is not involved with the planning and construction of the road, or if the road services more than one settlement or tribal group of residents. The concept of self-help is more likely to succeed for maintenance activities of paths and trails rather than tracks and roads, since these activities are simpler and in the direct interest of the users.

15.18 From the above description, it is clear that the last three maintenance methods (Paras 15.15–15.17) are the most suitable for rural access infrastructure and are certainly the most feasible for paths and trails. Since routine maintenance of all the rural access infrastructure considered in this handbook requires small but continuous inputs of resources over a rather large number of separate points, labour-based maintenance is particularly cost-effective where population densities are high enough so that workers living adjacent to the infrastructure obviate the need for expensive transport [22, 75].

15.19 Using labour-based maintenance methods in developing areas has the further advantage of supplying cash incomes in the non-cash environment of subsistence farming. These maintenance activities therefore help the transfer to a cash economy which better access and improved farming methods seek to encourage. However, routine maintenance is needed most during the planting and harvest seasons when traffic is heaviest and the weather is often wet. This is the time when the available labour supply is lowest. As development proceeds, the labour pool of workers available for maintenance activities during these critical seasons will continue to decline as the agricultural cash economy increases.

IV. MAINTENANCE BY INFRASTRUCTURE TYPE

15.20 Path and trail maintenance is frequently undertaken by the users who cut back the brush, replace or reshape the surface, and alter the alignment when short sections become difficult. When an outside agency participates in the construction of such infrastructure it can only be with the understanding that maintenance will be undertaken as a self-help activity. Therefore, the agency's input to the maintenance activity will necessarily be limited to introducing motivation and training in proper maintenance techniques.

15.21 In principle, track maintenance is also predominantly a self-help activity; however, since the added width requires a greater maintenance effort, communities without a self-help history may be reluctant to participate, or may assume some one else is going to provide the maintenance. Some sort of community contract system should be introduced in order to make one or several individuals responsible for maintenance activities. Motivation may consist of a promoter, a continuing supply of tools, and/or a token payment to ensure responsibility. Alternatively, the lengthman system may be introduced.

15.22 Earth and gravel road routine maintenance activities require a more serious and ongoing participation by some formal governmental organization. Sole dependency on self-help maintenance for these roads is an extremely optimistic approach. Community contract maintenance (Para 15.15) of earth and gravel roads is also not likely to be successful if there is no strong community organization with widespread support, or if the labourers are coerced [14]. However, community contract maintenance undertaken through an active, well-organized body can supply a maintenance capability that may not otherwise exist. The overseeing governmental agency should make every effort to support such community contractors, not only through prompt payments and proper monitoring, but also through training opportunities for both local supervisors and workers.

15.23 The lengthman system of low-volume rural road maintenance is again receiving increasing interest in developing countries. Since it involves contracts between individuals and a governmental agency, it can be utilized in areas where community organization activities are lacking, providing the population base is sufficient. The lengthman system is investigated in more detail in Technical Note 7.2, Lengthman Maintenance Contract System.

15.24 Low-volume earth and gravel road maintenance is, of necessity, a decentralized operation. As such, combinations of community contract and lengthman maintenance activities are appropriate providing there is no perceived difference in the remuneration of labourers working on adjacent sections of the same road.

15.25 In areas where population density or social strictures will not support labour-based maintenance of rural roads, the patrol gang method of maintenance (Para 15.14) is probably the most appropriate one [75]. The ethnic, religious or social factors which may affect the establishment and performance of such maintenance units should, however, be reviewed before the system is implemented [52].

V. ROAD SURFACE MAINTENANCE ACTIVITIES

15.26 Corrugations form in some degree on all earth and gravel road surfaces. They are most common in non-plastic materials and are a major factor in determining the level of effort required to maintain such surfaces in a rideable condition. While road roughness is normally associated with higher vehicle operating costs on more heavily trafficked roads, a more immediate concern on the type of low-volume rural access infrastructure described here is the water retained on the surface by corrugations. Retained water leads to structural degradation and pothole formation, both causes of early road failure. The rate of corrugation formation [79] is influenced by traffic type and speed, type of surface material, road crown condition, and traffic volume.

15.27 Surface corrugations are most easily removed systematically before a hard crust is formed, especially when such removal is done by hand labour. Otherwise the entire road surface will have to be reshaped, a much more difficult operation. Corrugations can be removed by hand labour with brooms or rakes; by drags consisting of piles of brush, chains, large tyres chained together, or weighted frames with blades; by mechanical brooms; or by road graders. The drags, mechanical brooms and small road graders may be towed by animals, tractors or trucks. Large graders are self-propelled, and are particularly necessary when corrugations have become densely packed and hard under the action of traffic.

15.28 The frequency of such maintenance procedures can best be determined by observation. However, since corrugation removal can be such a predominant factor in the amount of surface maintenance on earth and granular surfaced roads, the estimated number of vehicles passing over the roadway between corrugation removal operations is often used as an indicator to determine the frequency of corrugation removal. Technical Note 7.3, discusses the relationship between traffic volumes and corrugation-induced maintenance activities. Technical Note 7.4 summarizes various published data about labour-based maintenance worker output, and Technical Note 7.5 lists the tools and light equipment the labour-based maintenance worker requires.

VI. PLANNING FOR MAINTENANCE

15.29 As indicated in Para 14.04, no construction should begin before the method and funding for maintaining the infrastructure have been established. When community contract maintenance is planned, a specific maintenance agreement executed prior to construction will both take advantage of the existing mobilization of the community around the infrastructure project and make it easier for the local people to accept partial construction or spot improvements if their only other choice is no improvement at all [65].

15.30 Labour-based maintenance stresses supervision and prompt payment [43]. Therefore, there must be an ongoing training activity for supervisors, not only to replace existing personnel, but also to provide for maintenance of additional infrastructure. The financial resources to pay the maintenance workers must be properly allocated and be available as soon as construction is completed [16]. The allocation of maintenance funding before budgeting new capital projects [17] seems to be the only successful method for ensuring the prompt payments which keep a labour-based maintenance programme from collapsing.

15.31 The maintenance operation must begin with clear lines of responsibility. A maintenance supervisor should be responsible for all feeder road maintenance activities in an area. Foremen or overseers should be responsible for specific road links, or work tasks in the case of mobile gangs. If the lengthman maintenance system is adopted, roadmen are assigned individual road sections. Maintenance deficiencies can therefore be related to individuals and causes can be discussed with the persons responsible. This system of responsibility also encourages competitiveness and improved performance [46].

15.32 From the infrastructure condition inventory described in Chapter 12, it should be possible to assess the adequacy of any existing maintenance activities in a project area. Future maintenance needs can also be evaluated and the most appropriate solutions determined concerning maintenance technology, organization, and funding. The emphasis should, of course, be on developing maintenance systems and procedures which are sustainable. If the current maintenance system is functioning well, it may be possible to expand it. Otherwise, the existing maintenance system must be modified and improved, to include not only maintenance activities on the motorable roads but also on the other links that constitute the rural access infrastructure network.

15.33 The evaluation of a proposed maintenance programme must consider (i) the required maintenance needs and tasks, (ii) the appropriate maintenance technology, (iii) the required instruments and capabilities for implementing the maintenance programme, (iv) the form the maintenance organization will take, and (v) the provision of resources for the maintenance efforts. Technical Note 7.6, Maintenance Programme Checklist, itemizes the considerations that must be evaluated when planning a maintenance programme for rural access infrastructure.

15.34 The evaluation procedure (Para 15.32) may uncover existing or potential constraints to some or all of the maintenance programme options thought feasible for a specific RAI network. Technical Note 7.7, Reducing Maintenance Constraints, consists of a matrix of possible constraints and potential solutions. The elimination of such constraints must become an integral part of any maintenance option selected. A successful maintenance programme must function within the level of resources that are available on a sustainable basis. Therefore, when an analysis indicates an adequate maintenance system cannot be established for a potential RAI improvement programme, the programme should be modified so that it can be adequately maintained. If adequate maintenance is impossible, the improvement programme should be abandoned.

VII. CONCLUSIONS

15.35 The largest untapped or underutilized available maintenance resource in many developing countries is local labour. This is particularly true for the maintenance of rural access infrastructure in areas with high underemployment and/or unemployment. In such situations major tasks such as road realignment, regravelling, bridge building and repairing larger landslides and washouts will be beyond the capacity of small local labour-based RAI maintenance organizations. However routine maintenance and emergency repairs of minor landslips and washouts should be well within their capacity. The use of an available labour resource for maintenance, therefore, requires certain adjustments to conventional maintenance concepts, such as:

(i) accepting a construction methodology and implementation that will be conducive to labour-based maintenance;
(ii) developing an ability to think small;
(iii) foregoing self-contained maintenance units which have the capability and equipment to deal with any periodic or emergency maintenance activity without outside support;
(iv) accepting a limited routine maintenance capability that must be supplemented or assisted as soon as any increase in transport demand occurs;
(v) providing a steady supply of trained maintenance supervisors; and
(vi) maintaining a steady flow of cash to promptly meet all financial obligations.

15.36 A large dispersed labour-based maintenance operation is very fragile and without motivation, in the form of supervision, and a steady cash flow, it will disappear almost without being noticed. While self-help and low-cost community contract maintenance may ease the financial burden, it is unrealistic to believe that a viable national access infrastructure maintenance programme for earth and gravel feeder roads will function on goodwill or promises of future largesse.

15.37 The use of available labour resources for RAI maintenance is beneficial to both the local population and the government, because the funding is in local currency which stays within the local community, and because the employment of local labour shows the concern of the central authorities for the wellbeing of the local population by providing both cash and better access. A properly organized community contract or lengthman maintenance programme for low-volume feeder roads is;

(i) cost-effective;
(ii) relatively simple to administer;
(iii) immune to increasing fuel and equipment costs;
(iv) independent of externally-induced shortages;
(v) capable of producing visible accomplishments; and
(vi) a decentralized approach to the necessity of developing a local capability for maintenance, which is the most viable option for a continuing infrastructure maintenance programme in rural areas.

CHAPTER 16

Institutional Issues

16.01 This chapter explores institutional issues that frequently arise in rural roads projects or in RAI components of development projects [13]. Its objective is to provide guidance to project designers and programme planners on factors which should be taken into account in establishing the institutional framework for rural access infrastructure planning, construction and maintenance. Improved institutional design will result in projects that have a better chance of being sustainable, i.e., making an effective long-term contribution to rural development.

16.02 Rural access infrastructure (RAI) is fundamentally different from highways in ways that affect the choice of an appropriate institutional setting for each of the three stages of planning, construction and maintenance. The preparation of a RAI project requires inputs from several agencies and from the communities to be served. Frequently it requires rapid screening of many proposals rather than detailed study of a few alternatives. In construction, the need to keep costs down and the desire to benefit the rural poor place a premium on the use of labour-based methods wherever feasible. Maintenance responsibilities require the development of local government and/or community capabilities complemented by effective technical and financial support from the central or regional government.

I. THE POLICY ENVIRONMENT

16.03 In preparing RAI programmes, planners should first consider the overall country context, including the degree of political commitment to rural development objectives, the role of the private sector in development activities, the opportunities for introducing labour-based methods of RAI construction and maintenance, the legal and regulatory framework for rural development, central government planning and budgeting procedures, and modes of project formulation. Investment in RAI improvements may be identified either as part of an overall strategy for the transport sector, or as a supporting investment for agriculture, forestry, irrigation or rural development programmes. The first investment in RAI would normally be regarded as a pilot project with potential implications for the design of a larger RAI programme. Those agencies that are expected to participate in an expanded RAI programme should be involved in the planning and financing of the pilot project.

16.04 If a government wishes to implement a large-scale RAI programme using labour-based techniques, it should be aware of policies that indirectly affect the feasibility and profitability of such techniques. Among the policy issues that are likely to be important are the following:

(i) fiscal policies and use of shadow prices for labour or surcharges for equipment;
(ii) discriminatory tariffs on equipment intended to build roads with equipment-based techniques;
(iii) interest rate policies that encourage labour-based construction, i.e., do not subsidize the import of construction equipment;
(iv) tendering procedures which are not biased towards the use of equipment, e.g., minimum plant holding requirements; and
(v) equal status between staff of technical departments working on labour-based and equipment-based techniques.

16.05 Regardless of the choice of construction technology, policy makers need to consider support for the following strategies:

(i) special training programmes for engineers, technicians, local government officials, community leaders and others involved in RAI work;
(ii) simplified administrative standards and procedures;
(iii) simplified engineering preparation, contract documents, and bidding and disbursement procedures to be applied in RAI work; and
(iv) development of a domestic contracting industry.

16.06 In relating RAI investments to a more general rural development investment strategy, planners face three main options: (i) separate single-sector investments; (ii) integrated rural development projects focused at the local level; and (iii) integrated, time-phased investments at the regional level. Single sector investments are attractive because they are relatively easy to implement. Integrated rural development projects are designed to maximize the return on resources in the long run, but are expensive and complicated to plan and implement if a large beneficiary population is to be reached. Time-phased sector-specific investments permit flexibility in the allocation of resources, require less stringent intersectoral coordination, and will spread immediate benefits to a wider group of people. Single sector investments seem most appropriate under conditions of low institutional development. Integrated rural development requires a high level of institutional capacity, with time-phased programmes occupying an intermediate position.

16.07 In preparing a programme of RAI improvements, planners must look beyond the technical perspective of the public works ministry to consider the priorities and constraints of the finance ministry, other participating ministries, the national legislature, and local government, as well as those of potential donors. Prescreening criteria may reflect the non-quantifiable but very real benefits derived from assuring an equitable geographic distribution of investments, meeting the transport needs of agency personnel, or serving special population subgroups. Planned RAI construction should never be allowed to exceed the country's foreseeable maintenance capacity, unless adequate additional resources for maintenance will be mobilized through the project.

16.08 Planning and budgeting procedures may range from a rigid, formal, centrally controlled process to a flexible, informal approach which can be responsive to local variations in needs and resources. The first is sometimes referred to as a 'blueprint' approach, in which every staff action and decision takes place according to a set of predetermined rules. The second is called the 'learning process' approach. By allowing agency administrators to make their own decisions and to take independent action within broad general guidelines, the organization can make the most effective use of its physical, financial, and human resources and can respond rapidly to changes that may occur in the project environment. Use of the 'learning process' approach is recommended whenever there is uncertainty about the relationships of the project to broader rural development objectives.

II. ORGANIZATIONAL OPTIONS

16.09 When loan funding is employed, the project planner must select a set of implementing institutions to carry out the various tasks involved in the RAI component. These institutions may include the borrower itself, a branch of the borrower institution, an executing agency which will receive funds from the borrower, a special project unit, a parastatal organization, a non-governmental organization, or a community group. The more links in the chain from sources of funding to implementing institution, the more complex and cumbersome will be the implementation process. The situation will be further complicated by having several sub-borrowers, executing agencies, contractors and/or communities participating in the project. Such complex projects will require a large investment in administrative overhead in order to deal with the multiple demands and delays likely to arise in project implementation.

16.10 Responsibility for planning, building and maintaining RAI should be decentralized to the greatest extent possible, consistent with the availability of qualified staff and the degree of local and regional control over resources. In this way, RAI programmes can contribute to the broader objective of building self-sustaining institutions at the local level.

16.11 In general, it is better to modify existing institutions to meet new needs rather than to create new institutions, such as special project units, to execute RAI improvement programmes. However, existing institutions should not be limited to the traditional road-building agency. Other sectoral agencies such as agriculture, irrigation or forestry, regional development authorities, private voluntary organizations and community groups may all offer viable alternatives in particular circumstances and may be a more suitable choice than the traditional road-building agency.

16.12 Many World Bank-financed projects have supported the growth of a feeder roads unit within the traditional road-building agency. This approach has been effective in focusing attention on, and allocating resources to, the rural access problem in many developing countries. However, care must be taken to ensure that the growth of a feeder roads unit does not undermine the main mission of the parent agency. Furthermore, experience indicates that feeder road units are often not responsive to RAI below feeder roads, i.e., tracks, trails and paths, and in those cases it would appear necessary to also include other agencies in RAI programmes. The continuity of a feeder road unit depends upon the expansion of the rural economy and the gradual conversion of its capabilities from construction to recurrent tasks.

III. INTERAGENCY LINKAGES

16.13 Project planners must also pay attention to issues of interagency coordination and control. Coordination should begin at the point where possible RAI investments are identified in terms of their potential contribution to the rural development process. Coordination at the project preparation stage may be limited to cross-checking of facility location or may, in addition, call for a formal review of the RAI programme by an interagency planning committee or coordinating council, or may incorporate agency criteria and/or staff views in the infrastructure pre-screening process. The effectiveness of interagency coordinating committees appears to be inversely related to their organizational level and directly related to the resources which they control.

16.14 In addition to horizontal linkages between the road building agency and other agencies involved in the rural development process, vertical linkages need to be developed between the project and its sources of political support, including both the beneficiary communities and the central government. RAI projects require strong support from the central government in the form of financial commitments and expedited procurement procedures. Political support can be mobilized through action by regional authorities, community organizations, and other agencies whose programmes depend on the progress of the RAI component.

16.15 Projects envisaging feeder road construction through separate project units should specifically include support for the transfer of resources and skills needed for subsequent maintenance and repair to an appropriate institution, whether it be a district office of Public Works, an arm of the regional development authority, or local government. This institution must have legal responsibility for road maintenance and the authority to generate and handle funds. By the end of the project, it should be expected to control the fiscal, material and staff resources needed to prepare and carry out annual work programmes. Therefore, local communities must be made responsible in the case of maintenance of tracks and, in particular, trails and paths. Technical Note 8.1, Organizational Framework for Decentralized Road Maintenance, offers guidance for decentralizing national maintenance programmes to better accommodate feeder road maintenance activities.

IV. IMPLEMENTATION ISSUES

16.16 Use of the 'learning process' approach requires agency staff to re-think their goals and objectives, to take a fresh look at the potential contributions rural people can make to projects, and to learn to communicate more effectively with their clientèle. This process is called 'bureaucratic reorientation'. To be effective, it requires strong leadership and commitment from senior agency staff, supported by a national ideology favouring local involvement in development planning.

16.17 Expanding RAI activities usually means recruiting and training new staff. Unskilled labour should be hired locally on a temporary basis. Lower level management (crew chiefs and site supervisors) should be selected as much as possible from within the ranks and given the training needed for them to carry out their jobs effectively. Training programmes should be designed on the assumption that a certain amount of trained manpower will ultimately find opportunity to use their acquired skills elsewhere, for instance in the private sector.

16.18 Hiring local people as paraprofessionals is one way to improve communications between agency staff and beneficiary groups. Another way is to add socially trained outreach workers to the agency staff. In both cases, these persons assist the project by transmitting information concerning local needs and priorities to agency staff, and by presenting agency plans and proposals in terms which can be understood by the community [56].

16.19 Training should not be limited to agency staff, but should also support the development of local contractors and the maintenance management capabilities of local officials. Where community assistance is expected in carrying out the project, specific attention should be paid to the design of communication strategies to mobilize such support, to provide needed information, and to promote skill development among local residents. Broader training programmes build public support and provide another channel for feedback from project beneficiaries.

16.20 Local labour productivity is influenced by the payment method selected (Para 14.07); however, worker motivation is also influenced by non-monetary factors such as the quality of tools used, the relations among members of a work crew, relations between workers and supervisory staff, participation in work-related decision making, and observance of local customs and traditions. Local crew chiefs should be permitted to select their own workers to ensure compatibility on the job. Employment security and support services on field assignments are important for the morale of skilled workers and other civil service employees.

16.21 The motivation of higher level staff derives from values which have been constantly reinforced during their formative years, their professional training, and the organizational culture of their agency. Early value orientations are likely to be elitist, positivist and particularist. Professional training is likely to reinforce these values, adding to them a pride in the technical complexity of work performed. Consequently, these professionals prefer bridge or major road construction to the improvement of rural roads and maintenance activities. Normally, the organizational culture of a road-building agency will support the same value system. Only if the organizational culture is significantly changed (through 'bureaucratic reorientation') will higher level staff be motivated to participate effectively in the process of rural development. Their involvement will increase if they perceive their task as an organization and management challenge and as an effective means of promoting the well-being of the country's rural population.

16.22 Salary levels often hinder recruitment and retention of competent staff. Due to macro-economic constraints, salaries in the public sector may be fixed at levels that are not competitive with the private sector or with those in neighbouring countries. However, there are other mechanisms through which financial incentives can be offered to civil service staff. Chief among these are field day allowances, travel allowances and subsidized housing and schooling. Although expensive, such programmes represent a form of staff renumeration which is almost certainly less costly than the alternative of hiring expatriates and, in the long term, more effective.

16.23 In order for local staff to manage projects effectively, the responsibility for procurement should be decentralized to the lowest possible level, and off-shelf procurement of urgently needed spare parts, tools and materials without formal competitive bidding should be permitted at the discretion of the project manager. Local agency staff should also have access to the funds needed to provide for the prompt cash payment of wages. However, autonomy with respect to management of the project cash flow must be complemented with effective cost accounting and financial control procedures.

16.24 Monitoring and evaluation of RAI projects should recognize that such programmes are part of the broader rural development process. Monitoring will usually focus on the achievement of project progress or 'efficiency' objectives, while an evaluation of organizational matters conducted during the project period should focus on the achievement of desired behavioural change or 'instrumental' objectives. Evaluation of the socio-economic impact of RAI investments requires a longer time-frame, and it is difficult to sort out the effects of RAI investments from the effects of other changes that take place over this time period. Monitoring and evaluation activities may be qualitative as well as quantitative, and should include the participation of project beneficiaries.

16.25 As an individual country develops skills in RAI techniques, it should codify its experiences. Handbooks can be prepared giving instruction on the different features of planning and nurturing RAI which are specific to the environment of that country. Such handbooks, if clearly written, can be of value not only in training technical staff, but also in the local people's general education about their country's resources.

V. LOCAL PARTICIPATION

16.26 Local participation in project activities may occur at any time during the three stages of planning, construction and maintenance. By increasing the probability that local resources will be mobilized for subsequent maintenance, local participation in planning and construction of RAI projects can make these projects more cost-effective. The executing agency for RAI improvements should coordinate closely with local authorities and community organizations, as well as with other line agencies, in order to promote effective local participation in RAI projects.

16.27 World Bank-financed projects which have included feeder roads, but only few tracks and trails, illustrate a range of options for local participation in planning rural road projects. The key variable defining this range is local capacity to mobilize and manage resources. Where local institutions are well-established, a local government entity may retain control over the entire planning process, using consultants and contractors to deal with specialized design and implementation issues. A lesser degree of participation is achieved when local government or communities take part in decision making, including subproject selection, but do not have final responsibility for programme implementation. Communities may also make important contributions to project details, based on local knowledge of the physical and socio-economic environment. At a minimum, most national rural road programmes expect local government or communities to take the initiative in proposing specific RAI improvements.

16.28 Local participation in construction comes about when a beneficiary community supplies some of the resources needed to carry out a RAI improvement project. These may include land, labour, tools and equipment, construction materials, support services, and/or funds to pay for these resources. Such participation may occur in equipment-based as well as in labour-based projects.

16.29 Land acquisition for RAI improvements may lead to an inequitable distribution of costs and benefits in the community. Where financial compensation is planned, timely public notice concerning the new alignment may help to inhibit speculation by 'insiders'. Where road improvements will result in a significant increase in adjacent land values, it may be necessary to provide protection for the tenure rights of those owning or working the land.

16.30 The use of local labour to carry out RAI projects obviates the need for site camps and minimizes the employer's responsibility for the welfare of workers and their families. It also avoids the social and economic problems that may arise when a temporary work force is introduced into a rural community. Furthermore, it ensures that wages paid for RAI work circulate within the rural economy, contributing to its growth through a multiplier effect.

16.31 Many local government entities and rural communities have traditional methods of construction for minor civil works, using local materials and familiar forms of work organization. RAI projects designed to maximize use of these methods can achieve an acceptable standard of performance at considerably lower costs. The use of local materials and familiar forms of technology also makes it more likely that local government and communities will be able to assume future responsibility for RAI maintenance.

16.32 In choosing an appropriate technology for RAI construction, planners need to consider the quality and availability of local tools and equipment. If tools are to be provided by a road-building agency, communities can be asked to assume collective responsibility for their security. Local government may also be in a position to provide spare parts, tool maintenance and repair services. Locally owned vehicles can be used to haul select materials and to assist in simple earthworks, either as part of a community contribution or through individual contracts.

16.33 RAI programmes may not provide an adequate market to support the growth of a specialized hand tool manufacturing industry in each country. Therefore, procurement of internationally competitive tools is recommended in most cases. Specifications should be drawn up to ensure an adequate standard of performance on the job. However, cumbersome central procurement procedures should be relaxed, and field managers should be permitted to procure suitable tools from local suppliers.

16.34 Direct community participation in financing RAI construction is rare, except in cases where communities have substantial cash income from remittances or other sources. Ideally, the local government should have the ability to collect local revenues which would be supplemented by appropriations from the central government where needed. Local government may increase its revenues by expanding the types of taxes which it is authorized to collect and by stimulating development that will increase the value of the tax base. Financial resources may be transferred to local government by the central government through general revenue sharing, earmarked taxes, or lending for specific projects. Because local participation in project funding must place a significant part of the decision making power at the local level, it requires a certain level of technical and financial skill at this level in order to be successfully implemented.

16.35 Many projects have assumed that the maintenance of improved roads will be carried out by local governments or rural communities. However, there are few examples of successful implementation of this strategy. The poor past performance of local government is linked to the limited extent of authentic community participation in rural road planning and construction, combined with the top-down approaches prevailing in most public works departments. The approach outlined in this chapter is intended to improve the implementation of this strategy in future projects.

16.36 Chapter 15 described the following institutional options for local participation in maintenance: patrol gang (perhaps carried out by local contractors), community contract, lengthman, and self-help systems. Local contractor capacity can be developed through provision of credit, training, and supervised subcontracts. In some countries, parastatal firms may undertake road maintenance on a contractual basis. To be effective, this option requires a government commitment to support the private or parastatal sector.

16.37 The willingness of rural people to participate in RAI maintenance depends upon their perception of the costs and benefits which are brought to them by the RAI. An equitable distribution of costs and benefits within the community may be as important as their magnitude in determining community response. An improved understanding of the distribution of project costs and benefits can be gained by involving local people in project monitoring and evaluation activities.

VI. THE PARTICIPATION PROCESS

16.38 The initiative for local participation in RAI projects may come either from the central government or from the local government or community leaders. Centrally initiated or 'top-down' approaches have the advantage of greater resources, more consistent detailed planning, improved financial accountability, and transfer of skills to local leaders. However, plans may be too rigid or inappropriate, local resources may be ignored, and participation may be restricted to an elite group. Locally initiated or 'bottom-up' approaches tend to be more politically acceptable, help to relieve pressures on agency staff, and foster the development of self-confidence and skills at the local level. However, such initiatives are more difficult to coordinate and control and are subject to some of the same dangers of being co-opted by a local elite.

16.39 Local participation may be purely voluntary, induced by a system of rewards, or coerced by a system of punishments for non-compliance. Coerced participation is not usually very effective or lasting. Purely voluntary participation is preferable, but it, too, may be hard to sustain over the long run. Planned use of inducements such as cash payments or Food for Work may be the best way to ensure the continuity of cooperative efforts throughout the project period.

16.40 Participation may take place through formal organizations, informal groups or individual involvement in project activities. Formal organizations tend to have rules, standards and procedures that prevent them from reaching the rural population. Also, the goals of the project may to some extent conflict with other organizational goals. Local voluntary associations or informal groups are likely to be more flexible and responsive to the needs of rural people, and are therefore more likely to elicit effective participation.

16.41 While direct participation in project activities by individuals helps to develop local skills and may foster cooperation between different groups, it increases the complexity of project management and may add to the level of conflict in the community. Large projects that reach a diverse population may be more effectively managed through indirect participation, e.g., representation, while small projects with a homogeneous target population may accomplish more by adopting direct forms of participation, e.g., community meetings.

16.42 Three key variables which have a significant influence on patterns of participation are age, sex and land tenure status. Young persons are more likely to participate in project implementation, while older people have more influence in project decision-making. Women are often excluded, by custom or by decision, from direct participation in project activities, although they may exert considerable influence in an indirect role. Landless persons and others who depend on wage labour to support their families are also at a disadvantage in the participation process. Strategies for successful participation should include specific ways of reaching out to these disadvantaged groups, in order to make the most effective use of all available human resources.

16.43 The role of local leaders in the participation process must be carefully analyzed, since they can either facilitate community participation or help prevent it from taking place. Use of local people as agency paraprofessionals enhances the prospects for effective communication between agency staff and local leaders. Local paraprofessionals will be more effective in their roles if they are also supported by reliable training and technical supervision.

16.44 Other actors in the participation process include local line agency staff, central government officials, and foreign personnel. The participation process is also limited by the attitudes and behaviours adopted by each of these groups, reflecting in part the values and objectives of the organizations and cultures to which they belong. Some bureaucratic reorientation may therefore be needed within both donor and implementing institutions in order to make community participation a truly effective tool for development.

16.45 The following strategies for successful participation in rural roads projects are therefore recommended:

(i) seeking central government support;
(ii) promoting bureaucratic reorientation in executing agencies;
(iii) supporting local organizations;
(iv) reaching out to disadvantaged groups;
(v) strengthening local government; and
(vi) overcoming donor agency constraints.

16.46 Donors' role in designing RAI projects for local participation must be limited and indirect. It can be most effective in strengthening the legal and administrative structures that surround participatory projects, and in promoting the attitudes and behaviours of agency staff that will facilitate local participation.

16.47 The goal of institution-building components in RAI projects is to develop self-sustaining institutions at the local level which continue to function because they succeed in identifying and meeting community needs efficiently and economically. These institutions should be closely linked to the sources of political support for RAI programmes, including the central government, the line agency responsible for road building, other agencies involved in the rural development process, and the beneficiary communities. Institutional arrangements that work well for the planning, construction and maintenance of RAI may then serve as a prototype for similar structures intended to mobilize resources in order to meet community needs in other sectors as part of the broader rural development process.

VII. CONCLUSIONS

16.48 To be successful, an RAI programme must:

(i) receive strong political support and commitment;
(ii) be integrated within a national development strategy or policy that determines priorities and the optimal allocation of scarce available resources;
(iii) be coordinated with area agricultural and social service development;
(iv) be administered by an institution that will provide leadership, technical expertise, funding control, and an appropriate response to local needs; and
(v) have as a long-term objective the development of decentralized local institutional capability to continue and expand the national plan.

16.49 Several recommendations [54] can be made concerning the institutional issues that arise within the context of establishing effective organizations to meet the needs of the rural population and concurrently make a long-term contribution to national development. These include:

(i) modifying existing institutions to meet new needs rather than creating new institutions;
(ii) decentralizing responsibility for planning, building, and maintaining rural transport infrastructure to the greatest extent possible, consistent with the availability of qualified staff and the degree of local and regional control over resources;
(iii) developing strong links at the working level between the road building organization and other agencies involved in rural development;
(iv) placing the greatest possible reliance on the local sector, including community groups, contractors, and other non-government organizations, consistent with their absorptive capacity constraints; and
(v) actively encouraging beneficiary participation in planning and construction to ensure that the rural access infrastructure programme meets their needs, thereby encouraging continued local participation in maintenance activities and strengthening local decision-making and development capabilities.

CHAPTER 17

Cost Evaluation and Control

17.01 Cost evaluations are necessary for determining the economic feasibility of rural access infrastructure (RAI) projects, as well as for selecting the appropriate technology to be used in their execution. The evaluations may be in financial costs or economic costs. Financial costs are the actual outlay of funds for the project. When these costs are adjusted by deleting taxes and/or subsidies, and/or by applying shadow prices to foreign exchange costs and wages, they become economic costs.

17.02 The use of economic costs in feasibility studies is desirable to reflect the true costs to a country. For instance, taxes and subsidies are merely intergovernmental shifts of resources. In many countries minimum wage laws result in paying wages that do not reflect the real cost of labour. In situations of extensive under- or un-employment, the real labour costs may be less than actual wage rates; consequently shadow prices rather than actual labour wages are used in economic analyses. Economic costs are important when a country decides how to allocate its limited resources. However construction and maintenance organizations, either privately- or government-owned, must use financial costs when preparing the budgets for their activities. The costs used in the technical notes supporting this chapter are financial costs.

17.03 Financial cost control is normally considered as an attempt to monitor project costs to determine if the actual costs of the project works are within budget. Cost control further implies added efficiencies in executing the project activities through assuring the timely availability of inputs such as materials, spare parts, tools, fuel and manpower during the life of the project. Cost evaluation and cost control are inevitably intertwined. Cost evaluations are ideally based on cost data derived from cost control documentation, while cost control uses cost evaluations to benchmark progress. Unfortunately, cost control techniques are not uniformly practised in many developing countries and, in the case of RAI, they may not be directly applicable to a technology which is not currently being used. In such circumstances, the cost evaluators must include intelligent estimates of all the factors which will determine the actual costs of project execution. The failure to properly account for these costs, either by overestimating unit productivity or by neglecting all but the obvious costs for production units, can have serious effects on donor agency project cost evaluations [2].

17.04 A study of per kilometre costs, submitted by different countries for proposed low-volume rural road construction projects under similar topographical and environmental constraints, using similar design standards, indicated that uneven accounting practices can frustrate any analysis of other variables such as terrain or design differences. This chapter therefore addresses some of the components necessary for both cost evaluation and cost control, since unit costing is the basis for both activities.

I. COST ELEMENTS

17.05 Unit costs are the foundation for all cost estimates and controls. Unit costs can be divided into the main categories: direct costs and indirect costs. These costs are usually further divided into local and foreign exchange costs in donor-supported projects. Direct costs are those costs directly associated with the particular project [25] and are indispensable to the execution of the work [55]. Direct costs include material, tool, labour and equipment costs that can be attributed to specific measurable quantities of work. Direct costs may include overhead costs attributed to a project but not included in productive unit costs, such as electricity, site administrative costs and mobilization. Technical Note 9.01 contains an itemized list of direct costs.

17.06 Indirect costs are costs not directly associated with a particular project site but allocated on a percentage basis to specific projects [96]. Such headquarters costs include: all headquarters expatriate and local staff expenses; office buildings and land costs, personnel and delivery transport; staff training activities; and support activities such as health and nutrition programmes for employees, tool development, local contractor development activities, programme preparation overheads, demonstration project overheads and funding agency expenses.

17.07 Accounting methods vary in their assignment of particular charges. In some systems an equipment operator's costs are included in the unit cost of equipment, while other methods include operator's cost as a labour charge. Some accounting methods include labour housing in camp sites as a site overhead, while other methods prorate camp housing as a labour cost. Since there is no worldwide standardized costing system, the first step in developing a project costing procedure is to identify and include all direct and indirect costs and format their application as simply and directly as possible in some type of a logical accounting system.

17.08 Many overhead direct costs and most indirect costs are designated as percentages of more obvious costs. A supervisor who spends 20% of his time on a particular project should have 20% of his salary charged to that project. However, fire insurance on a training complex may also be a cost of doing business which must be budgeted and funded, either as (i) a general line budget cost, (ii) an indirect cost only to those projects receiving some direct benefit from the training centre, or (iii) an indirect cost to all projects.

II. ELEMENTS OF EQUIPMENT OWNING AND OPERATING COSTS [100]

17.09 Equipment owning and operating costs are made up of two types of charges. Time-dependent charges are incurred irrespective of machine usage, and are therefore fixed on a calendar basis. Usage-dependent charges are incurred only when a piece of equipment is actually in use and are determined using machine clock hours or mileage.

17.10 The time-dependent costs include:

(i) interest charges on capital. Even though equipment may be purchased from government budgets and not by borrowing money or on supplier credits, it is necessary to include a charge for interest payments which reflects the 'value' of the capital investment tied up in the equipment;
(ii) fixed maintenance costs, including the cost of providing repair and service facilities for routine maintenance as well as for repairs;
(iii) insurance, to protect the investment in the equipment and to protect the owner against claims for damages caused by the equipment; and
(iv) tax payments where appropriate, for example when a contractor owns the equipment.

17.11 The usage-dependent costs include:

(i) depreciation, which is the reduction in capital value of the machine due to age and wear in service. A distinction should be made between the accounting methods of depreciation and the actual reduction in the value of a machine. The former represents merely a method of calculating the time dependent 'book value' of the equipment and should not be followed by construction authorities in calculating usage charges. True depreciation should be assessed by the actual reduction in the life of the machine because of usage;
(ii) variable maintenance, which is generally considered to include the cost of spare parts, replacement tyres and labour costs involved in operating the maintenance facilities for routine maintenance and repairs; and
(iii) running costs, which consist of the cost of fuel, lubricants, and on-site routine service.

17.12 Even though some of the equipment costs are defined as fixed, implicit in these costs is an assumed usage rate. For example, the interest charge is a function of the average value of the equipment during the operating period. A higher rate of usage would therefore imply a lower total interest charge. Similarly, the provision of maintenance facilities is generally determined by the equipment usage rate. Technical Note 9.02 discusses the derivation of these different equipment cost elements.

III. ELEMENTS OF LABOUR-BASED COSTS [16]

17.13 The basic direct costs per labourer are described in Technical Note 9.01. section (iii). The overhead direct costs i.e., logistical and supervision costs, are based on the size of the nominal labour force rather than the actual or 'effective' labour force. The effective labour force is the number of labourers required to produce the project work activities at the anticipated output per man-day. The nominal labour force is the number of labourers that must be recruited to assure an effective work force will be available to complete the project within the specified time frame. Technical Note 9.03, Forecast of the Effective Labour Force and Working Days for a Project, illustrates the methodology used to forecast these values.

17.14 Logistical costs may be quite modest for individual items but their aggregate total cost can be significant. They include transportation of workers, materials, tools and equipment; and labour camps and storage facilities for materials, tools and equipment. Camps to house labourers who cannot live at home, and storage compounds for tools, equipment and materials, all generate costs for construction, utility services and upkeep. All of these logistical costs must be based on the size of the nominal labour force rather than the effective labour force. Technical Note 9.04 cites some examples of camp requirements.

17.15 Daily transport to deliver workers is a costly undertaking. It should be avoided whenever possible by hiring persons from the immediate vicinity of the work site. If such a practice is not feasible, the daily cost of transporting labourers from distant locations should be compared with the cost of setting up additional labour camps. Technical Note 9.05, Sample Calculation of Requirements for Daily Transport of Labour to a Work Site, illustrates the cost not only of transporting labourers over selected distances, but also the labourers' lost productive time.

IV. UNIT PRODUCTIVITY

17.16 Once the cost of a production unit like a piece of equipment, a fleet of trucks, a work gang, an individual labourer or a loaded donkey is known, its productivity must be determined. Unit costs – the costs per lineal metre, square metre, cubic metre, or ton-kilometre – are found by dividing costs per time unit by productive output in the same time unit. Unit costs become the basis for both estimating project costs and evaluating cost control.

17.17 Equipment

Manufacturers' publications and many engineering textbooks depict curves showing equipment productivity. The curves show maximum production under specified conditions such as distance travelled, material density and swell, and coefficient of traction. The production figures must be modified by job correction factors for such specific conditions as job efficiency, i.e., number of minutes per hour actually worked, operator skill, condition and weight of material handled, visibility, transmission type, type and size of blade or bucket being used, working gradients and elevation. The product of the individual correction factors is used to determine actual production. For example, if a track-type bulldozer with a production curve rating of 400 loose cubic metres (LCM) per hour is being driven by an average operator (factor 0.75) moving dry non-cohesive material (factor 0.80) on flat ground (factor 1.00) and is working 50 minutes per hour (factor 0.83), the actual production is $400 \times 0.75 \times 0.80 \times 1.00 \times 0.83$ or 200 LCM per hour, one half of the maximum production.

17.18 The above value is a measure of efficiency. A second equally important factor is the utilization of a piece of equipment. If equipment is not fully utilized, its immediate usage-dependent costs reduce (Para 17.09) but its total time-dependent costs increase due to the interest, insurance and other charges over a longer economic life. In order to determine the true cost of equipment underutilization it is necessary to recalculate both the time-dependent and user-dependent costs to determine the true usage costs per hour.

17.19 A more common, but unsound, approach to underutilization is to reduce the output factor to account for idle equipment time. Technical Note 9.06, Evaluation of Equipment Utilization, begins with an example (Table 23) of a calculation using underutilization as an output factor. No allowance is made for reduced fuel costs when the equipment is not working since fuel costs in the example are included in the direct hire hourly charge for the equipment. Table 24 of the same Technical Note illustrates the impact of underutilization, i.e., the increase in time dependent costs per unit of output because the underutilized equipment has a longer economic life than fully utilized equipment. In the example used in Technical Note 9.06, the reduction in user-dependent costs (fuel, lubricants and oil) is greater than the increase in time-dependent costs.

17.20 **Labour**

Labour productivity has not been documented to the same degree as equipment productivity. For this reason any new labour-based programme is usually preceded by a pilot project, not only to determine the workers productivity, but also to train the supervising personnel in the proper usage and control of that productivity. If a country's experience with labour-based technology is limited, its implementation unit will have to base any estimates of direct costs on the experience of foreign countries. The accuracy of the estimates depends on the ability of the implementation unit to modify the experience in foreign countries to fit local conditions. The specific costs involved in creating support activities for labour-based technology during pilot projects should be justified on the basis of the projected long-term benefits from all similar operations which may follow [16].

17.21 All labour productivity measurements should refer to a single task or operation. In the example shown in Para 17.17, one bulldozer excavated, moved, and deposited (or spread) a specified quantity of material per hour. This operation must be divided into smaller tasks, each undertaken by a group of labourers, in a labour-based project. These tasks include: (i) excavation, (ii) loading, (iii) hauling and dumping, and (iv) spreading. The specific conditions for each job site must be evaluated, e.g., the labourer's strength; type of tools; distance travelled; material hardness, density, and swell; traction or footing; gradients traversed; and elevation. However, supervision expertise replaces proper equipment selection, since the supervisor must adroitly juggle the number of labourers engaged in each task to achieve maximum efficiency. When equipment size remains constant, differences in operating difficulty often show up as reduced ouput or increased equipment repair costs; labour efficiency is much more sensitive and its cost much more direct and readily correctable, because people can be added, removed, or shifted around to achieve gang balance and optimum productivity.

17.22 Animals

Animal productivity is measured in two ways: ability to pull (draught animal), and ability to carry (pack animal). Assuming all work conditions remain constant, the maximum daily output for an individual animal remains constant. This output consists of three interrelated variables: (i) force or effort exerted, (ii) velocity, and (iii) effective working hours. Maximum productivity occurs at a specific combination of these three variables. A good average draught horse produces its greatest day's work when applying a tractive power of about 68 kg at a speed of 4 km/hr for eight hours [90]. This is the equivalent of one horsepower. A bull of the same weight, pulling the same load, will assume a normal pace of about 2.5 km/hr. At the end of an eight-hour day, the horse will have moved the load further, or produced more work than the bull, and is thus said to be more powerful. Tests show that light horses, bulls, buffalos, mules and camels all provide about three-quarter horsepower, cows about half horsepower and donkeys one-third horsepower [90]. These are the rates at which the animals normally deliver force, not the maximum. In pulling tests, horses and donkeys have, for several seconds, pulled up to twice their weight and oxen have pulled up to their actual weight. Such efforts rapidly use up an animal's strength and reduce the time it is able to work. More details about animal production are included in Technical Note 9.07, Animal Productivity.

V. PERFORMANCE BUDGETING [77]

17.23 Performance budgeting is the allocation of cost needs to specific work accomplishment. It requres a definition of the work to be done; the equipment, animal, and/or manpower requirement in numbers and time to be expended; the materials required; and a calculation of costs in terms of equipment, labour, materials, and any contractual services required. Unit costs are found by dividing the calculated costs by the proposed output.

17.24 Performance budgeting is useful in both construction and maintenance activities. For construction tasks, it presents the unit costs in a formalized structure which is easily understood by both planners and field supervisors. It allows quick evaluation for cost control purposes, and is especially useful when the work is to be undertaken by local contractor, under a community contract, or in decentralized organizations; because it not only describes the anticipated output, but also indicates the technology desired and the material requirements.

17.25 Technical Note 9.08 is a copy of the descriptive material used in the Philippines [3] to outline a specific work accomplishment. Technical Note 9.09 is a copy of the standard costing form used to estimate the unit cost of the surface course described in Technical Note 9.08.

17.26 Performance budgeting for maintenance activities is done on a yearly basis. It need not be a complex task. If current cost data is maintained for equipment, labour and materials, these can be readily applied to the resource needs identified in the annual work programme. Since the cost needs are related to specific work accomplishment, objective evaluations can be made of the consequences when budget reductions are necessary, or of budgetary increases required to increase the level of maintenance service [77].

17.27 Maintenance activities must be catalogued, defined, described, and assigned productivity rates before performance budgeting is introduced. Technical Note 9.10, Maintenance Performance Budgeting, consists of a sample list of maintenance activity definitions and a sample description of a specific activity taken from the definition list [77].

VI. INVENTORY CONTROL

17.28 Inventory control is the accounting of capital assets, manpower and consumables. Capital assets are plant and equipment which, because of the investment represented, should be subjected to an individual accounting. Management must know how equipment is being used, but more importantly, management should know when and why equipment is not being used [77].

17.29 One source of equipment utilization data, for both construction and maintenance activities, is the daily/weekly crew activity reports showing by week the work performed each day, the material consumed and the specific units of manpower and equipment employed. The major inadequacy of many work reports is their identification of equipment availability, which is the time the equipment is on the job site in working condition rather than actual equipment usage in 'clock hours' or kilometres travelled. Reports showing equipment usage should also explain non-usage on particular days in terms such as; no work required, down for servicing and repair, or waiting for spare parts. These data, properly reviewed and analyzed, can help management make proper decisions about [77]:

(i) the types and quantity of equipment to be included in future acquisition programmes;
(ii) more effective allocation of equipment units among field work forces;
(iii) sharing of lightly used equipment among field work forces;
(iv) upgrading spare parts inventory to reduce the amount of down time; and
(v) improving and scheduling of equipment servicing and repair operations.

17.30 The most reliable method of gathering this type of data is to use an equipment log which remains with each piece of equipment [34]. All the details necessary to determine the usage of the vehicle should be recorded in the equipment log as and when they occur. Such details include fuel, oil, and lubricant consumption; hours or kilometres of use; servicing activities; repairs, including identification of spare parts used; and names, dates, and signatures of the authorized operator or mechanic accountable for such activities.

17.31 A store ledger should also be maintained on each project [34]. All materials, hand tools, oil, fuel, spare parts, etc., that are received, issued, or returned should be itemized in this ledger as the transactions occur (Figures 28, 29). The ledger should be formatted so that it is always possible to:

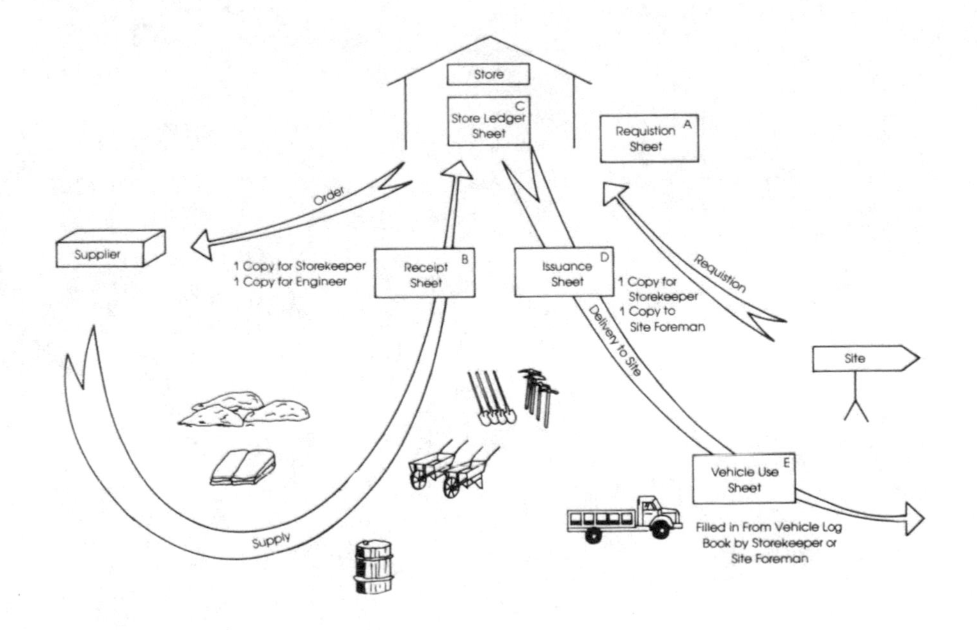

Fig. 28 *Inventory control sheet routing diagram*

(i) find out when and to whom material, tools, and fuel have been issued;
(ii) verify the information on all equipment logs; and
(iii) verify any physical check (inventory) of the balance of usable and unusable items on hand by comparing such counts with the balance recorded in the ledger.

17.32 Muster rolls should also be maintained daily [34]. Usually separate muster rolls are kept for the regular (monthly employed) personnel who receive government benefits, and for the unskilled casual, i.e., daily employed, labour. In practice duplicates of muster rolls are also maintained so that a record is always available on site even when the original muster rolls are required at headquarters for preparing and checking payrolls.

17.33 Daily/weekly reports, i.e, reports that are filled in daily and submitted weekly, are made out by foremen or gang bosses. These reports describe how many and which type of people are working on each activity; the number and type of vehicles and equipment working; the work accomplishments such as loads hauled, culverts built, quantity of walls built; and other relevant information including actual equipment usage [34].

17.34 Each week a supervisor should summarize all the information on each daily/weekly report in a project weekly report [34]. The stores ledger and muster rolls should also be summarized in this weekly report. All three sources of data should be compared for obvious inaccuracies or unusual occurrences as the weekly report is being compiled.

17.35 A formal monthly report should be completed by a senior supervisor in the presence of the supervisor who prepared the project weekly reports [34].

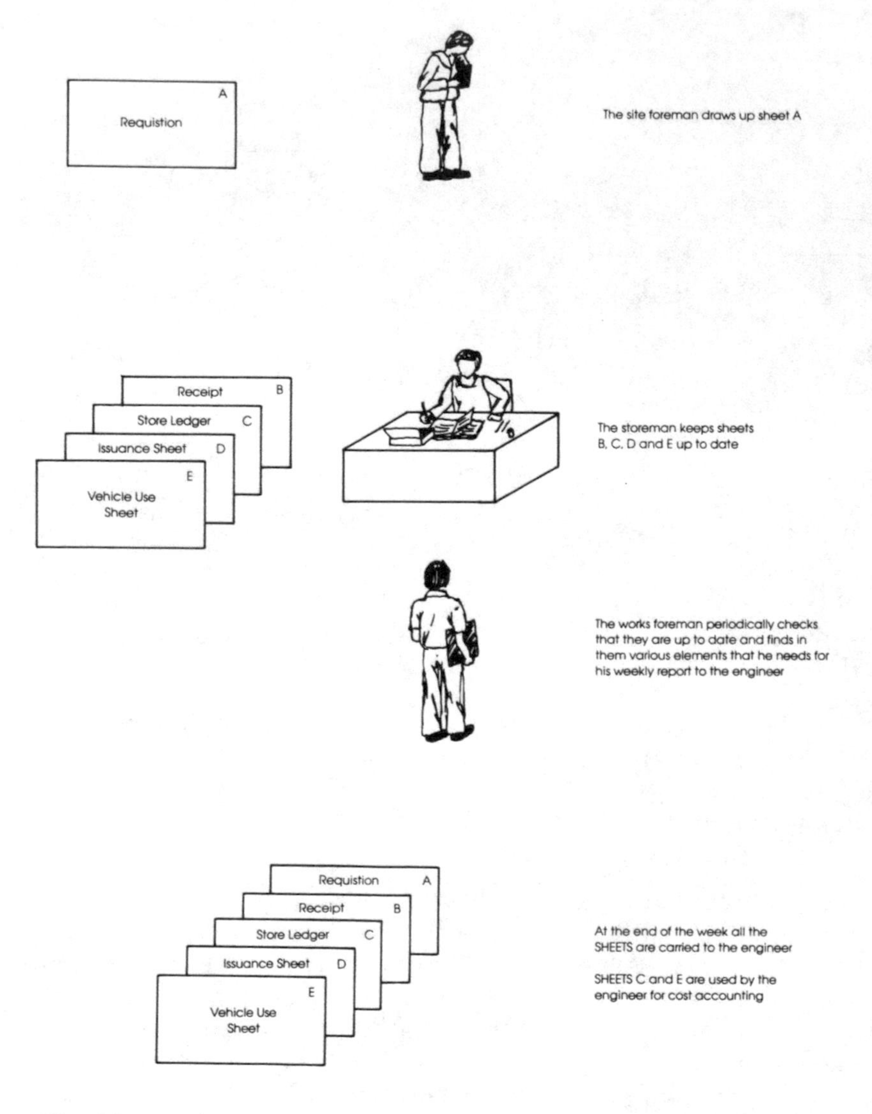

Fig. 29 *Inventory control sheet routing*

This report should summarize the weekly reports and the information in the equipment logs. All information such as progress, equipment clock hours, vehicle mileage, fuel consumption, store balances and man-days, should be physically checked and compared to the information taken from the weekly reports before the monthly report is submitted to headquarters.

VII. CONCLUSIONS

17.36 Cost evaluations, to be realistic, must be based on cost factors developed by the concerned construction or maintenance authority. They must include all the direct and indirect costs pertinent to the activities being evaluated. Costs per kilometre or other broad-based costs, are an easy method of expressing anticipated expenditures that should only be applied after the elements that make up the costs are properly documented and their suitability tested against the circumstances of the specific project being evaluated.

17.37 Equipment owning and operating costs are made up of two elements: time-dependent costs and user-dependent costs. Actual per hour equipment costs are therefore not known until the equipment has been used up or otherwise disposed of, consequently estimated hourly costs are used in equipment cost accounting.

17.38 Estimated equipment output per hour is based on optimum equipment productivity as modified by selected efficiency factors. These factors are job-specific and based on site conditions, task requirements, operator skills and job equipment balance. Their product is applied directly to the optimum equipment production rates. Usage factors, on the other hand, are included in the time-dependent costs of the overall hourly unit cost. Changing usage factors therefore requires a recalculation of the equipment's time-dependent costs.

17.39 Labour construction productivity is based on the output of an effective labour force. Man-day outputs vary by task, site conditions, supervision, worker motivation and the individual labourer's physical strength and training. Total labour force output is very sensitive to the proper balance of task work both within and between labour gangs.

17.40 A nominal construction labour force must be engaged to assure the presence of an effective labour force. Logistical support such as housing and/or transportation, supplies and materials, and supervision capacity must be provided for the nominal labour force rather than the effective labour force. This requirement increases the overhead costs for the effective labour force.

17.41 When using animals, it is important not only to have some idea of their productive capability under actual work conditions, but also to have someone available who can advise on possible changes in harnessing, feeding, work programme, or other factors which may result in increased productivity. Since different work conditions like climate and type of work, suit different species and breeds of animals, it is very important to use the correct type of animal, bearing in mind factors such as cost and food availability. A knowledge of the amount of work which can be expected from an individual animal is necessary to prevent overworking or underutilizing animals, to assure a sufficient supply of animals is available, and to allow the correct sizing of panniers, carts, rollers and other implements [101].

17.42 Performance budgeting, the allocation of cost needs to specific work accomplishment, is a method of relating (i) the costs of the indiviudal elements required to perform a certain activity, (ii) the anticipated output of those elements in accomplishing the activity, and (iii) the total cost of the performance of that activity for the project under evaluation. Properly documented performance budgeting permits compilation of more general cost parameters such as cost per kilometre. It also provides a monitoring tool to identify areas of cost overruns and to pinpoint the causes of each overrun such as substandard productivity, increased actual costs, inefficient use of resources, lack of resources and changes in the magnitude of task requirements.

17.43 Actual costs of doing work must be monitored not only to determine the status of projects underway, but also to build a data base for estimating future projects. Such monitoring should include built-in checking mechanisms to minimize improper accounting procedures, mathematical mistakes, and theft; and to flag areas of improper or inefficient resource expenditures. Common monitoring tools include equipment and vehicle logs, store ledgers, muster rolls, daily/weekly reports, weekly summary reports and monthly progress reports. These tools should be used not only to monitor progress, but also to prepare bi-weekly or monthly work schedules to improve resource allocation and increase project productivity.

CHAPTER 18

Storage Facilities

18.01 The automatic assumption that a costly rural road improvement is the only solution to a transport problem must always be avoided. All-weather access to remote agricultural areas, proposed on the grounds that cash crops, such as yams or maize, deteriorate before they can be delivered to the market, should be proven as a better solution than the sometimes much less expensive provision of adequate crop storage facilities for periods when existing infrastruture is impassable [12].

I. RELATIONSHIP TO TRANSPORT ISSUES

18.02 The suitability of a storage alternative is linked to the perishability of the specific crop being produced. Some crops, such as pineapples grown in certain parts of Bangladesh, have to be at the point of consumption within two days of harvesting, while in Kenya only the morning surplus milk yield is marketed because there is no evening collection to transport the milk to cooling facilities [29]. Ethiopian coffee must be processed in a washing facility within five hours of being picked. Each of these crops require a different transport analysis.

18.03 The pineapple example indicates a seasonal transport demand, requiring all-weather access only if the harvest and rainy seasons coincide. Milk production is a year-long herd output even though individual cows do not produce constantly, therefore all-weather transport is necessary to get the milk to the processing plant although short delays between the processing plant and the consumer may be acceptable. The coffee example indicates the need for all-weather transport between the farm and the processing plant, since the coffee must be picked at a specific time in its ripening process, whether the rains have stopped or not. However, sufficient storage for the washed coffee can obviate the need for immediate transport to market.

18.04 Current headloading practices may limit the volume transported during a harvest period when the farmer must budget his transport activities to allow enough time to harvest his crop. Alternative transport aids and on-farm storage both encourage increased individual farmer productivity, either by increasing the area under cultivation or by the use of fertilizer, because the farmer's transport time requirements are reduced during the harvest season.

18.05 All farmers benefit from proper on-farm storage, not only for the preservation of subsistence crops, but also for the reduction of losses, which permit an increased portion of the surplus crops harvested to reach the marketplace. Improved storage also allows the additional marketing of some of the agricultural produce previously saved to replace spoilage of the subsistence portion of the harvest. Therefore proper on-farm storage and cooperative village storage, if such storage is acceptable to the farmer, increase his or her total cash crop output even if no additional land is placed under cultivation and no improved agricultural techniques are practised.

18.06 Local storage facilities also dampen maximum transport demand peaks and consolidate local transport demand. This allows a larger volume of produce to be hauled with the existing transport fleet, thereby increasing utilization with its attendant savings in capital expenditures. In a competitive market some of these transport savings are passed on to the farmer in addition to the increased price he/she may receive for properly stored produce withheld during the harvest period when prices are depressed.

II. CURRENT PROJECTS INCLUDING STORAGE

18.07 Storage considerations in transport-oriented projects are usually limited to storage needs between high volume modes of transport, such as storage at ports, or at the focal points of transport systems like truck terminals, railroad yards and airports. Little consideration has been given to the incorporation of local storage in rural areas in most current transport projects.

18.08 Rural development projects, on the other hand, being multidiscipIined, frequently consider rural storage as a means of distributing inputs in conjunction with agricultural extension programmes; and sometimes include storage of outputs as a vehicle to improve farmers' income capabilities and the nutritional value of the rural food supply. However, seldom is the inclusion of such storage facilities reflected in a cost reduction of transport infrastructure, even though the provision of more complex storage and processing facilities – such as milk processing plants – required for specific agricultural produce almost always correctly identifies the need for all-weather access.

III. STORAGE IN FUTURE TRANSPORT PROJECTS

18.09 It is unreasonable to expect transport planners to become agricultural experts. However it is not unreasonable for them to acquire a working knowledge of agriculture and the way the agricultural community operates. Without such knowledge, evaluations of accessibility (Chapter 4), transport aids (Chapter 5), and infrastructure requirements (Chapter 6) would be undertaken in a vacuum.

18.10 It is equally unreasonable to expect agricultural experts to evaluate the possible tradeoffs between storage costs and rural transport infrastructure costs. When a direct linkage between storage and infrastructure is anticipated in an agricultural or rural development project, the investigative effort should be instigated by the transport planners and agreed to by the agricultural representatives. If, on the other hand, the storage component is part of a transport project, the transport planners should seek the cooperation of extension services in the project area or encourage their introduction in conjunction with the transport project. As a last resort, the transport agency should fund the seconding of an agricultural extension agent to help implement the storage segment of the transport project.

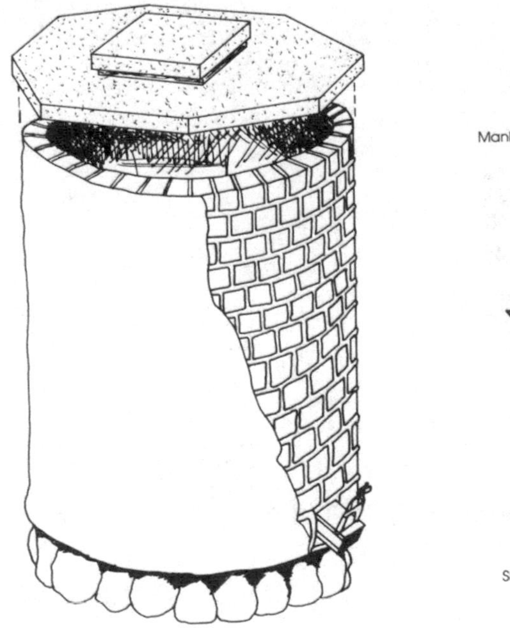

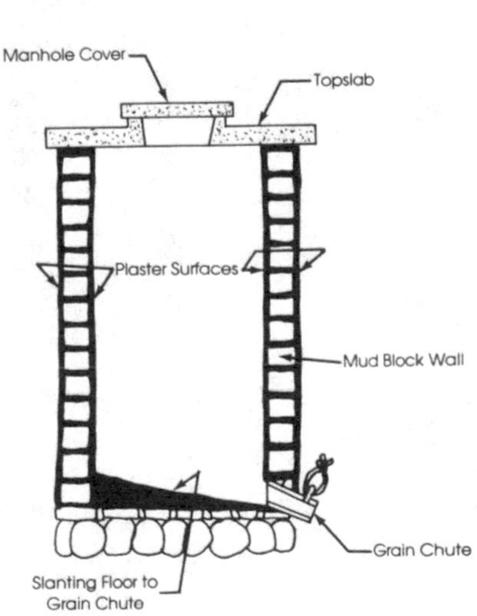

Fig. 30 *Improved mud block silo*

18.11 The successful introduction of improved storage facilities (Figures 30, 31 and 32) into rural communities requires an effort or programme which recognizes local social, cultural and behavioural factors, relies and builds on local skills and knowledge, utilizes local resources, and responds to local needs and goals. Such programmes are best carried out by agricultural extension agents who have, or can develop, first-hand knowledge of [39]:

(i) the local farmers' recognition of storage problems and awareness of storage quality;
(ii) the level of farmer concern for specific storage problems;
(iii) the degree of farmer sophistication in understanding storage problems and principles; and
(iv) the resources the farmers have at their disposal.

18.12 Storage facility construction which is justified as a less costly alternative to a higher stage of rural access infrastructure (RAI) construction, is a legitimate transport sector investment. The investment may be limited financially to technology transfer to the local population if a strong awareness of the benefits of storage exists. Other levels of investment may include both technology transfer and the establishment of a line of credit for on-farm storage; the additional supply and delivery of non-local material to be purchased with project-supplied credit, or as a project cost if credit is not supplied; or the

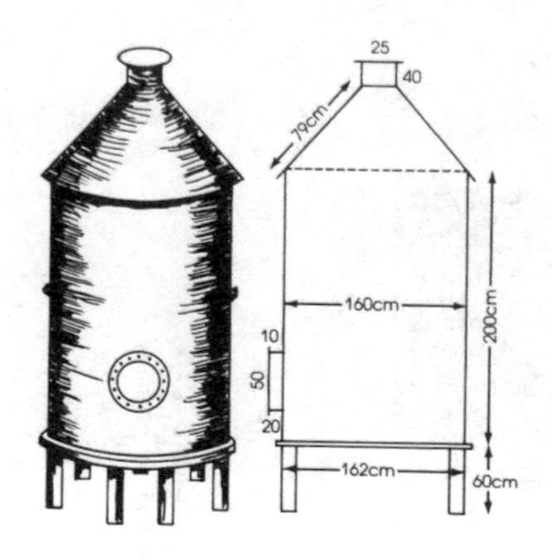

Fig. 31 *3-ton sheet metal silo*

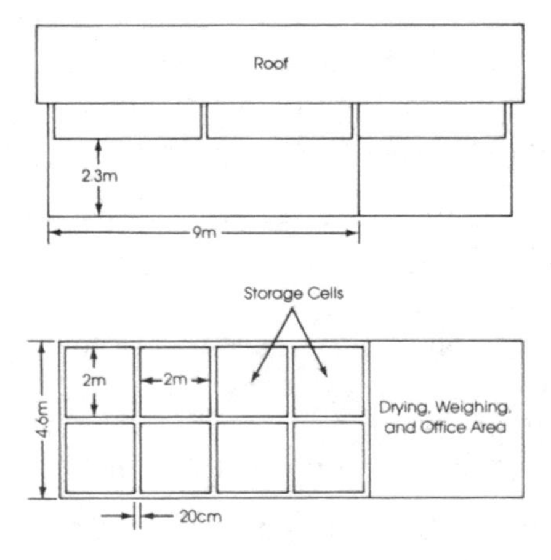

Fig. 32 *30-ton concrete block cooperative silo*

construction of village or community storage facilities by local contractors or under a community contract system.

IV. STORAGE PRINCIPLES [38]

18.13 All methods of agricultural produce storage try to follow three storage principles:

(i) keep the stored material cool and dry (Figure 33);

Fig. 33 *Wide thatch roof to keep rain and sun off a mud crib*

(ii) protect the stored material from insects; and
(iii) protect the stored material from rodents (Figure 34).

18.14 Stored produce is kept cool by building storage structures of low conduction building materials, by keeping or building storage containers away from direct sunlight, and by painting containers white. It is kept dry by waterproofing buildings and containers. Insect protection is increased by correctly drying and cleaning the material before storing, by applying insecticide, and/or using airtight storage. Rodent control includes not only clean, tight buildings or storage containers, but also clean, clear areas around the storage units and an active trapping and/or poisoning campaign at the first sign of any rodent activity.

18.15 Grain should be dried before it is stored since heat builds up more quickly in stored wet grain. Heat in turn causes rapid mould formation, insect multiplication, and grain germination (sprouting) in the storage container (Figure 35), while well-dried grain deteriorates slowly even at fairly high temperatures. Technical Note 10.1 contains further information about crop drying.

18.16 It is impractical to dry grain to a low moisture level and then store it in storage facilities which will not keep the grain dry, such as cribs (Figure 36), unsealed gourds or baskets (Figure 37), sacks, most kinds of earthen pits, or mud-walled structures which do not have extra protection against moisture. In general, only further airtight storage warrants the use of specially constructed dryers.

V. SPECIFIC STORAGE FACILITIES

18.17 Short-term storage usually implies that some deterioration losses are acceptable, although heat, insect and rodent losses can be minimized by proper precautions such as cleaning and some degree of drying (Figure 38). In crib storage, suitable grain is left in its covering and continued drying occurs during

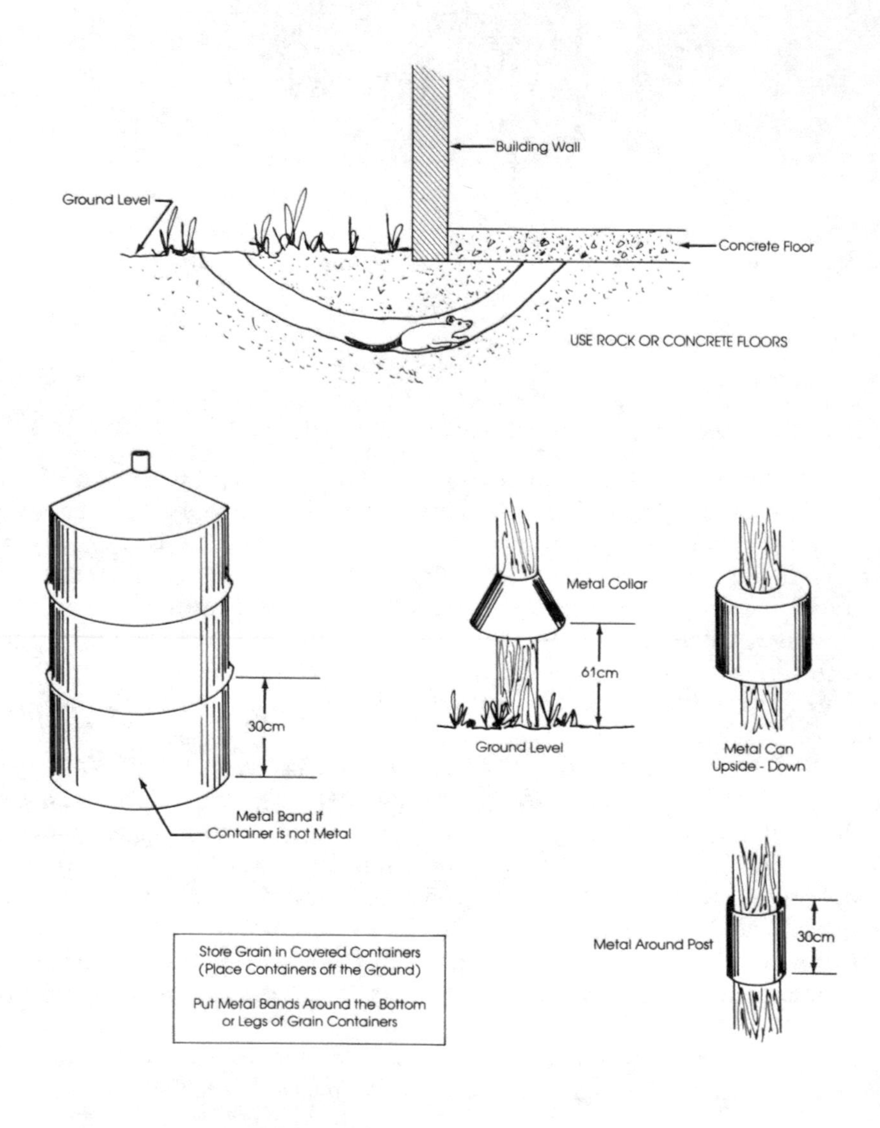

Fig. 34 *Rodent proofing*

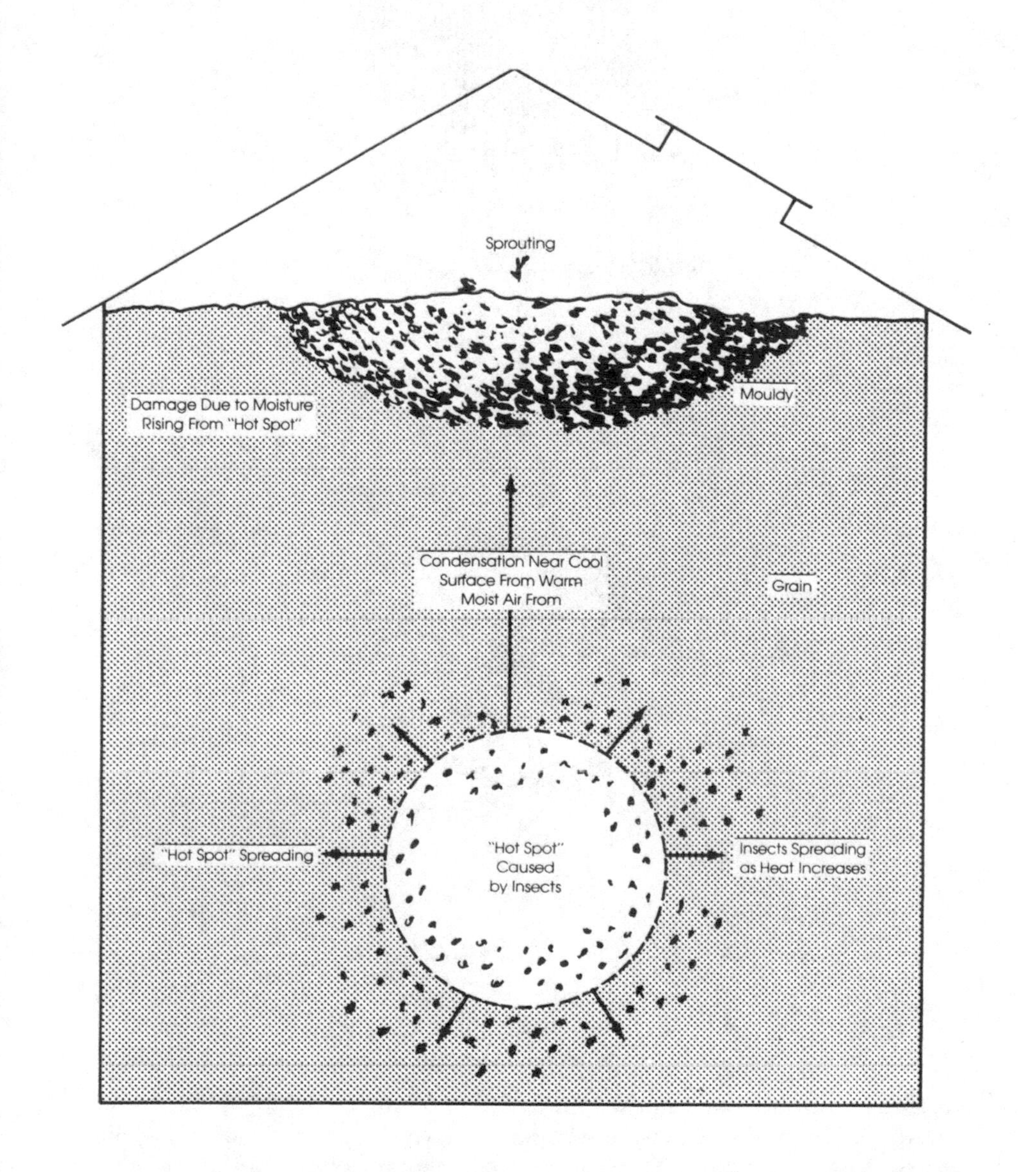

Fig. 35 *Moisture- and heat-induced spoilage*

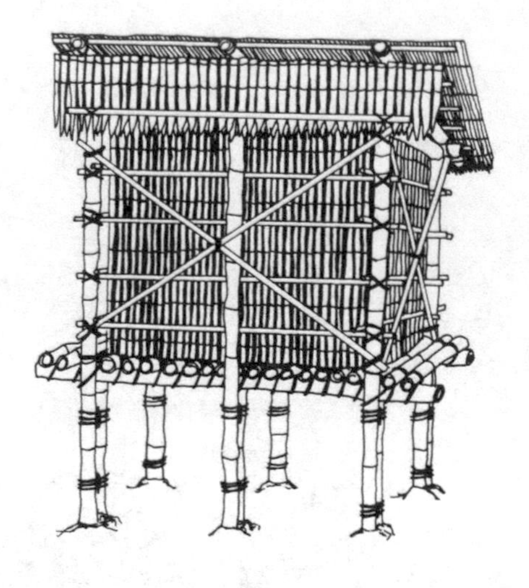

Fig. 36 *Crib for drying maize*

Fig. 37 *Basket granary*

storage. Long-term grain storage, intended to retain the quality of the produce stored, requires both cleaning and complete drying of the grain and takes place only in airtight containers. However, airtight containers markedly increase the rate of deterioration of improperly dried grains, due to rapid heat buildup.

18.18 Size is an important storage facility criterion. Small capacity storage facilities are considered acceptable for on-farm storage, while larger capacity storage facilities are commonly considered as village, community or cooperative storage activities. Capacities of commercial and government-run storage

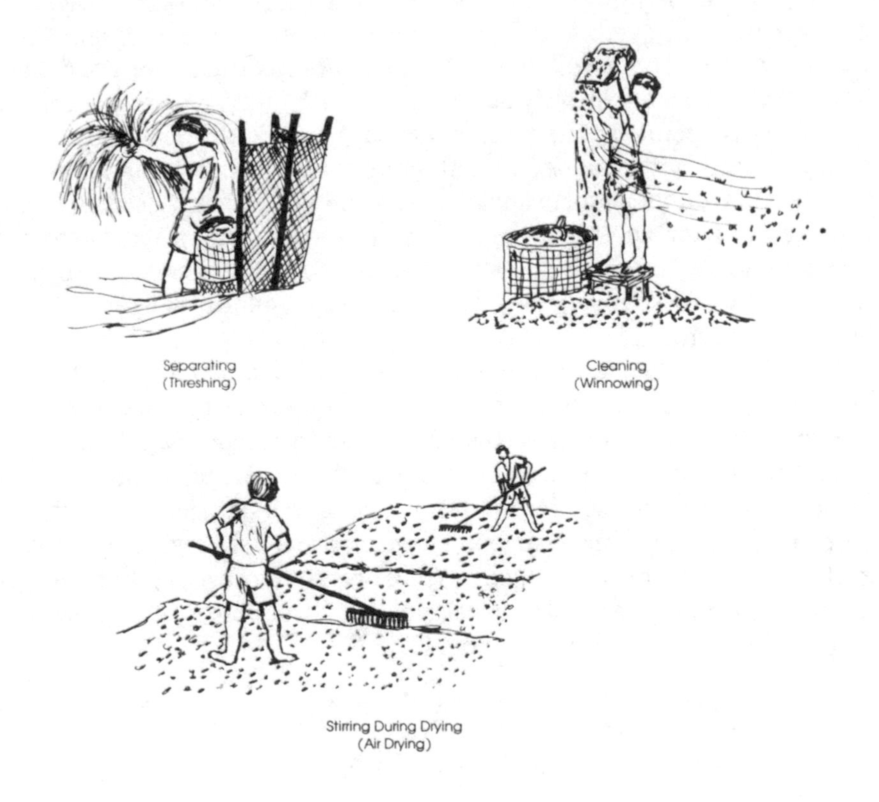

Fig. 38 *Typical operations prior to grain storage*

facilities are carefully calculated to take advantage of the economies of scale which may be substantial in larger storage facilities. Large granaries are seldom employed by marginal cash crop farmers; small granaries are common among the rural landless, especially those who labour in the fields and receive part or all of their wages in grain. The predisposition of the target population towards small storage facilities may be due to the following factors:

(i) many indigenous storage facilities are short-lived, such as the Indian puri, a paddy straw rope-woven storage facilities which is unwrapped as the stored material is consumed;
(ii) local storage facilities are often built by one or two people, which physically limits their structure size;
(iii) many farmers are reluctant to use community storage facilities if their own grain can be contaminated by other people's grain;
(iv) large granaries often attract thieves and tax collectors; and
(v) economies of scale are often less important than the commitment of resources before they are required, leading small farmers to increase storage capacity as required by building more small traditional storage units.

18.19 In-transit storage facilities such as a village godown or riverside barge landing shed, tend to be larger. Small producer shipments usually travel in sacks, which are best suited to short-term storage at transfer points, and allow each producer to identify his own crop. Such in-transit storage facilities are frequently operated by merchants who buy the sacked produce on delivery and hold it until a full load is amassed; or by cooperatives which perform the same function, although some hold the produce until it is purchased by a truck or barge operator and then disperse the payments among the farmers in proportion to their share of the produce sold. In other economies, cash crops must be sold to the government and the village godown is operated by a government agent.

18.20 Technical Note 10.2, Storage Facilities [38], describes some types of storge containers and structures used for on-farm grain storage, and one 30-ton individually compartmentalized community granary (Figure 32). Waterproofing and/or airproofing methods for some of these storage facilities include plastic membranes; plaster of cement, sand and lime; paint; and local materials processed from tree saps, bamboo resin, coconut oil, gum from leaves, ox-blood, and oils from berries, nuts, seeds or fruit. Storage capacities for the various storage facilities in Technical Note 10.2 can be roughly determined using the following conversion factors [37]:

(i) paddy, approximately 550 kg per m^3;
(ii) maize, approximately 710 kg per m^3;
(iii) wheat, approximately 750 kg per m^3.

VI. CONCLUSIONS

18.21 Increased storage facilities can sometimes substitute for complete accessibility in remote areas where all-weather access is expensive, storage requirements during periods of unfavourable conditions are reasonable, and area residents realize the personal benefits to be derived from good storage facilities. The evaluation of storage substitution for all-weather access should include the following items:

(i) the effect of delay on the area's produce;
(ii) the volume of produce involved;
(iii) the type of storage required;
(iv) the population's awareness of storage needs and techniques;
(v) the implementation requirements of providing additional storage;
(vi) the cost difference between providing additional storage facilities and improving access.

18.22 Improving storage capability requires an extension service input to explain and demonstrate the principles of storage technology and its benefits. These services are best provided by agricultural agencies. However, if no other sponsor can be found, the transport authorities should underwrite the educational effort themselves to insure the anticipated benefits of any tradeoff between improved access and increased or improved storage facilities.

18.23 The transport planner must be aware that successful storage facilities should meet local needs, utilize local resources, and rely on local operational expertise. The information provided in this text is not intended to expand the rural transport organization's role from rural infrastructure development into agricultural activities. It is presented to apprise transportation officials of the various low-cost storage options available to meet local requirements, so those officials can present a rational case to the agricultural sector for cooperation in satisfying their common goals.

CHAPTER 19

Economic Analysis of Rural Access Infrastructure

19.01 This chapter provides (i) simplified operational procedures for the appraisal of feeder roads and other rural access infrastructure (RAI) in transportation, rural development and agricultural projects, and (ii) clarification on issues often encountered in the screening and appraisal of RAI. It does not claim an exhaustive set of methods for any conditions prevailing in these projects. The principles of analysis are described for feeder roads and motor vehicle operating costs, but also apply to other levels of RAI when the operating costs of other transport aids are substituted and the RAI improvements suit those transport aids.

I. APPRAISAL OF RURAL ACCESS INFRASTRUCTURE

19.02 In general, the benefits from feeder road improvement or construction relate to normal and generated traffic. Normal traffic is the expected traffic growth that would have taken place on the existing road in any case, even without the new investment. Generated traffic is traffic arising from economic and social activity which would not develop without the lowering of transport costs and the improvement in transport services caused by the new investment. Sometimes road improvements result in diverted traffic, which is traffic diverted from existing infrastructure to the new or improved infrastructure.

19.03 Figure 39 considers a linear relationship between volume of transport and unit transport cost between a farmgate x and local market y. Farmgate x is representative of all farmgates in the area of influence of a feeder road considered for improvement or construction; it is normally the road's midpoint. Vehicle operating costs (VOC) savings on normal traffic are represented by the shaded rectangle in this figure. They are measured by the quantity q_1, transported from x to y in the without-project situation multiplied by the difference between the unit transport costs between x and y in the without- (C_1) and with- (C_2) project situation, or q_1 $(C_1–C_2)$.

19.04 Benefits stemming from generated road traffic or new transport induced by the transport cost reduction are represented by the shaded triangle in Figure 39 and measured by the product ½ $(q_2–q_1)$ $(C_1–C_2)$, where q_2 is the quantity transported from x to y in the with-project situation. Quantifying the benefits of generated traffic can be difficult since it requires full specification of the road demand function, which may not be a linear relationship. In other words, the benefits related to this traffic are ½ $(q_2–q_1)$ $(C_1–C_2) + e$, where e is an error term if, in fact, the demand function is not linear. For feeder roads, the error term can be significant, particularly if the cost change $(C_1–C_2)$ is large and/or the normal benefits are small. Establishment of demand elasticity is

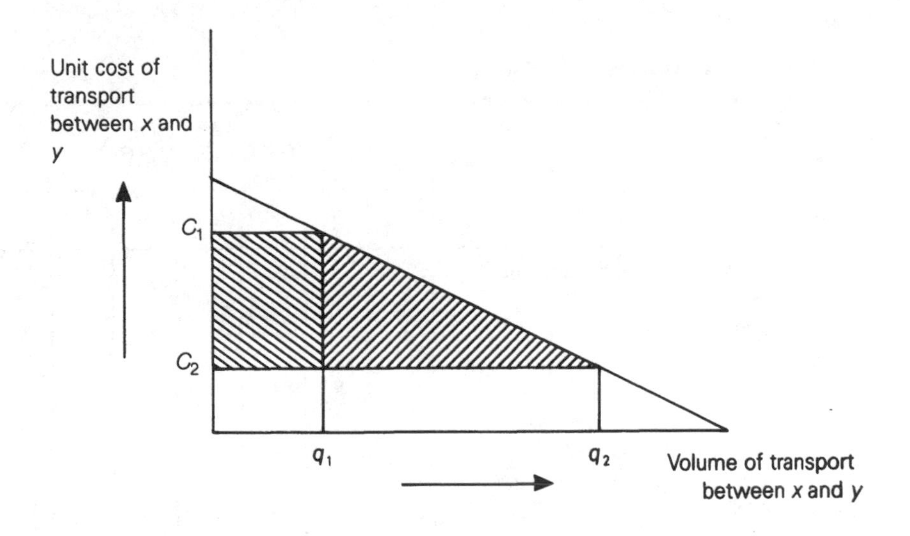

Fig. 39 *Transport demand function*

further complicated if feeder road improvements are complemented by investments in agricultural extension services, storage facilities, irrigation works, etc., in the road's zone of influence.

19.05 Benefits stemming from diverted traffic are measured by the volume of transport directed from an existing feeder road to a new or improved road multiplied by the difference between the unit cost of transport along the existing route and the unit cost of transport along the new route. In projects with feeder road components, diverted traffic seldom takes place. In other words, the benefits stemming from a feeder road improvement or construction are in most cases determined by $q_1(C_1-C_2) + \frac{1}{2}(q_2-q_1)(C_1-C_2)+e$.

19.06 Eight simplified operational procedures are available for the appraisal of feeder road components. In order to save time, these procedures imply the determination of a simple economic return (ER). A simple ER is defined as an ER based on a significant portion of quantifiable benefits rather than all. The extent to which benefits are considered depends on the underlying assumptions, which are summarized in Table 10. None of the procedures considers diverted traffic.

19.07 Simple ERs are underestimates of actual ERs. In general, the higher the number of simplifying assumptions made in the procedures, the stronger the underestimation becomes. For instance, if the application of procedures IV or V does not result in an acceptable simple ER, one may want to try Procedures VI, VII. or VIII.[1] The difference between some of these proce-

[1] If one is reluctant to make any of the simplifying assumptions of Table 10, the use of a Rural Roads Computer Programme [5] may be considered.

Table 10. Procedures' Assumptions

(1)	(2)	(3)	(4)	(5) PROCEDURE	(6)	(7)	(8)
Assumptions[1]	*I*	*II*	*III*	*IV*	*V*	*VI*	*VII*
1. Benefits stemming from generated traffic are insignificant	X						
2. Constant growth rate in ADT	X						
3. Constant growth rate in overall agricultural production		X	X				
4. Local market prices in the with- and without-project situations remain the same		X	X	X	X	X	X
5. VOC savings related to non-agricultural traffic insignificant		X	X	X	X		
6. On-farm consumption in the with- and without-project situations negligible		X	X	X			

[1] A cross in one of the columns (2)–(8), means that the assumption of column (1) is made; none of these assumptions apply to procedure VIII.

Table 11. Input Data of Procedures

(1)	(2)	(3)	(4)	(5)	(6) PROCEDURE	(7)	(8)	(9)
Input Data[1]	*I*	*II*	*III*	*IV*	*V*	*VI*	*VII*	*VIII*
1. Prevailing opportunity cost of capital	X	X	X	X	X	X	X	X
2. Costs of RAI improvements/ construction	X	X	X	X	X	X	X	X
3. Expected life of improved/ constructed RAI	X	X	X	X	X	X	X	X
4. Average daily traffic (ADT) on existing (unimproved) RAI	X							
5. Growth rate of ADT on existing RAI	X							
6. Vehicle operating cost (VOC) savings expected from RAI improvement during first year[2] after completion of construction works	X	X						
7. Growth rate of above VOC savings	X							
8. Average annual tonnage of marketable agricultural production travelling over the existing (unimproved) RAI		X	X					
9. Growth rate of agricultural production		X	X					
10. Local market prices of these products in the without-project situation		X	X	X	X	X		

Table 11. (*contd.*)

(1)	*(2)*	*(3)*	*(4)*	*(5)*	*(6)*	*(7)*	*(8)*	*(9)*
	PROCEDURE							
Input Data[1]	*I*	*II*	*III*	*IV*	*V*	*VI*	*VII*	*VIII*
11. Size of cultivated areas in zone of influence in the with-and without-project situations				X	X	X	X	X
12. Yields of agricultural products produced in these areas in the with- and without-project situations				X	X	X	X	X
13. Production costs of these products in the with-and without-project situations				X	X	X	X	X
14. Farmgate prices of these products in the without-project situations							X	
15. Local market prices of these products in the with- and without-project situations								X
16. Home comsumption of these products in the with- and without-project situations					X	X	X	X
17. VOC savings related to non-agricultural traffic							X	X
18. Transport costs in the with- and without-project situations				X	X	X	X	X

[1] A cross in one of the columns (2)–(9), means that the imput data of column (1) is required.

[2] The following references give information on VOC:
Staff Working Paper 610, Appendix K [6]; S. W. Abaynayaka et al., Tables for *Estimating Vehicle Operating Costs in Developing Countries*, TRRL Laboratory Report 723, Transport and Road Research Laboratory, 1976 [1]; Jan de Weille, *Quantification of Road User Savings*, World Bank Occasional Paper No. 2, The John Hopkins Press, 1966 [20]; R. S. P. Bonney and N. F. Stevens, *Vehicle Operating Costs on Bituminous, Gravel and Earth Roads in East and Central Africa*, Road Research Technical Paper No. 76, Road research Laboratory, Ministry of Transport, London, Her Majesty's Stationery Office, 1976, p. 22 [8].

dures is the choice of required data. Table 11 summarizes the data needed to use each of the procedures.

19.08 As indicated in Table 10, Procedure I ignores the benefits related to generated traffic or those represented by the shaded triangle of Figure 39. These benefits are generally insignificant if the shaded rectangle of Figure 39 is large compared to the shaded triangle, because q_1 is relatively large. They are normally significant if the shaded rectangle is large due to a significant difference between C_1 and C_2. Thus, the use of Procedure I is recommended wherever existing traffic is significant, benefits from additional agricultural value added in the with-project situation are insignificant, and traffic is expected to grow at a given compound rate. The procedure recognizes that the use of aggregate growth rates for various means of transport may be preferable in situations where detailed data for making separate estimates of growth rates are unreliable. Procedure I can be used to assess the economic viability of feeder road improvements when accessibility does not change.

19.09 Procedures II and III are based on an assessment of agricultural production increasing at an anticipated growth rate. They are best suited for analyzing areas that already have agricultural activity capable of being accelerated through development activities which lead to a progressively increasing marketable surplus. Procedure II can be used to assess the economic viability of feeder road improvements resulting in reduced per ton-km transport costs. Feeder road improvements include activities which make passage easier, such as proper shaping and compaction of the road's surface. While such activities may permit passenger cars to travel over a feeder road that was too rough for comfortable passage previously, the overall accessibility of the feeder road is not considered for improvement in this evaluation because any obstacles such as stream crossings, unclimbable grades, or other localized obstacles, are not corrected. Feeder road improvements are therefore activities that are carried out over the length of a feeder road resulting primarily in ton-km cost reductions; their costs may be expressed as an average investment per km of road length.

19.10 Procedure III may be used to evaluate the impact of eliminating localized obstacles such as stream crossings and unclimbable grades. Eliminating obstacles decreases delays thereby improving access as described throughout this handbook. While obstacle elimination may also be considered as a means of reducing per ton-km costs, its primary purpose is to minimize the time the road is impassable. Since obstacle reduction costs depend on discrete activities undertaken at identifiable locations, they are best expressed as a total investment on a route rather than per km costs. In this simplified procedure, obstacle elimination is evaluated as a crop loss reduction; agricultural produce previously lost to spoilage because of access problems is saved when the spoilage inducing delays are eliminated. Eliminating obstacles can also be an inducement for the local farmers to immediately increase the total amount of crops under cultivation. In such a scenario, the increased production can be considered as a portion of the new total output that was previously lost because the farmers found the risk of planting the crops unacceptable. The cost of producing this additional output must be deducted from the market price used in this simplified evaluation procedure.

19.11 All three procedures use weighted average vehicle operating costs. Procedure I is based on per km vehicle operating costs (VOC) while Procedures II and III are based on per ton-km costs (voc) which are found by dividing the vehicle operating costs (VOC) by the average load carried in tons. The following example illustrates how a weighted average VOC is computed:

$$VOC_w = q \cdot (VOC_A) + (1 - q) \cdot (VOC_B)$$

where

VOC_w = per km weighted average operating cost,
q = percentage of vehicle A in the traffic stream,
VOC_A = per km average vehicle operating cost for vehicle A, and
VOC_B = per km average vehicle operating cost for vehicle B.

If vehicle A is a seven ton truck with a VOC of \$1.05 and vehicle B is a three-quarter ton oxcart with a VOC of \$0.20 and the average daily traffic consists of two trucks and 10 oxcarts, the weighted average vehicle operating cost is:

$$VOC_w = (2/12) \cdot (\$1.05) + (10/12) \cdot (\$0.20) = \$0.34$$

19.12 The weighted per ton-km vehicle operating cost (voc) for the above example, as applied in Procedures II and III, is found as follows:

$$voc_w = \frac{VOC_w (t_A + t_B)}{t_A \cdot f_A \cdot L_A + t_B \cdot f_B \cdot L_B}$$

where: voc_W = per ton-km weighted average vehicle operating cost,
t_A = number of daily trips for vehicle A,
f_A = load factor for vehicle A,
L_A = load capacity for vehicle A,
t_B = number of daily trips for vehicle B,
f_B = load factor for vehicle B,
L_B = load capacity for vehicle B, and
q, VOC_A, VOC_B are as defined previously.

When in the aforementioned example, vehicle A only carried full loads of produce from the field to the market its load factor is 0.50 because it travels empty half the time. Assuming vehicle B always carries a full load from the field but also carried two full loads back in every five trips, its load factor is 0.70. The per ton-km weighted average vehicle operating cost becomes:

$$voc_W = \frac{(2 + 10) \cdot (0.34)}{(2) \cdot (0.50) \cdot (7 \text{ ton}) + (10) \cdot (0.70) \cdot (3/4 \text{ ton})} = \$0.33$$

Vehicle A evacuates 7 tons of crop in one round trip while vehicle B is evacuating 3.75 tons in five round trips, therefore vehicle A, which constitutes 17% of the ADT, evacuates 65% of the crop while vehicle B carries 35% of the crop and an additional 1.5 tons of goods back from the market.

19.13 Procedure I consists of determining the critical value of average daily traffic or ADT^* with the following expression:[2]

$$ADT^* = \frac{I}{VOC \cdot p \cdot 365 \cdot A\,(N, i_o)}$$

where: I = per kilometre cost of the feeder road improvement,
VOC = per kilometre weighted average vehicle operating cost before improvement,
p = percentage of VOC savings resulting from the feeder road improvement (as a decimal),
$A(N, i_o)$ = present worth of an annuity factor, as a function of N and i_o,
N = expected feeder road life in years,

[2] The derivation of this expression is presented in Annexe 11, Technical Note 11.1.

$$i_o = \left(\frac{1+i}{1+g}\right) - 1, \text{ where}$$

i = the prevailing opportunity cost of capital, and

q = the constant per annum traffic growth in the with-project situation.

Values of the present worth factor, $A(N,i_o)$ are given in Technical Note 11.2 for N from one to 20 years and for i_o for one to 30%. The critical value of average daily traffic ADT^* is the traffic needed on the feeder road for the projected VOC savings to equal the cost of the improvement. If present actual traffic equals or exceeds the ADT^* value, the simple ER is acceptable, or equal to or larger than the prevailing opportunity cost of capital. This formula cannot be used in combination with Technical Note 11.2 if the opportunity cost of capital is less than the per annum traffic growth rate since that note does not include negative percentages. Maintenance costs may be ignored if their present value is less than 15% of the present value of construction costs.

19.14 The following example illustrates the use of the above ADT^* concept. A decision is to be made about the economic viability of the proposed improvement of a feeder road with the following characteristics:

I = \$25,000 per km.

VOC = \$0.50, a weighted average of the vehicle operating cost of all vehicles such as passenger cars, vans, trucks of various sizes, buses, motorcyles and animal drawn vehicles, observed on the existing road.

p = 25%

q = 5% per annum

i = 12% per annum

N = 10 years,

ADT = 95 vehicles, actual at evaluation time.

The computations are as follows:

(i) Calculate $i_o = \left(\frac{1-i}{1+g}\right) - 1 = \left(\frac{1{\cdot}12}{1.05}\right) - 1 = 0.0667$

or 6.67%

(ii) Look up the present worth factor for 10 years for 6% and 7% in Technical Note 11.2, and interpolate

6% = 7.3601		7.3601
	$0.3365 \times 0.67 = -0.2255$
7% = 7.0236		$A\ (N,i_o) = 7.1346$

(iii) Solver for $ADT^* = \dfrac{I}{VOC{\cdot}p\ 365{\cdot}A\ (N,i_o)}$

$$\underline{ADT^*} = \frac{25{,}000}{(0.50){\cdot}(0.25){\cdot}(365){\cdot}(7.135)}$$

ADT^* = 77 vehicles

Since the actual ADT (95) is greater than the critical ADT* (77), the proposed improvement is economically viable.

19.15 Procedure II consists of determining the critical value of annual marketable agricultural production, or Q^*, with the following expression:[3]

$$Q^* = \frac{I}{voc \cdot p \cdot A(N, i_p)}$$

where I and p are as defined as before and

voc = per ton-km pre-investment weighted average vehicle operating cost,
$A(N,i_p)$ = present worth of an annuity factor, as a function of N and i_p (using Technical Note 11.2),
N = expected RAI life in years, and
$i_p = \left(\frac{1+i}{1+r}\right) - 1$ where:
i = the prevailing opportunity cost of capital,
r = the annual growth rate in marketable agricultural production in the with project situation.

The critical value of the current annual marketable agricultural production in tons, Q^*, is the marketable production quantity that must be transported over the feeder road for the projected ton-km cost savings to equal the per kilometre investment of the improvement. If the total present marketable agricultural production in tons, equals or exceeds the Q^* value, the simple ER is acceptable.

19.16 The following example illustrates the use of the above Q^* concept. A decision is to be made about the economic viability of a proposed improvement of a track with the following characteristics:

I = \$1,500 per kilometre,
voc = \$0.75 per ton-kilometre (weighted average vehicle operating cost of vehicles using the track),
p = 30%,
r = 10% per annum,
i = 15% per annum,
N = 5 years, and
Q = 300 tons (total marketable production at time of evaluation).

The computation is as follows:

(i) Calculate $i_p = \left(\frac{1+i}{1+r}\right) - 1 = 0.0455$ or 4.55%

(ii) Look up the present worth factor for 5 years for 4% and 5% in Technical Note 11.2

4% = 4.4518   4.4518
$0.1223 \times 0.55 = -\ 0.0672$
5% = 4.3295   4.3846

[3] The derivations of expressions using Q^* and Q^{**} are also presented in Technical Note 11.1.

(iii) Solve for $Q^* = \dfrac{I}{voc \cdot p \cdot A(N,i_p)}$

$$Q^* = \frac{1500}{(0.75)\cdot(0.30)\cdot(4.385)} = 1{,}520 \text{ tons}$$

Since the actual Q (300 tons) is less than the critical Q^* (1,520 tons), the improvement is not economically viable.

19.17 Procedure III can be used to determine the critical value of annual marketable agricultural production saved, or Q^{**}, through spot improvements. Spot improvements increase accessibility. Therefore production that would otherwise spoil can reach the market. Since the cost of spot improvements is not a constant per km cost, the length of the trip as well as the market price of the lost production must be considered. Procedure II may be modified to include these values in the determination of the critical annual marketable agricultural production saved, Q^{**}, using the following expression:

$$Q^{**} = \frac{IE}{(P_m - voc \cdot D) \cdot A(N, i_p)}$$

where: IE = total cost of eliminating one or more obstacles on the feeder road under consideration,

P_m = weighted average of crops' local market prices per ton,

voc = weighted average per ton-km pre-improvement vehicle operating cost,

D = average distance travelled, round trip from mid-point of area benefiting from the obstacle elimination to the destination, in km, and

$A(N, i_p)$ = present worth of an annuity factor, see Para 19.15.

The critical value of the current annual marketable agricultural production saved, Q^{**}, is the quantity needed for the net excess market sales to equal the cost of eliminating the obstacle. If the quantity saved equals or exceeds the Q^{**} value, the simple ER is acceptable.

19.18 The following example illustrates the use of the Q^{**} concept. A decision is to be made about the economic viability of building a ford at the entrance to the track investigated in Para 19.16. Since the ford is located at the beginning of the track, the location of the additional marketable production is assumed to be the track's mid-point. Available data indicate that:

IE = \$6,000 for the ford
P_m = \$148.00
voc = \$0.75
D = 10 km
$A(N, i_p)$ = 4.385
Q = 100 tons, actual loss at evaluation time.

Thus, $Q^{**} = \dfrac{IE}{(P_m - voc \cdot D) \cdot A(N,i_p)}$

$$Q^{**} = \frac{6{,}000}{[148-(0.75)\cdot 10]\cdot(4.385)} = 9.74 \text{ tons}$$

If the ford is built 6 km from the end of the road, the average location of the lost production becomes km 8 and D becomes 16 km. In this case the Q^{**} value is as follows:

$$Q^{**} = \frac{6{,}000}{[148.00-(0.75)\cdot(16)]\cdot(4.385)} = 10.06 \text{ tons}$$

In other words, the exact location of the ford in this example has small impact on the critical Q^{**} value because the ton-km transport cost is a small portion of the market value of a ton of salvaged production.

19.19 When the transport cost to market value ratio is low, the exact location of the obstacle to be eliminated is not critical, but as that ratio increases, the significance of the locations of the individual obstacles to be eliminated increases. For example, if the cost of the ford increases four-fold to \$24,000, Q^{**} becomes 38.96 tons when the ford is at km 0 and 40.24 tons when the ford is at km 6. Note that the Q^{**} increases at the same rate as the investment increases. However, if the market price is reduced to a quarter (\$37) while the obstacle elimination cost remains at \$6,000, Q^{**} becomes 46.38 tons when the ford is at km 0 and 54.73 tons when the ford is at km 6, or approximately 12% more because of the difference in location.

19.20 It is noted that Procedures II and III cannot be combined by adding I and IE and dividing by the sum of the respective denominators of the Q expressions. If the above two examples (Para 19.16 and Para 19.18) are combined to show a total area production of 400 tons, their combined critical Q^{*} would be 130 tons, in part, due to a *voc* savings realized when transporting the additional 100 tons of salvaged produce. However, in reality, the low investment required for obstacle elimination masks the high investment required to produce the *voc* savings. Unless each component results in a viable option, their combination is not acceptable.

19.21 Procedures I–III do not determine a simple ER *per se*. While they are basically feasible–unfeasible determinations, the more the present traffic volume in Procedure I or the transport volume in Procedures II or III exceed the critical or break-even volume, the higher the simple ER should be. Procedures IV–VIII consist of determining benefits and cost streams related to investments for the improvement or construction of RAI and agricultural investments. With this information, one can easily determine the simple ER. Since techniques for calculating ERs are well known, they are not discussed here.

19.22 Procedures IV and V are based on an assessment of economic activity, particularly agricultural production, in the feeder road's zone of influence. They are best suited for situations where there are reasonably accurate data regarding boundaries of zone of influence, prices and yields of agricultural products produced in this area, and the agricultural potential. These procedures ignore *VOC* savings related to non-agricultural traffic such as passenger traffic.

19.23 Boundaries of zones of influence are determined by examining (i) the road network around the feeder road being analyzed, (ii) the distance between farms and local markets, (iii) the terrain and (iv) the means of transport used, such as headloading, pack animals, animal-drawn carts, agricultural pickups, trucks, passenger cars and buses. The annual benefits B stemming from investments in feeder road and agricultural components in a zone of influence with one agricultural crop are given by:[4]

$$B = P_m(Q_2-Q_1)-K_2-Q_2k_2+K_1+Q_1k_1+H_2k_2-H_1k_1 \qquad \text{................ (1)}$$

where: P_m = crop's local market price ($ per ton),
Q_2,Q_1 = annual production of crop in with- and without-project situation, respectively (tons),
K_2,K_1 = economic cost of producing Q_2 and Q_1, respectively ($),
k_2,k_1 = economic costs of transport over the feeder road in the with- and without-project situation, respectively ($ per ton), and
H_2,H_1 = on-farm (home) consumption of an agricultural crop produced in the area of influence, in the with- and without-project situation, respectively (tons).

The difference between Procedures IV and V is that IV ignores home consumption or the terms $H_2k_2- H_1k_1$. Normally, the effect of ignoring these terms on the ER is insignificant.

19.24 The step-by-step procedures for solving expression (1) consist of determining the annual benefit B over the project's expected life for each of the principal products in the zone of influence. They require the collection of data on cultivated areas in a road's zone of influence, yields and prices of the agricultural products grown in those areas, and costs of agricultural production such as cost of labour, fertilizers, pesticides, farm implements and extension services. If these data are not available at the Ministry of Agriculture, a survey is required. Such a survey normally starts with the identification of representative farms, followed by an analysis of yields and costs of production related to these farms. When a number of RAI links are located in a region where representative farms are evenly distributed, one survey for these RAI normally suffices. A further simplification may be introduced in carrying out the procedures by applying them first to the major crops in the zone of influence. If an acceptable simple ER is obtained, computations related to other minor crops can be omitted; otherwise, they have to be carried out to check the resulting ER.

19.25 Procedures VI, VII and VIII are also based on expression (1), or a derivation thereof, for the assessment of economic activity in the zone of influence. In addition, they do not ignore *VOC* savings related to non-agricultural traffic or the portion of the shaded rectangle of Figure 39 pertaining to this traffic. The quantification of these benefits is as described in Para 19.03. The difference between Procedures VI and VII is that VI uses local market prices while VII uses farmgate prices in the without-project situation.

[1] References [6] and [11] discuss and give derivations for expressions used in this section.

The choice of using either Procedure VI or VII should be based on which information is more readily available.

19.26 Unlike Procedures IV to VII, Procedure VIII allows for a difference in local market prices in the without- and with-project situations. This difference may occur when the quality of agricultural products in the with-project situation is better than that in the without-project situation. The difference may also occur when increases in production due to significant agricultural investments in RAI's zones of influence are so substantial that they may affect the prices of agricultural products in a country. Price effects caused by a project, which are often included in the definition of externalities, are the exception.

19.27 Expressions (2), (3) and (4) represent the annual benefits B stemming from investments in RAI and agricultural components in a zone of influence with one agricultural crop, and established with Procedure VI, VII, or VIII, respectively.

$$B = P_m(Q_2-Q_1)-K_2-Q_2k_2+K_1+Q_1k_1+H_2k_2-H_1k_1+B_t \qquad \text{.........(2)}$$

$$B = P_1(Q_2-Q_1)-K_2-Q_2k_2+K_1+Q_1k_1+H_2k_2-H_1k_1+F_1(V_2-V_1)+B_t \qquad \text{........(3)}$$

and

$$B = P_mQ_2-P_{m1}Q_1-C_2-Q_2k_2+C_1+Q_1k_1+H_2k_2-H_1k_1+B_t \qquad \text{........(4)}$$

where B_t = VOC savings from the feeder road improvement and related to non-agricultural traffic ($)

P_1 = farmgate price of an agricultural product in the without-project situation ($ per ton),

F_1 = freight rate for transporting one ton of agricultural products over the RAI in the without-project situation ($ per ton),

V_2, V_1 = quantities of an agricultural product 'exported' from the zone of influence in the with- and without-project situation, respectively, or $V_2 = Q_2-H_2$ and $V_1 = Q_1-H_1$ (tons),

P_{m2}, P_{m1} = local market price of an agricultural product in the with and without-project situation, respectively ($ per ton), and all other terms are as defined above.

19.28 Procedures IV to VIII make no distinction between production patterns on farms of different size. Ignoring this distinction may affect the ER if yields of products from large farmers are significantly better than yields of products from small farms. In this situation, a weighted average yield of the small and large farms should be used in the application of these procedures. The procedures do not recognize differences between farmgate prices of on-farm consumption and prices of exports from a road's zone of influence since these differences are rarely significant enough to affect the ER.

19.29 Procedure IV consists of carrying out Steps 1 through 6 of the following algorithm; Procedure V calls for carrying out Steps 1 through 7, while Procedure VI calls for carrying out Steps 1 through 8.

Algorithm:

Step 1: determine for each crop in the zone of influence in the without-project situation, the annual production during the expected life by computing the product of 'area cultivated with this crop' times 'its yield';

Step 2: determine for each crop in the zone of influence in the with-project situation, the annual production during the expected life by computing the sum of 'unimproved area cultivated with the crop' times 'its yield' and 'improved area cultivated with the crop' times 'its yield';

Step 3: with the prevailing local market price and annual production of Steps 1 and 2, calculate annual production values in the without- and with-project situations and consequent annual incremental production values;

Step 4: for each crop in the zone of influence, determine annual incremental agricultural production costs during the expected life;

Step 5: for each crop in the zone of influence, determine the annual incremental transport costs during the expected life;

Step 6: estimate annual incremental RAI maintenance costs (routine and periodic);

Step 7: for each crop in the zone of influence, determine the difference between the products of 'annual home consumption' times 'costs of transport' in the without- and with-project situations; and

Step 8: determine *VOC* savings related to non-agricultural traffic.

Results of Steps 3 through 6, together with costs of investments of RAI and agricultural components, are all the data necessart to establish the simple ER with Procedure IV. Results of Steps 3 through 7, together with these investment costs, suffice to determine the simple ER with Procedure V, while Procedure VI calls for the results of Steps 3 through 8.

19.30 Procedure VII consists of Steps 1 through 8 of the aforementioned algorithm, with the following alteration and addition:

- replace 'the prevailing local market price' of Step 3 by 'the farmgate price in the without-project-situation' ($ per ton), and add the following step to Steps 1 through 8 –

Step 9: for each crop in the zone of influence, determine the product of 'freight rate for transporting one ton of agricultural products on the feeder road link in the without-project situation' times 'annual incremental exports from the zone of influence'.

19.31 Procedure VIII consists of Steps 1 through 8 of the aforementioned algorithm where Step 3 is replaced by the following Step 3:

Step 3: with the prevailing local market prices P_{m1} and P_{m2} and annual production of Steps 1 and 2, determine annual production values in the without- and with-project situations and consequent annual incremental production values. Para 19.27 gives the definitions of P_{m1} and P_{m2}.

19.32 Other procedures may be applied. For instance P_1 of Procedure VII may be replaced by (P_m-F_1), $(P_2+F_2-F_1)$, $(P_m-k_1-r_1)$, or $(P_m+k_2+r_2-k_1-r_1)$, where

r_1, r_2 = transporter's profit related to his transporting one ton over the feeder road in the without- and with-project situation, respectively ($ per ton),

F_2 = freight rate for transporting one ton of agricultural products over the feeder road in the with-project situation ($ per ton), and

all other symbols are as defined in Paras 19.23 and 19.27. Procedure VII has been described in terms of P_1 since information about P_1 is generally more readily available than information about F_2, r_1 and r_2. Procedure IV may be used with Step 3 of Procedure VIII, if home consumption and *VOC* savings related to non-agricultural traffic are negligible.

19.33 The following example illustrates the use of Procedure IV. The construction of a 30 km gravel feeder road and wells together with a strengthening of extension services is proposed in a zone of influence of 6500 ha. Costs of constructing the road and wells amount to $100 and $0.20 million, respectively. The expected life of these components is 10 years. Table 12 gives the crop area breakdown and the crops' yields, while Table 13 shows the crops' annual production costs, including costs of extension services, costs per ton of transport, and road maintenance costs in the with- and without-project situations. The costs of per ton of transport in columns (8) and (9) have been arrived at by multiplying the length of the road (30 km) by the per ton-km transport costs in the without- and with-project situations ($0.29 and $0.11, respectively). The results of steps 1–6 of Para 19.29 are located in Tables 11 and 12 as follows: Specifically, the results of:

Step 1 are shown in columns (7) and (14) of Table 12,
Step 2 are shown in columns (8) and (15) of Table 12,
Step 3 are shown in columns (20), (21) and (22) of Table 12,
Step 4 are shown in columns (6) and (7) of Table 13,
Step 5 are shown in columns (14) and (15) of Table 13,
Step 6 are shown in column (18) of Table 13.

The investment costs of $1.00 and $0.20 million and cost and benefit streams obtained from carrying out Steps 3 through 6 result in an ER of 30.99%

19.34 Procedure V calls for the determination of home consumption. To do so, the population in the zone of influence is first determined. Referring to the example of Para 19.33, suppose the area of influence has a base population of 60,000 which is growing at a rate of 2.5% per annum and has a per capita home consumption of 15 kg and 12 kg of wheat and 0.5 kg and 0.3 kg of tomatoes in the without- and with-project situation, respectively. Table 14 shows the results of Step 7 (Para 19.29) applied to this example. The notations (k_1, k_2, H_1 and H_2) of Table 14 are defined in Para 19.23. Columns (9) and (12) of Table 14 indicate that unless the population in the zone of influence is large in relation to that total production, i.e., the area consists of small farms inhabited by marginal cash crop farmers, the effect of on-farm consumption on the simple ER is small. In this example the ER of 30.17% is slightly lower than the one computed in Para 19.33.

Table 12: Crop Area Breakdown, Yields and Production Values

	WHEAT									TOMATOES											
Year	w/o Proj. Areas (000ha)	w/Project Areas (000 ha) Unlmp	Imp	Yields (ton/ha) Unlmp	Imp	Production (000tons) w/o	With	Production Value (000$) w/o	With	w/o Project Areas (000ha)	w/Project Areas (000 ha) Unlmp	Imp	Yields (ton/ha) Unlmp	Imp	Production (000 tons) w/o	With	Production Value (000$) w/o	With	Prod. value (all crops) (000$) w/o	With	Value Added (all crops) (000$)
(1)	(2)	(3)	(4)	(5)	(6)	(7)	(8)	(9)	(10)	(11)	(12)	(13)	(14)	(15)	(16)	(17)	(18)	(19)	(20)	(21)	(22)
1	4.00	4.00	1.00	0.56	0.90	2.240	3.140	16	61	0.30	0.30	0.16	9.11	15.0	2.733	5.133	8	64	24	125	101
2	4.50	3.50	2.00	0.59	0.90	2.655	8.865	38	120	0.30	0.20	0.18	9.55	20.0	2.865	5.980	16	130	54	248	194
3	4.50	3.10	2.40	0.61	1.00	2.750	4.291	53	180	0.30	0.20	0.28	10.00	20.0	3.000	7.600	24	198	77	378	301
4	4.85	2.00	3.50	0.63	1.00	3.055	4.760	70	239	0.40	0.10	0.39	10.00	20.0	4.000	8.800	32	261	102	500	398
5	4.90	1.00	4.70	0.65	1.00	3.190	5.350	87	300	0.50	0.00	0.50	10.00	20.0	5.000	10.000	40	325	127	625	498
6	5.00	0.00	6.00	0.70	1.00	3.500	6.000	125	360	0.50	0.00	0.50	10.00	20.0	5.000	10.000	40	325	165	685	520
7	5.00	0.00	6.00	0.70	1.00	3.500	6.000	125	360	0.50	0.00	0.50	10.00	20.0	5.000	10.000	40	325	165	685	520
8	5.00	0.00	6.00	0.70	1.00	3.500	6.000	125	360	0.50	0.00	0.50	10.00	20.0	5.000	10.000	40	325	165	685	520
9	5.00	0.00	6.00	0.70	1.00	3.500	6.000	125	360	0.50	0.00	0.50	10.00	20.0	5.000	10.000	40	325	165	685	520
10	5.00	0.00	6.00	0.70	1.00	3.500	6.000	125	360	0.50	0.00	0.50	10.00	20.0	5.000	10.000	40	325	165	685	520

Note: Column 7 = Column (2) × Column 5
" 8 = Column (3) × Column (5) + Column (4) × Column (6)
" 9 = Column (7) × \$150 − Column (2) × \$80
" 10 = Column (8) × \$150 − Column (3) × \$80 − Column (4) × \$90
" 16 = Column (11) × Column (14)
" 17 = Column (12) × Column (14) + Column (13) × Column (15)
" 18 = Column (16) × 60 − Column (11) × \$520
" 19 = Column (17) × 60 − Column (12) × \$520 − Column (13) × \$550
" 20 = Column (9) + Column (18)
" 21 = Column (10) + Column (19)
" 22 = Column (21) − Column (20)

Table 13: Annual Agricultural Production, Transport and Road Maintenance Costs

Year	Agricultural Production Costs ('000$)		Agricultural Production Costs ('000$)		Incremental Agricultural ('000$)		Transport Costs $ per ton km		Agricultural Transport Costs ('000$)		Agricultural Transport Costs ('000$)		Incremental Agricultural Transport Costs ('000$)		Road Maintenance Costs ('000$)		Incremental Road Maintenance Costs ('000$)
	Without-Situation		With-Situation						Without-Situation		With-Situation						
	Wheat	Tomatoes	Wheat	Tomatoes	Wheat	Tomatoes	Without	With	Wheat	Tomatoes	Wheat	Tomatoes	Wheat	Tomatoes	Without	With	
(1)	(2)[a]	(3)	(4)	(5)	(6)	(7)	(8)	(9)	(10)	(11)	(12)	(13)	(14)	(15)	(16)	(17)	(18)
1	328.00	156.60	423.00	245.40	95.00	88.80	8.70	3.30	19.49	23.77	10.36	16.94	−9.13	−6.83	0	0	0
2	369.00	156.60	477.00	230.40	108.00	73.80	8.70	3.30	23.10	24.93	12.75	19.73	−10.13	−5.20	0	0	0
3	369.00	156.60	482.20	259.80	113.20	103.20	8.70	3.30	23.93	26.20	14.16	25.08	−9.77	−1.02	0	10	10
4	397.70	208.80	496.50	268.65	98.80	59.85	8.70	3.30	26.58	34.80	15.71	29.04	−10.87	−5.76	0	10	10
5	401.80	261.00	528.50	277.50	126.70	16.50	8.70	3.30	27.75	42.50	17.66	33.00	−10.09	−10.50	0	10	10
6	410.00	261.00	570.00	277.50	160.00	16.50	8.70	3.30	30.45	43.50	19.80	33.00	−9.78	−10.50	0	25	25
7	410.00	261.00	570.00	277.50	160.00	16.50	8.70	3.30	30.45	43.50	19.80	33.00	−10.65	10.50	0	10	10
8	410.00	261.00	570.00	277.50	160.00	16.50	8.70	3.30	30.45	43.50	19.80	33.00	−10.65	10.50	0	10	10
9	410.00	261.00	570.00	277.50	160.00	16.50	8.70	3.30	30.45	43.50	19.80	33.00	−10.65	10.50	0	10	10
10	410.00	261.00	570.00	277.50	160.00	16.50	8.70	3.30	30.45	43.50	19.80	33.00	−19.80	−10.50	0	25	25

Notes: Column (6) = Column (4) − Column (2)
Column (7) = Column (5) − Column (3)
Column (10) = Column (7) of Table 3 times Column (8)
Column (11) = Column (16) of Table 3 times Column (8)
Column (12) = Column (8) of Table 3 times Column (9)
Column (13) = Column (17) of Table 3 times Column (9)
Column (14) = Column (12) − Column (10)
Column (15) = Column (13) − Column (11)
Column (18) = Column (17) − Column (16)

[a] The agricultural production costs include both the agricultural production costs of Table 12 and the annual cost of the extension services. The annual costs of extension services per hectare are $2 for unimproved production and $5 for improved production technology.

Table 14. Annual On-Farm (Home) Consumption, H_1k_1, and H_2k_2

Year	Population ('000)	Wheat on Farm Consumption ('000 tons)		Tomatoes on Farm Consumption ('000 tons)		H_2k_1 ('000$)	H_2k_2 ('000$)	$H_1k_1-H_2k_2$ ('000$)	H_1k_1 ('000$)	H_2k_2 ('000$)	$H_1k_1-H_2k_2$ ('000$)
		Without (H_1)	With (H_2)	Without (H_1)	With (H_2)						
(1)	(2)	(3)	(4)	(5)	(6)	(7)	(8)	(9)	(10)	(11)	(12)
1	60.00	0.90	0.90	0.03	0.03	7.83	2.97	4.86	0.26	0.10	0.16
2	61.50	0.92	0.92	0.03	0.03	8.00	3.04	4.96	0.26	0.10	0.16
3	63.00	0.95	0.76	0.03	0.02	8.26	2.51	5.75	0.26	0.07	0.19
4	66.20	0.99	0.79	0.03	0.02	8.61	2.61	6.00	0.26	0.07	0.19
5	67.80	1.02	0.81	0.03	0.02	8.87	2.67	6.20	0.26	0.07	0.19
6	69.50	1.04	0.83	0.03	0.02	9.05	2.74	6.31	0.26	0.07	0.19
7	71.30	1.07	0.36	0.04	0.02	9.31	2.84	6.47	0.35	0.07	0.28
8	73.10	1.10	0.88	0.04	0.02	9.57	2.90	6.67	0.35	0.07	0.28
9	74.90	1.12	0.90	0.04	0.02	9.74	2.97	6.77	0.35	0.07	0.28
10	76.80	1.15	0.92	0.04	0.02	10.01	3.04	6.97	0.35	0.07	0.28

[1] k_1 and k_2 values are given in Columns (8) and (9) of Table 13, respectively.

Note: Column 7 = Column (3) times Column (8) of Table 13
Column 8 = Column (4) times Column (9) of Table 13
Column 10 = Column (5) times Column (8) of Table 13
Column 11 = Column (6) times Column (9) of Table 13
Column (9) = Column (7) – Column (8)
Column (12) = Column (10) – Column (11)

II. SPECIAL ISSUES

19.35 Screening.

Prior to the economic appraisal of RAI, one may wish to screen out candidates to be considered for such appraisal, if the total costs of improving existing RAI far exceeds the available budget. Screening techniques for RAI components are sometimes based on one composite index of economic, social and political factors. This practice is not recommended because the screening may accept a large number of candidates which are subsequently rejected in the economic appraisal. A more efficient procedure, if social and political factors are to be considered, starts with screening based on these criteria only. The RAI links remaining after this screening process may subsequently be subjected to a screening on the basis of economic criteria.

19.36 State of Project Preparation.

The implementation of RAI improvements is often behind schedule, resulting in disbursement delays, because engineering work for the first year's programme is not completed before negotiation with a project donor. This engineering work should be completed before negotiation in order to ensure a timely start of construction works and to provide an adequate basis for the cost estimate of RAI components.

19.37 For many simple RAI improvements, it may suffice to complete limited engineering work, provided design standards, construction procedures and related historical data (updated) for representative items are available at the time of appraisal. 'Final' engineering, as appropriate to such cases, would normally be performed only shortly before construction starts.

19.38 Given the limited financial resources of most developing countries, project preparation should consider a redirection from routine application of predetermined rural feeder road design standards to specific application of selected design features necessary to increase accessibility to the level required for initial development. This can best be accomplished by spot improving the existing tracks and roads which represent a significant bottleneck. Such an incremental improvement approach will not only require a minimum investment, but will also permit further RAI upgrading to parallel actual development demand, maximize future construction options, and minimize unnecessary construction and maintenance. Spot improvements concentrate all resource expenditures on identified existing environmental and operational constraints, require minimal engineering input in formal paper work and drawings, can easily be tailored to the specific types of transport found in the project area, and are frequently individually small operations which can maximize the use of low technology, local materials, and local capabilities.

19.39 Consideration of incremental improvements should not be done to the exclusion of long-term investment needs. Early attention to the balance between highway and RAI construction works, their maintenance and spot improvements, is called for in the project preparation phase. In other words, the size and composition of a government's current road investment programme should be examined. Such an examination includes the assessment of tradeoffs between long-lived capital structures which offer economies of scale and incremental improvements.

19.40 **Construction Costs.**

RAI construction costs should be fully accounted for in the economic appraisal. This applies also to situations where not all benefits of RAI links to be improved or constructed can be quantified, due to lack of data on the future use of these links by passengers in or outside the RAI's zone of influence, government officials and/or farmers who are not living in the project's zone of influence. Taking into account all construction costs may result in an ER below the prevailing opportunity cost of capital. Para 19.50 below describes conditions under which the World Bank may still consider the financing of a RAI with such an ER.

19.41 **Maintenance.**

Review of project completion reports reveals that in a large portion of completed projects with rural road components, insufficient attention was paid to their maintenance during project preparation and/or implementation. It is, therefore, important that during project preparation particular attention be given to the estimation of required maintenance costs and corresponding annual budget allocations, together with an assessment of institutional and manpower capabilities to carry out the maintenance. Detailed training programmes are to be formulated if these capabilities are insufficient. Estimation of maintenance costs should, *inter alia*, be based on an assessment of standards for periodic and routine maintenance, equipment and workshops.

19.42 **Complementary Needs.**

The justification for investments in RAI construction and improvements often hinges on improved conditions for agricultural production. Although good rural roads are important for agricultural production, farmers are unlikely to produce more if transport savings mainly accrue to truckers and final consumers, and/or other obstacles to their production are not dealt with. Questions to be raised during project preparation include:

- are transport services fully competitive or hampered by regulation?
- do crop prices provide adequate incentives to farmers to produce more?
- what is the land tenure situation?
- how is access to credit and what are its terms and conditions?
- are agricultural extension services adequate?
- is there a need for complementary agricultural investments in, for instance, storage and processing facilities, and wells?
- are inputs such as fertilizers and pesticides adequately available?

19.43 Complementary agricultural investments are sometimes needed to fully capture benefits to be had from RAI construction or improvements. These investments may either be included in the project with the RAI components in preparation or be secured by an agricultural project simultaneously being prepared. If a proposed RAI project without agricultural components complements ongoing agricultural projects, the costs of agricultural investments are sunk costs and therefore should not be considered in the RAI appraisal. The benefits from these agricultural projects should only be taken into account in the economic evaluation of a RAI link if they would not materialize without it. Costs and benefits of *planned* (future) agricultural projects, which are dependent on RAI components of a proposed project, should be considered together with the costs and benefits of these links, regardless of potential source of finance. This situation calls for the formulation of investment packages consisting of interdependent RAI and agricultural components.

19.44 **Interdependencies.**

During the appraisal of rural development or agricultural projects, only one ER is sometimes established for the entire investment, including all RAI and agricultural components such as irrigation works, supply of inputs and extension services in all project areas. The reason for this approach is based on the argument that the establishment of separate ERs for the RAI components of these projects requires arbitrary decisions on how to attribute essentially joint results to different investment components.

19.45 Rural access infrastructure links included in a rural development or agricultural project should not, however, *a priori*, be thought of as being interdependent among each other and agricultural components as well. Neither is it necessarily true that the RAI component in an investment package which is part of a rural development project is dependent on that package's agricultural components. Project elements, including RAI, should constitute an economically viable package of investment and production techniques through their appropriate design, standards and synergism. A project may consist of one or more of such packages. In addition, a project may itself be an economically viable package in a set of other ongoing projects within a larger programme. Thus, ERs should be determined for economically viable investment packages, which may consist of one or more RAI and agricultural investments, agricultural components, or one or more rural roads.

19.46 Clear cases of interdependence among RAI and agricultural components are (i) the construction of irrigation canals and dams together with RAI to maintain the irrigation works, (ii) the improvement of RAI in areas where agricultural investments are proposed and where the present condition of such infrastructure would impede development operations during the rainy season, and (iii) the improvement of RAI in areas with proposed investments for the production of perishable products such as tomatoes, peaches, pears, etc. Assuming interdependency without testing it, may lead to the 'mixing in' of productive elements with elements which by themselves would not offer an acceptable ER. The identification of feasible project combinations of proposed projects calls for the consideration of possible interdependency among RAI links, among agricultural components and among RAI and agricultural components. For projects covering large areas, such consideration should start with a breakdown of subareas.

19.47 This handbook proposes an ER determination using procedures IV–VIII for groups of independent, individual RAI and agricultural components to save time, if the following two conditions are satisfied:

(i) The per km cost of construction or improvement of each of the rural roads belonging to the group is equal to or less than US$15,000 equivalent in 1984 prices; and
(ii) The RAI links which constitute a group are located in areas where proposed agricultural investments are interdependent with each other.

Although RAI links satisfying these conditions are not necessarily interdependent RAI links, they may be treated in the same way as interdependent groups of RAI when applying procedures IV–VIII. Spot improvements (Para 19.38) may also be grouped to determine their economic viability.

19.48 One ER for groups of independent RAI may also be acceptable provided the areas in which these RAI are located have similar farming patterns and variations in costs of RAI improvements or construction are less than about 20%. Finally, when a RAI component of a project is relatively small, i.e., its costs constitute less than 10% of total project costs, the determination of a separate ER of the RAI component may not be necessary, provided the components have a reasonable, general justification on project or regional grounds and their design standards are acceptable.

19.49 **Social Aspects.**

RAI construction or improvement is usually suggested to enhance agricultural production and/or to alleviate social problems such as poor access to hospitals and/or schools. The World Bank's policy is that investment packages containing RAI to enhance agricultural production must have an ER equal to or higher than the prevailing opportunity cost of capital in order to be considered for Bank financing ('acceptable ER').

19.50 The World Bank may consider the financing of RAI to alleviate social problems although the determination of an acceptable ER is difficult or impossible with available data, provided:

(i) the majority of the population in the RAI's zone of influence consists of the target population or the population with relative poverty;
(ii) the equivalent annual construction cost per person in the RAI's zone of influence does not exceed a limit to be determined on the basis of household expenditures of the target population;
(iii) the absence of the proposed RAI results in virtual inaccessibility to facilities such as schools and health clinics; and
(iv) total construction costs of these RAI do not exceed 20% of the costs of the total construction programme of the proposed project.

19.51 One crucial issue to be addressed when deciding to include RAI links without determining an ER is their design standards. That is, if people in the RAIs' zones of influence are too poor to buy motorized vehicles or to make significant use of a future bus service, a simple bicycle track or good footpath will suffice. In other words, absolute minimum design criteria have to be employed for these RAI links.

19.52 Paras 19.35–19.47 already imply errors made in the appraisal of RAI components. Other errors observed during a review of rural road appraisals during the past five years pertain to:

- double counting of benefits or the adding of VOC savings related to agricultural traffic to the above given expressions (1) through (4) for B in Paras 19.23 and 19.27;
- insufficient attention to variations in average daily traffic (ADT) due to seasons, weekend versus non-weekend activities, and/or market-day versus non-market-day activities;
- insufficient attention to alternative RAI design standards;
- the use of shadow wage rather than market wage rates although unemployment or underemployment does not prevail; and
- insufficient number of representative farms when analyzing agricultural production potential (Para 19.24).

III. CONCLUSIONS

19.53 All of the benefits from improving rural acccss infrastructure are not quantifiable. Furthermore, the costs of the RAI component in rural development and agriculture projects are often a relatively small portion of the overall project costs. Simplified operational procedures are available to reduce the time and effort involved in appraising these RAI project components. They imply the determination of a simple economic return, i.e., an ER based on a significant portion of quantifiable benefits rather than all. The extent to which benefits are considered depends on the underlying assumptions and the type and amount of data collected. All simple ERs are underestimates of actual ERs.

19.54 This chapter presents eight appraisal procedures in increasing order of complexity. There is not much difference in the degree of complexity of the last three procedures presented; the difference between these three methods pertains to different input requirements. If a proposed RAI link, group of interdependent RAI links or investment packages of interdependent RAI links and agricultural components passes the test of a simple appraisal method, no further analysis is required unless the analyst wishes to know the percentage points by which an ER exceeds the opportunity cost of capital. In other words, to save time one accepts a project component or group of interdependent components if its 'simple' ER exceeds the opportunity cost of capital.

19.55 While RAI links should not, *a priori*, be assumed interdependent among each other or with agricultural project components, they may be treated in the same way as interdependent groups of RAI links if their per km cost is below US$15,000 equivalent (in 1984 prices) and they are located in areas where proposed agricultural investments are interdependent with each other. Spot improvements may be also grouped to determine their economic viability.

19.56 One ER for groups of independent RAI links may also be acceptable in areas where farming patterns, terrain and soil conditions are similar and improvement costs vary by less than about 20%. The World Bank may consider financing RAI networks to alleviate social problems although the determination of an acceptable ER is difficult or impossible with available data when the conditions outlined in Para 19.50 are applicable.

PART IV
TECHNICAL NOTES

1. The Technical Notes in Part IV were developed in conjunction with Part III and are referenced at appropriate locations throughout the previous text. These Technical Notes are either expanded technical details of specific subject matter, or suggestions for a practical method of solving specific problems identified within the presentation. They are intended primarily as a reference collection for readers of the previous material. In order to access easily related information the Technical Notes are grouped in a series of eleven annexes, each annexe covering a specific field (see table of contents).

2. A list of cross references is given at the beginning of each annexe. It identifies, by paragraph or technical note number, direct references to the technical notes which make up each annexe.

ANNEXE 1

Accessibility

This annexe contains two Technical Notes:
T.N. 1.1., *Rainfall and Slippery Soils*, is referenced in Para 9.19.
T.N. 1.2, *Evaluation of a Farm Family's Transport Requirements*, is referenced in Paras 2.20, 8.33, 8.36 and 9.21.

Technical Note 1.1
Rainfall and Slippery Soils

1. Observations in Kenya [49] indicate that 'red coffee' clay soils become impassable because they become too slippery for normal traffic after less than one hour of moderate rainfall. One hour of moderate rainfall, 5.0 mm, will also cause roads built on other types of clay soil such as black cotton soil, and on more silty soils, to become impassable. Generally, however, providing the road is well shaped and possesses adequate drainage, the road will become passable at some time in the succeeding 24 hours provided there is no further rain. The tentative conclusion is that on any day in Kenya when rainfall exceeds 5.0 mm, clay and some silty soil roads will certainly be impassable for up to 24 hours, and if the rainfall lies between 2.5 mm and 5.0 mm, then it is still highly likely that they will be impassable for some period of the day [49].
2. Road surface drying is greatly accelerated by exposure to the sun. Thus in the tropics roads with an east-west orientation have an advantage in this respect. In wet climates it is important, particularly on roads with a north-south orientation, to make sure that the vegetation at the roadside is kept trimmed back in order to expose the road surface to the sun. (See Technical Note 4.04, Para 3, and Figure 60.)

Technical Note 1.2
Evaluation of a Farm Family's Transportation Requirements

1. The following data, taken from a World Bank report for a project in Benin [93], describe the agricultural activities of a typical farm family of eight that produces a small surplus of agricultural produce for the cash economy:

Crop	*Production (kg/yr)*	*Subsistence (kg/yr)*	*Subsistence (ha required)*	*Surplus (kg/yr)*
Cereals	1550	1200	2.00	350
Yams/Cassava	3300	1400	0.40	900
Groundnuts	270	20	0.03	250
Rice	130			130
Cotton	20			20
	5270	3620	2.43	1650

The subsistence quantities, from comparison with another World Bank report [98] for verification of caloric content, were found to include 2% of total production for seeds and 15% of the remainder as loss due to inefficient storage, rodents and birds. While it is estimated that 15% of a person's caloric needs must come from other foods such as pulses, oil, sugar, meat and vegetables, the above surplus was evaluated to determine the transport needs without considering the need for acquiring other foods or any agricultural inputs.

2. The family in Para 1 sells 31% by weight of their production. Their yearly total surplus output represents less than one-quarter of the capacity of a single 7-ton truck of the type normally operating in the area. The family itself is faced with a minimum off-road agricultural transport demand of 5270 kg rather than the 1650 kg indicated in the road transport requirements. Considering the small quantity of surplus in each category, the family would probably move the surplus twice, once to their home and again to the local market or collection point at some designated timc, increasing their total agricultural transport demand to 5270 kg + 1650 kg = 6920 kg.

3. The major additional transport demand for subsistence is water and firewood. Studies in Kenya [91] indicate a farm family's water requirement at 50 kg/day and a firewood requirement of 30 kg/day. This additional transport demand is 29,200 kg per year or eight times the weight of the produce required for subsistence. In this example, the total farmer yearly transport demand includes, at a minimum – excluding any farm inputs, personal acquisitions, or double movement of produce – subsistence foodstuffs (3620 kg), water and fuel (29,200 kg), and surplus (1650 kg), or a total of 34,470 kg; more than 20 times the surplus evaluated for possible road transport demand.

ANNEXE 2

Transport Aids

This annexe contains nine technical notes:

T.N. 2.1, *Transport Aids – Excluding Conventional Motor Vehicles*, is referenced in Paras 5.03, 6.01 and 10.02.

T.N. 2.2, *Feature of Transport Aids*, is referenced in Paras 10.04 and 11.06 and Technical Note 9.07.

T.N. 2.3, *Suggested Infrastructure Requirements for Transport Aids*, is referenced in Paras 5.03, 6.01, 10.04, 10.05 and 11.05 and Technical Note 4.13.

T.N. 2.4, *Suggested Criteria for Vacious Types of Infrastructure*, is referenced in Paras 10.04, 10.05 and 11.05, and Technical Note 4.01.

T.N. 2.5, *Improved Animal-Drawn Vehicles (ADV)*, is referenced in Para 10.10.

T.N. 2.6, *The Supply and Quality of Rural Transport Services in Developing Countries: A Comparative Review*, is referenced in Para 10.12.

T.N. 2.7, *Overcoming Constraints to Existing or New Transport Aid Usage*, is referenced in Paras 5.03 and 10.13.

T.N. 2.8, *Requirements for Developing the Use of Alternative Transport Aids*, is referenced in Para 10.13.

T.N. 2.9, *Illustrative Comparison of Transport Costs Using Optional Aids*, is referenced in Para 10.20.

Fig. 40 *Headloading*

Technical Note 2.1
Transport Aids – Excluding Conventional Motor Vehicles

Headloading
- head basket (Fig. 40)
- head pan
- sack

Shoulderloading
- pickul bar (Fig. 41)
- shoulder yoke
- shoulder pan

Backloading:
- chee-kee (Fig. 1)
- rucksack
- sack with forehead strap

Stretcher (Litter), 2 Person:
- single shoulder pole
- two pole

Wheelbarrow
- Western (Fig. 42)
- Chinese (Fig. 43)

Single Track Cycles
- bicycle (Fig. 3)
- motorized bicycle
- motorscooter
- motorcycle

Pack Animals: (Fig. 44)
- donkey
- pony
- mule
- horse
- camel
- llama
- elephant

Two-Track Cycles:
- bicycle with sidecar (Fig. 45)
- motorcycle with sidecar (Figs. 46 & 47)

Three-Track Carriers:
- tricycle, 2 rear wheels (Fig. 4)
- tricycle, 2 front wheels (Fig. 48)
- motorized tricycle
- bicycle with trailer (Fig. 49)
- tricycle with trailer
- motor-scooter with trailer
- motorcycle with trailer
- three-wheel motocycle
- auto rickshaw (Fig. 50)

Handcarts
- two-wheel (Fig. 51)
- four-wheel (Fig. 52)

Rickshaws

Animal Sledges: (Fig. 53)
- ox
- bullock
- buffalo
- donkey
- pony
- mule
- horse
- elephant

Two-Wheel Animal Carts: (Fig. 54)
- ox
- bullock (Fig. 55)
- buffalo
- donkey
- pony
- mule
- horse
- dog
- goat

Four-Wheel Animal Carts:
- ox
- bullock
- buffalo
- mule
- horse

Tractor-Trailer Combinations:
- two-wheel tractor (Fig. 5)
- three-wheel tractor
- four-wheel tractor

Fig. 41 *Indonesian pickul bar*

Fig. 42 *Western wheelbarrow*

Fig. 43 *Chinese wheelbarrow*

Fig. 44 *Typical pack animal*

Fig. 45 *Bicycle with sidecar*

Fig. 46 *Motorcycle with goods sidecar*

Fig. 47 *Motorcycle with passenger sidecar*

Fig. 48 *Tricycle with two front wheels*

Fig. 49 *Bicycle and trailer*

Fig. 50 Auto rickshaw

Fig. 51 *Two-wheel handcart*

Fig. 52 *Four-wheel handcart*

Technical Note 2.2
Features of Transport Aids

Transport mode	*Reported load capacity (kg)*	*Reported average speed (km/h)*	*Reported maximum trip length (km)*	*Reported load width (cm)*
Headload	40% of body weight [29] 25–35 [29]	3 [47][1]	20 [47][1] 40 [62][2]	80 [47][1]
	men – 30, women – 20 [62][2] 50 [67][2][3]	1.5–3 [91]	20 [47][1]	80 [47][1]
Shoulderload	40% of body weight [29] 25–35 [29]	3.5 [47][1]	40 [47][1]	190 [47][1]
	62 (men) [42][1] 35 [27]	3–5 [27]	10 [27]	
Backload	35 (women) [42][1] 60–80 [29][3][4] 70 [27]	3.5 [47][1] 3–5 [27]	10 [47][1] 10 [27]	70 [47][1]
Wheelbarrow, Western	120 [29] 60–100 [91]	3–5 [27] 1.5–3 [91]	1 [27]	
Wheelbarrow, Chinese	180 [29][5]	3–5 [27]	3–5 [27]	
Four-wheel handcart	2000 [29][5]			
Bicycle, pass.			30 [47][1]	55 [47][1]
Bicycle, agric. product	150 [47] [67][1]	5 [47][1]	15 [47][1]	150 [47][1]
Bicycle, goods transport	105 [47][1] >100 [42] 60–100 [91] 80 [27] 40–80 [62]	8 [47][1] 6–12 [91] 10–15 [27]	30 [47][1] 30–40 [42] 40 [27]	120 [47][1]
Motorcycle	30 [47][1] 150–200 [27] 100–120 [63][7]	30–60 [27] 25 [47][1]	100 [27] 100 [47][1]	80 [47][1]
Animal Sled	100–150 [63][7] 70–150 [27]	3–5 [27]	20 [27]	
Donkey, pack	70–120 [29] 50 [62] 60–100 [91]	2.5–3 [29]	20 [27]	
Pony, pack	100–150 [29]	3–4.5 [29]	20 [27]	
Mule, pack	75–150 s29] 100–150 [62]	3–4.5 [29]	20 [27]	
Horse, pack	100 [47][1] 100–150 [62]	4 [47][1]	15 [47][1] 20 [27]	90 [47][1]
Camel, pack	200–300 [62] 120–680 [29]	3–5 [29]		
Bicycle, with sidecar	150 [27]	10–15 [27]	40 [27]	
Motorcycle, with sidecar	6-pass. [50][7] 250–400 [29][7]	30–60 [27]	100 [27]	
Tricycle, two front wheels	200 or 2 pass. [47][1] 150 or 2 pass. [92]	12 [47][1]	12 [47][1]	
Tricycle, two rear wheels	150 [91] 150–200 or 4 pass. [29] 250 or 3 pass. [42][1]	6–12 [91] 10–15 [27]	40 [27]	

Bicycle trailer	80–130 [91] 150 [27]	6–12 [91] 10–15 [27]	40 [27]	
Motorcycle trailer	200 or 5 pass. [29][8] 200–300 [27]	30–60 [27]	100 [27]	
Three-wheel motorcycle	310–700 [92] 250–400 [29] 200–300 [27] 225–275 or 2 pass. [92] 600 [63]	30–60 [27]	100 [27]	
Auto rickshaw	7 pass. [92] 750 [62][1]			
Hand cart	180 [27]	3–5 [27]	3–5 [27]	
Ox-cart	800 [91] 1200 w/1 ox [42][1] 1500 w/2 ox [42][1] 1000 w/2 ox [29] 500–1000 [62][2] 400 [29][1] 200–300 [63][8]	3 [91] 5 [42][1] 3–5 [27]	15 [47][1] 13 [48][2] 50 [27]	190 [47][1]
Donkey-cart	250–400 [91] 400–500 w/2 [92]	1.5–3 [91]		
Horse/mule cart	1200 or 8 pass. with 2 animals [47][1] 1000 w/2 [92]	10 [47][1]	15 [47][1]	160 [47][1]
Pony carriage (four-wheel)	300 or 6 pass. [47][1] 400–600 or 5–7 pass. [42][1]	12 [42][1]	12 [47][1]	140 [47][1]
Ox-cart, with pneumatic tyres	500–550 [29][1] 2500 w/2 ox [29] 1500–1700 w/2 [92] 2000–3000 [62][2]		30 [48][2]	
Two-wheel tractor-trailer	600 [91] 400–1500 [92] 1000 [95][2] 1200 [29]	8 [91] 10–15 [27]	50 [27]	60–110 [92]
Three-wheel tractor-trailer	800 [92] 720 [29]			
Four-wheel tractor trailer	1500–3500 [91] 3500–9000 [92] 3000–5000 [95][2]	5–12 [91]	20 [48][2]	70–180 [92]

[1] Indonesia
[2] India
[3] Nepal
[4] Korea
[5] China
[6] Bangladesh
[7] Philippines
[8] Vietnam

Technical Note 2.3
Suggested Infrastructure Requirements for Transport Aids

Transport mode	Minimum tread width (cm)	Minimum cleared width (m)	Minimum clearing for passing (m)	Maximum sustained grade[1] (%)	Maximum grade (%)	Minumum bridge width (m)	Maximum ford depth (cm)	Type of surface required. 1 = excellent 5 = poor (scale)	Remarks
Headloading	45	1.0	2.0	20	30	0.3	100	5	For maximum load. loading assistance required.
Shoulderloading	45	2.0	2.0	20	30	0.3	100	5	Loads less for long or difficult trip
Backloading	45	1.0	2.0	20	30	0.3	100	5	Loads less for long or difficult trip
Stretcher	45	1.0	2.0	20	20	0.6	200	4	Load capacity small per man
Wheelbarrow. Western	60	1.2	2.5	4	12	1.0	10	3	Best in flat, smooth areas: unstable
Wheelbarrow. Chinese	60	1.2	2.5	4	12	1.0	25	3	Best in flat, smooth areas: unstable
Ridden bicycle	45	1.0	1.5	4	8	1.0	10	3	May have to push sometimes
Transport bicycle	100	1.5	3.0	4	6	1.0	10	3	May have to push often: unstable
Motor scooter	45	1.0	1.5	10	15	1.0	10	3	Small wheels a handicap
Motorcycle	45	1.0	2.0	12	30	1.2	15	4	Different types. different requirements
Ridden animal	60	1.2	2.0	15	25	1.2	100	5	All animals require periodic watering
Donkey, pack	60	1.8	2.5	15	25	1.2	45	5	Donkeys will refuse heavy loads
Pony, pack	60	1.8	2.5	15	25	1.2	45	5	Smaller animals carry smaller loads
Mule, pack	60	1.8	2.5	15	25	1.2	60	5	More endurance than a horse
Horse, pack	60	1.8	2.5	15	25	1.2	60	5	Fastest of pack animals
Camel, pack	60	1.8	2.5					3	Limited to areas of light friable soils
Llama, pack	60	1.8	2.5	15	25	1.2	45	5	Surefooted
Elephant, pack	100	2.0	3.0					5	Feeding a problem
Bicycle, with sidecar	100	2.0	2.5	4	4	1.5	10	2	Best on smooth flat surfaces
Motorcycle, with sidecar	150	2.5	3.0	10	12	2.0	10	2	Reasonable speed requres good road
Tricycle	100	1.0	2.0	4	6	1.0	10	3	Most common in urban areas
Bicycle, with trailer	100	1.0	2.0	4	6	1.0	10	2	Best on smooth flat surfaces
Tricycle, with trailer	100	1.0	2.0	4	6	1.0	10	2	Best on smooth flat surfaces
Motor-scooter, w/trailer	100	1.0	2.0	6	10	1.0	10	2	Best on smooth flat surfaces
Motorcycle, w/trailer	100	1.0	2.0	10	12	1.2	15	3	Best on smooth flat surfaces
3-wheel motorcycle	100	1.0	2.0	10	12	1.2	15	4	Stable on rought surfaces
Auto rickshaw	100	1.0	2.0	8	8	1.2	10	3	Best on smooth flat surfaces
Handcart, 2-wheel	100	1.2	2.5	4	10	1.5	15	2	Needs flat and smooth surface to steer
Handcart, 4-wheel	150	1.5	3.0	4	6	1.5	15	1	Usually found only in urban areas
Rickshaw	100	1.0	2.0	4	6	1.0	15	2	Best on smooth flat surfaces
Ox sledge	150	1.5	3.0	12	15	1.5	10	3	Needs large tractive effort for

Technical Note 2.3 (cont.)

Bullock sledge	150	1.5	3.0	12	15	1.5	10	3	Needs large tractive effort for capacity
Buffalo sledge	150	1.5	3.0	12	15	1.5	10	3	Needs large tractive effort for capacity
Donkey sledge	150	1.5	3.0	12	15	1.5	10	3	Needs large tractive effort for capacity
Pony sledge	150	1.5	3.0	12	15	1.5	10	3	Needs large tractive effort for capacity
Mule sledge	150	1.5	3.0	12	15	1.5	10	3	Needs large tractive effort for capacity
Horse sledge	150	1.5	3.0	12	15	1.5	10	3	Needs large tractive effort for capacity
Elephant sledge	200	2.0	4.0	12	15			5	Needs large tractive effort for capacity
Ox-cart	200	2.5	4.0	4	6	2.0	45	4	Gradients significant, both directions
Bullock cart	200	2.5	4.0	4	6	2.0	45	4	Gradients significant, both directions
Buffalo cart	200	2.5	4.0	4	6	2.0	45	4	Gradients significant, both directions
Donkey cart	200	2.5	4.0	4	6	2.0	30	4	Gradients significant, both directions
Pony cart	200	2.5	4.0	4	6	2.0	30	4	Gradients significant, both directions
Mule cart	200	2.5	4.0	4	6	2.0	45	4	Gradients significant, both directions
Horse cart	200	2.5	4.0	4	6	2.0	45	4	Gradients significant, both directions
Dog cart	100	1.0	2.0	6	8	1.0	15	3	Suitable only for small loads
Goat cart	100	1.0	2.0	6	8	1.0	15	3	Suitable only for small loads
Pneumatic-tyred animal cart	250	3.0	5.0	4	6	2.5	45	2	Unsatisfactory in muddy conditions
2-wheel tractor/trailer	100	1.5	2.5	6	10	1.2	15	3	Many sizes; multipurpose investment
3-wheel tractor/trailer	150	1.5	3.0	6	10	1.5	15	3	Uncommon means of locomotion
4-wheel tractor/trailer	200	2.0	4.0	6	10	2.0	30	4	Many sizes; multipurpose investment
4-wheel drive vehicle	200	2.5	4.0	20	30	2.0	60	4	High fuel and maintenance costs
Pickup truck	250	2.5	4.0	12	18	2.5	60	3	High ton per km costs
Trucks, < 5-tons	250	2.5	5.0	12	18	3.0	60	3	Suitable for rural areas
Trucks, > 5-tons	300	3.0	6.0	10	15	3.0	75	3	Best serves intermediate collection points

[1] Maximum grades require frequent rest areas for humans and animals.
[2] Maximum fording depths are suggested maximums only. Actual maximum depths depend on size of person or animal and actual load being moved. As with short slopes, fords can be crossed by partial unloading and making several trips.
[3] The description of these routes can be found in Technical Note 9.07, Table 25.

Fig. 53 *Animal sledge*

Fig. 54 *Two-wheel animal cart*

Fig. 55 *Indonesian bullock cart*

Technical Note 2.4
Suggested Criteria for Various Types of Infrastructure

Infrastructure type · *Mode usage*	*Tread width* (cm)	*Min. clear* (m)	*Passing clear* (m)	*Max. sus-tained gr.* (%)	*Max. grade* (%)	*Bridge width* (m)	*Ford depth* (cm)[1]	*Surface type*
PATH	*50*[2]	*1.0*	*2.0*	*20*	*30*	*0.3*	*100*	*EARTH*
· Headload								
· Shoulderload		(2.0)[3]						
· Backload								
· Stretcher					(20)			
· Bicycle rider				(4)	(8)	(1.0)	(10)	(Good earth)
· Motorscooter rider				(10)	(15)	(1.0)	(10)	(Good earth)
· Motorcycle rider				(12)		(1.2)	(15)	
TRAIL	*100*	*2.0*	*2.5*	*15*	*25*	*1.2*	*45*	*EARTH*
· Wheelbarrow, Western				(4)	(12)		(10)	(Good earth)
· Wheelbarrow, Chinese				(4)	(12)		(15)	(Good earth)
· Transport bicycle			(3.0)	(4)	(6)		(10)	(Good earth)
· Animal rider							(100)	
· Donkey, pack								
· Pony, pack								
· Mule, pack							(60)	
· Horse, pack							(60)	
· Camel, pack								(Friable)
· Llama, pack								
· Elephant, pack			(3.0)					
· Bicycle, w/sidecar				(4)	(4)	(1.5)	(10)	(Smooth compacted)
· Tricycle				(4)	(6)		(10)	(Good earth)
· Bicycle w/trailer				(4)	(6)		(10)	(Smooth compacted)
· Tricycle w/trailer				(4)	(6)		(10)	(Smooth compacted)
· Motorschooler w/trailer				(4)	(6)		(10)	(Smooth compacted)
· Motorcycle w/trailer				(10)	(12)		(15)	(Good earth)
· Three-wheel motorcycle				(10)	(12)		(15)	
· Auto rickshaw				(8)	(8)		(10)	(Good earth)
· Handcart, two-wheel				(4)	(10)	(1.5)	(15)	(Smooth compacted)
· Rickshaw				(4)	(6)		(15)	(Smooth compacted)
· Dog cart				(6)	(8)		(15)	(Good earth)
· Goat cart				(6)	(8)		(15)	(Good earth)
· Two-wheel tractor/ trailer				(6)	(10)		(15)	(Good earth)

Technical Note 2.4
(*contd.*)

Infrastructure type · *Mode usage*	*Tread width* (cm)	*Min. clear* (m)	*Passing clear* (m)	*Max. sustained gr.* (%)	*Max. grade* (%)	*Bridge width* (m)	*Ford depth* (cm)[1]	*Surface type*
TRACK	*200*	*2.5*	*4.0*	*4*	*6*	*2.0*	*10*	*EARTH*
· Motorcycle w/sidecar				(10)	(15)			(Good earth)
· Handcart, four-wheel							(15)	(Smooth compacted)
· Ox sledge				(12)	(15)			(Good earth)
· Bullock sledge				(12)	(15)			(Good earth)
· Buffalo sledge				(12)	(15)			(Good earth)
· Donkey sledge				(12)	(15)			(Good earth)
· Pony sledge				(12)	(15)			(Good earth)
· Mule sledge				(12)	(15)			(Good earth)
· Horse sledge				(12)	(15)			(Good earth)
· Elephant sledge				(12)	(15)			(Good earth)
· Ox-cart							(45)	
· Bullock cart							(45)	
· Buffalo cart							(45)	
· Donkey cart							(30)	
· Pony cart							(30)	
· Mule cart							(45)	
· Horse cart							(45)	
· Three-wheel tractor/ trailer				(6)	(10)		(15)	(Good earth)
· Four-wheel tractor/ trailer				(6)	(10)		(30)	
· Four wheel drive vehicle				(20)	(30)		(60)	

[1] Water velocity <0.9 m/sec.
[2] Italicised values represent typical values.
[3] Parentheses indicate modifications for specific transport aids.

COMMENTS CONCERNING CRITERIA DEVELOPED FOR TECHNICAL NOTE 2.4

1. The travelled surface, called the treadway, is modified natural terrain in the sense that stumps, roots, large rocks and other safety hazards have been removed. The treadway width should be minimal to reduce erosion problems. The cleared width is a traffic function. Clearing allows additional width for overhanging loads, e.g., pickal bars and yokes, and includes, at a minimum, intermittent passing locations. Passing may involve one party stopping within

the cleared area while the second moves by, also utilizing the cleared area. In general, the cleared height is governed by the height an axe or machete will reach to permit passage of horseback riders. Where path and trail treadway soils are free draining (Para 3), branches should be trimmed flush with the tree trunks rather than removing trees beyond the required clearing width. This will maintain a leaf canopy which will exclude the sunlight, thereby reducing growth in the cleared area and decreasing maintenance frequency [88].

2. Maximum slopes are governed by two parameters, the maximum slope the user can negotiate in both directions and the soil's erosive properties. Many soils erode badly on treadways over 15%. Terraced or stepped treadways can be utilized on steeper slopes but they deter utilization by many wheeled transport aids. Foot traffic can operate on log or stone steps in slopes up to 30%.

3. Limited data collected in the United States indicate the following order of magnitude of resource investments in trail construction [66]. The information relates to trails with a tread width of 60 cm in a cleared area averaging 1.3 m wide by 2.6 m high:

(i) level terrain (0–2%) at 150 manhours/kilometre,
(ii) rolling terrain (2–5%) at 225 manhours/kilometre.
(iii) hilly terrain (5–10%) at 300 manhours/kilometre, and
(iv) rugged terrain (>10%) at 375 manhours/kilometre.

The trails were built on sandy or loamy soils which are very well suited for trail construction and do not require granular surfacing. The labour costs amounted to 65% of the total construction costs in the U.S. If the tread had to be surfaced, it was anticipated that the construction costs per kilometre could be increased by up to 300%.

Technical Note 2.5
Improved Animal-Drawn Vehicles (ADV)

India produces an alternative animal-drawn vehicle wheel with pneumatic tyres and ball bearings which can increase the bullock cart capacity to 2.5 tons on firm surfaces. Since ground conditions affect the tractive effort of bullock carts [62], the larger pneumatic-tyred carts have been found best on wet or dry sandy tracks and dry clayey and silty tracks, followed by standard size carts with wooden and all-steel wheels. However, on muddy tracks (wet clayey and silty soils) wooden wheels are the most efficient, followed by all-steel wheels, with the larger carts using pneumatic tyres the least efficient.

Technical Note 2.6
The Supply and Quality of Rural Transport Services in Developing Countries: A Comparative Review

1. Available evidence suggests that changes in the use of basic vehicles and travel and transport aids cannot be expected to take place just because the technology exists and is being used successfully elsewhere. Farmers and other entrepreneurs would introduce, adapt and use these items if they were available and if the potential user could afford to buy them or could get credit for their purchase. Credit is often available for agricultural equipment and supplies, but *not* for transport aids, which are equally important for agricultural production.

2. Reference [10] concludes that:

If government transport policies in developing countries are to be changed so as to explictly include measures to promote the provision of lower-cost vehicles for those currently denied transport services, then it will be necessary to:

(i) change official perceptions of the transport needs of small farmers;
(ii) make influential decision-makers more aware of the potential range of proven low-cost vehicles;
(iii) broaden planning procedures so that they reflect a transport system with commensurate attention to the supply of vehicles, and a less road oriented viewpoint; and
(iv) reconsider, as a consequene of items (i) to (iii), what changes might be required in the role that government and the international lending agencies play in the provision of transport services.

Technical Note 2.7
Overcoming Constraints to Existing or New Transport Usage

The matrix on the next page shows recommended actions to overcome these constraints.

Technical Note 2.8
Requirements for Developing the Use of Alternative Transport Aids

The International Labour Organization [30] identifies the following requirements to achieve improved use of alternative transport aids:

(i) facilities for the production, repair, and maintenance of improved non-motorized, low-cost forms of transport (e.g., backloading devices, simple pedal-powered or animal-powdered vehicles);
(ii) local artisans trained in the production, repair and maintenance of these devices or vehicles;
(iii) a supply system for the items/parts which cannot be manufactured locally;

Technical Note 2.7

Overcoming Constraints to Existing or New Transport Aid Usage[1]

	Limit usable locations to those meeting proper infra-structure require-ments	*Improve infra-structure*	*Limit to tasks not ham-pered by constraints*	*Redesign aid*	*Training/ educa-tion*	*Demon-stration promo-tion*	*Introduce veterin-ary service*	*Provide credit*	*Create local work-shops*	*Use local material*	*Create co-operatives*	*Introduce storage*	*Encour-age rental*	*Introd. new job opports.*	*Modify/ remove regulations*
Technical Constraints															
Maximum slope	X	X													
Maximum speed			X												
Maximum capacity			X	X											
Maximum width	X	X		X											
Loading difficulty			X	X	X										
Unloading difficulty			X	X	X										
Travel surface condition	X	X	X												
Tractive power requred			X	X											
Fording capability	X	X		X											
Loaded stability			X	X	X										
Practical range		X	X												
Manufacturing difficulties					X			X	X	X					
Maintenance difficulties					X			X	X	X					
Repair difficulties					X			X	X	X					
Animal diseases					X		X	X							
User Constraints															
Religious/cultural taboos					X										
Socially unacceptable					X	X									
Aid not known					X	X									
High purchase cost				X				X	X	X	X		X		
High operating cost	X	X	X	X	X			X			X		X		
High repair cost	X	X		X	X			X	X	X	X				
Repair difficulties				X	X	X			X	X	X				
Operation difficult	X	X	X		X	X					X				
Infrastructure unsuitable	X	X		X											
Too large for task				X							X	X	X		
Limited need for use					X	X					X		X		
Limited type of use					X	X					X		X		
Reduces employment opportunities					X									X	
Institution Constraints															
Resistance by government decisionmakers					X	X									X
Regulations/restrictions hamper effective use					X										

[1] An 'X' relates the constraint to usage shown in the first column to the potential actions for relieving or eliminating a constraint shown across the top of the chart.

(iv) construction of a network of improved tracks, trails and/or paths;
(v) an operational system for the maintenance of the infrastructure upon which the simple transport aids pass;
(vi) a self-sufficient local transport cooperative or subsidized purchase system; and
(vii) a local credit system for the manufacture and repair of the simple transport aids, and for their acquisition.

Technical Note 2.9
Illustrative Comparison of Transport Costs Using Optional Aids

1. The basic cost data (net of tax) used to develop the following graph were derived from a study made in Tunisia [60]. Ownership and operating costs were available for motor vehicles on flat earth roads and flat asphalt roads, both in fair condition. The costs of the trucks began at an assumed use of 15,000 km per year, while the costs of a pickup truck began at 10,000 km per year. These costs were averaged to determine the cost of running one-half of the distance on asphalt roads and one-half the distance on earth roads. The load factors were reduced as shown in the graph below, to represent the poor usage on earth roads. The load factors were applied to the total distance, that is, the cost of running 10,000 km on paved roads and 10,000 km on earth roads, to find the 20,000 km per year cost at a specific load factor. Animal cart costs were based on cost data for operating 2000 km per year.
2. The graph illustrates that as conditions become less favourable for motor vehicles, for example, poorer roads and shorter hauls which reduce the km that can be travelled per year, their ton-km costs increase to the point where a traditional transport aid such as a mule/horse cart efficiently used is more cost-effective than a motor vehicle inefficiently used under adverse conditions. In this example only the costs relating directly to the transport aids are considered. If the additional costs of improving the infrastructure to an appropriate level to accommodate each of the transport aids were considered, the cost effectiveness of the animal carts would be further enhanced.
3. When the animal cart is pulled by oxen or buffalo, the marginal cost for transport may be considerably less than indicated in this chart. Not only may the oxen be used for agricultural purposes, but if oxen are replaced by vehicles, the farmer's losses due to the non-availability of milk, droppings for fuel or fertilizer, and other animal products such as meat and hides, may be quite significant. If the farmer owns the oxen and cart, the cost of transport is not a cash expense, further enhancing the use of animal carts to the farmer even when the real cost is higher.

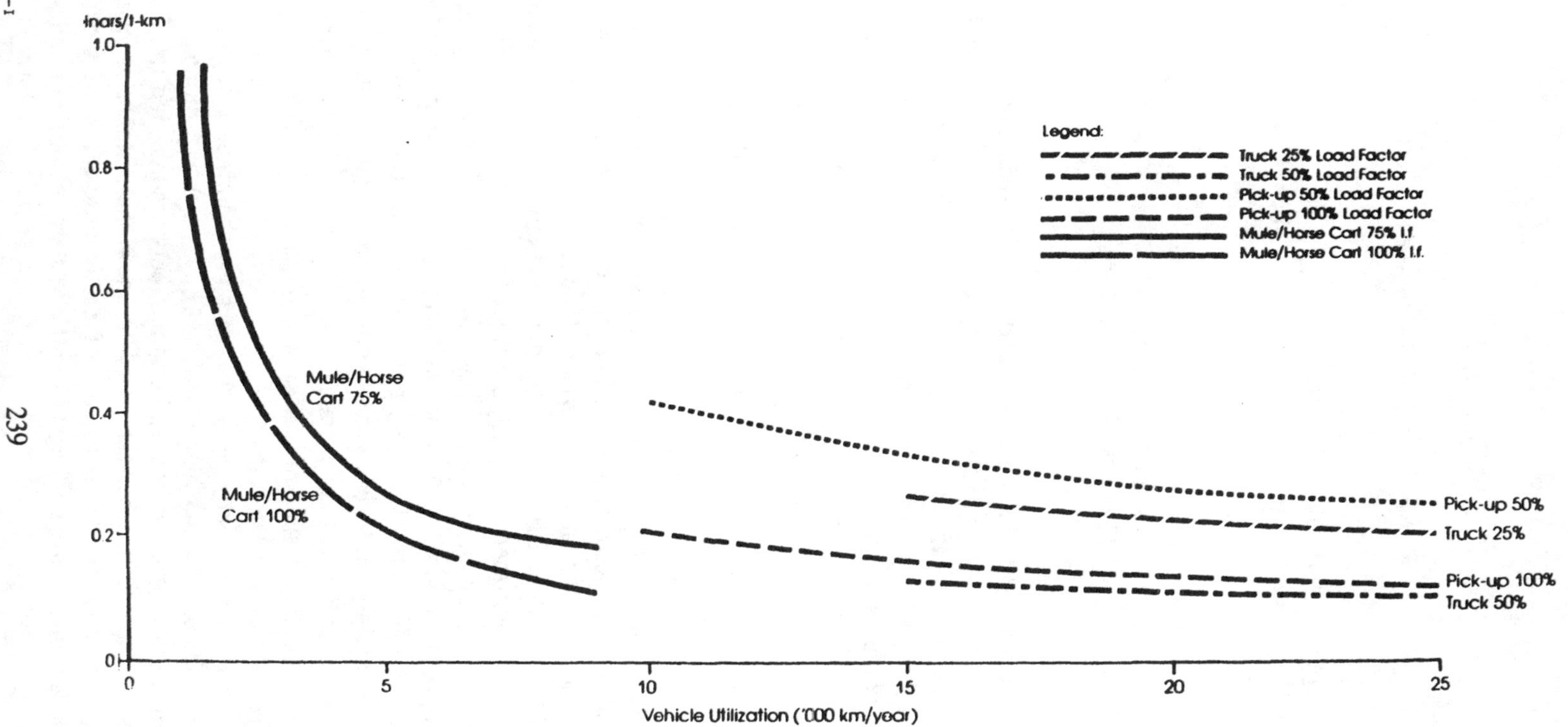

Fig. 56 *Transportation costs, Tunisia 1977*

RT-I

ANNEXE 3

Infrastructure Aspects

This annexe contains one Technical Note:
T.N. 3.1, *The Plan Book System*, is referenced in Paras 4.06, 4.11, 11.04, 11.09 and 12.20.

Technical Note 3.1
The Plan Book System

1. The basic requirement for this and similar planning techniques for rural access systems is a base map showing the existing transport infrastructure. Base maps developed for road inventories are usually suitable for rural physical planning after they are upgraded by overlays to include the other physical data development planners need.
2. Road inventory base maps [78] are reproducible scale drawings or air photos of an area with the existing roads coded by types, such as paved roads, gravel roads and single-lane roads. Maps for rural areas can be at a scale of 1:50,000, but if an area is densely populated and the tracks, trails and paths that are frequently used or represent routings that may be improved are to be recorded, a scale of 1:15,000 or 1:10,000 may be warranted.
3. The road system on a fully developed road inventory base map is divided into individual road sections, each section having reasonably uniform characteristics of surface type, width, condition and traffic volume. Road sections are also divided at each political or maintenance management organization boundary because road inventories are usually developed as a maintenance planning tool. Such road maintenance responsibility information is necessary to rural development planners since maintenance capability is a controlling consideration in rural access planning (Chapters 7 and 15). Tracks, trails and paths, if shown, are only designated by category. Consequently, road inventory maps are suitable only for conceptual planning of the lower RAI levels, i.e., to determine which paths, trails and tracks will require physical inventories before detailed planning is undertaken.
4. Route numbers are normally assigned to each long continuous segment of the area's regional or national highway network. Even route numbers are usually assigned to east-west routes and odd numbers are assigned to north-south routes. Kilometre numbers are used to show the distance from the beginning of the route to the beginning of each particular road section. Kilometre measurements usually begin from the western or southern end of a

route. Therefore, in the shorthand of a road inventory base map, the figure 17–34.1 represents a location which is 34.1 km north of the beginning of north-south route number 17 [78].

5. Physical road inventories record the locations of road features such as drainage structures, surface widths, platform widths, alignment changes, ditch locations, intersection locations and land usage on each side of the road. Physical inventories for tracks, trails, and paths are usually less detailed as noted in Paras 12.23 and 12.24; however the inventory of any type of infrastructure should include information on the location and nature of obvious weak points where traffic can be interrupted. Often, separate inventory sheets are prepared for each major drainage structure to describe its features in greater detail. Physical road inventories are usually the first activity undertaken after the basic road inventory base maps are prepared, but, because of the time involved, it is sometimes possible to obtain a road inventory base map for areas where no physical inventory has been taken and therefore no road section boundaries are noted.

6. The second activity in a road inventory process is the road condition inventory. This inventory consists of an evaluation of the then current state of the infrastructure. It identifies the weaknesses in the infrastructure such as failing culverts and poor road surfaces. Road physical inventories and road condition inventories can be made concurrently. Physical and condition inventories are usually combined for all lower RAI categories.

7. The last step in the road inventory process is a road needs inventory. This inventory activity involves an engineering assessment of the resource expenditures required to bring the system up to some predetermined level of service for high-volume through roads. It would be analogous to the rural transport infrastructure accessibility requirements described in Chapter 9. The predetermined accessibility defines the needs, i.e., the resource investments, for the RAI links.

8. From the description of the various inventories described above, it is obvious that inventory information at any level, such as raw inventory data sheets, condition inventories, or needs inventories, would be useful to area planners. Chapter 12 discusses in more detail simplified inventory techniques for collecting pertinent data when no inventories exist. However, a base map similar to the type described above is the foundation for all planning activities.

9. Road inventory base maps frequently show road-related features such as weigh stations, traffic count locations and resource locations such as borrow pits. They sometimes show off-the-road physical and cultural features such as important landmarks, streams, swamps and rivers, existing drainage and irrigation schemes, flood control facilities and population centres. If these items are not shown on the available base map, they should be added for area planning purposes. The base map for planning purposes should also include existing towns, villages, settlements, plantations, market locations, administrative centres, health centres, schools, social centres, religious places including burial areas, and recreational facilities. If the base map becomes too crowded when these cultural locations are added, some of the cultural features

can be shown on an overlay of the same scale as the base map, but all existing RAI must be included on the base map itself.

10. A recurrent difficulty in preparing road inventories lies in the accurate location of particular road features and in subsequently transferring this information to base maps. Technical Note 4.12 addresses this problem when inventories are based on odometer readings. Failing this, locations may be identified by references to bridges and culverts which are sequentially numbered as described in Para 12.24.

11. Planning overlays to the base map include tribal or other unofficial boundaries, population figures for the various settlements, areas under cultivation, areas suitable for expanded cultivation, grazing land, forests, obstructions currently limiting the existing infrastructure's zone of influence, proposed development activities by others, and all pertinent information required for the planning process. This information is placed on overlays because the base map represents existing permanent physical features while overlays contain transitory information.

12. The next overlay, or series of overlays, depict various development schemes or development phases. These overlays are often made in different colours so overlapping activities or conflicts between phases stand out. The last overlay, or series of overlays, consist of infrastructure modifications.

13. When a long-term development scheme has been determined, the access infrastructure shown on the previous overlays is frequently transferred to another series of overlays showing infrastructure development for different planning horizons, since the access planned for the immediate future development activities has the most reasonable certainty of being built. However, no infrastructure improvements should be marked on the base map until they are completed since the purpose of the base map is to show the transport infrastructure as it is on the ground.

14. All of the overlays and the base map for a study area are collected in a loose-leaf book for safe-keeping, mobility, and discussion purposes. This collection of a base map and its related overlays is frequently called a 'Plan Book', hence the name of this planning system.

ANNEXE 4

Infrastructure — Individual Links

This annexe contains fifteen technical notes:

T.N. 4.01, *Modified Geometric Design Criteria*, is referenced in Paras 10.06, 12.02 and 13.19, and Technical Note 6.7.
T.N. 4.02, *Simplified Soil Sampling Procedures*, is referenced in Paras 12.03 and 14.42, and Technical Notes 4.03 and 4.13.
T.N. 4.03, *Size and Type of Surface Material*, is referenced in Para 12.04 and Technical Note 9.08.
T.N. 4.04, *Surfacing Material Thickness*, is referenced in Para 12.04 and Technical Note 1.1.
T.N. 4.05, *Rainfall and Soil Formation*, is referenced in Para 12.05
T.N. 4.06, *Field Evaluation for Drainage Structures*, is referenced in Para 12.08.
T.N. 4.07, *Design Storm Definition*, is referenced in Para 12.10.
T.N. 4.08, *Controlling Travelway Erosion*, is referenced in Para 12.13.
T.N. 4.09, *Surface Water Diversion Spacing*, is referenced in Para 12.13.
T.N. 4.10, *Drainage Relief Systems*, is referenced in Para 12.14 and Technical Note 4.08.
T.N. 4.11, *Stage Construction of Water Crossings*, is referenced in Paras 6.03 and 12.17.
T.N. 4.12, *Recording Road Physical Inventory Notes*, is referenced in Paras 6.10 and 12.21, and Technical Note 3.1.
T.N. 4.13, *Recording Conditions and Needs Road Inventories*, is referenced in Paras 6.10, 12.22, 12.23 and 13.14.
T.N. 4.14, *Level 2 RAI Inventory Form*, is referenced in Paras 6.10 and 12.23.
T.N. 4.15, *Field Engineering for Realignment*, is referenced in Para 12.29.

Technical Note 4.01
Modified Geometric Design Criteria

1. Maximum slope, as defined in road design standards, is usually related to the length and gradient a truck can climb without its speed being reduced by more than 25 kph [73]. Maximum slope for motorable access relates to the slope and length that is negotiable at any speed by the specific trucks in the project area. Maximum slopes for other transport modes are also slopes that can be physically traversed, both up and down. Given the low operating costs

of alternative transport aids, their infrastructure can include slopes on which transport aids such as bicycles must be pushed up or walked down, or slopes which require partial or complete unloading to negotiate.

2. Maximum degree of curvature, or minimum curve radius, has two definitions in highway engineering [73]. One is the maximum degree of curvature that the specific vehicle can negotiate at a given speed, width and superelevation. The second is the minimum turning radius for specific motor vehicles. However, the minimum negotiable radius for access is any curve that is wide enough so that the specific vehicle in question can pass around it, even if the vehicle must go forward, back-up and go forward again until it has completed the turn. Therefore, curvature and width combine to determine accessibility, rather than vehicle speed or a fixed radius. While not advocating the design of motorable roads containing such curves, the widening of a tight curve to make such an operation possible may be an economically viable alternative to the construction of several kilometres of new road on a new alignment to provide initial access.

3. Minimum width in standardized highway design is a function of design speed; the higher the design speed, the wider the traffic lane must be to allow time for the small corrections in vehicle direction that are necessary to keep a vehicle in its own travel lane. The adoption of a 3.65 m lane allows a vehicle to travel in comfort at 100 kph in its own lane on any road tangent. Research [12, 73] has shown that as lane widths are reduced on low-volume rural roads, no related increase in accidents occurs. This suggests that narrow lanes may be a cost-effective means of reducing motorable access infrastructure investments even with higher traffic volumes than are being considered in this book.

4. In areas where low-volume roads enter or pass villages, there is usually a large increase in foot traffic which greatly increases the risk of accidents. In such cases it is often desirable to provide a pedestrian track paralleling, but separate from, those sections of roadway. Within such villages, it is often appropriate to widen the road to provide space for parked vehicles and for the many other roadside activities.

5. At traffic volumes below about 50 ADT, single-lane motorable infrastructure is quite sufficient if passing locations are included [73]. At the traffic volumes representing initial access requirements, any wide places along the roadway can provide sufficient access. However, if other transport modes also use the same facility, the dual usage can affect the width requirements, and must be evaluated. Formal passing bay design calls for passing bays to be located so two vehicles travelling in opposite directions will see each other before they are both between adjacent bays. The distance between passing bays when not controlled by sight distance depends on their cost. Accepted distances vary from 100 m [58] to 400 m [73]. Sharp curves that prevent the application of the above sight distance passing bay criteria should be widened to 6.0 m, which is also the recommended width of the road at passing bay locations (Figure 2).

6. Widths of paths, trails, and tracks given in Technical Note 2.4, Suggested Criteria for Various Types of Infrastructure, include three significant dimen-

sions: the tread width which represents the travelled way or surface width, the cleared width which allows for overhanging loads, and a passing cleared width which allows meeting units to move aside into the cleared areas to pass each other.

7. Path, trail and track designs are less formalized than those for low-volume roads. The applications of sound principles of travelway location, alignment and grade will eliminate many future operation and maintenance problems as well as produce a facility which meets the planned objectives of resource conservation and user desires. Factors affecting location include terrain, vegetation, soil types, moisture conditions, drainage patterns, cultural developments along and adjacent to the corridor, and land ownership patterns. A proper alignment for paths, trails, or tracks should follow contours, angle across natural slopes rather than run up and down to minimize treadway erosion, and take advantage of natural drainage to minimize the need for major drainage modifications [88]. It should avoid highly erosive, wet, shallow soils; water level areas subject to flooding; rock or landslide areas; lightning-prone areas; and flat area locations where drainage is difficult.

Technical Note 4.02
Simplified Soil Sampling Procedures

1. The Unified Soil Classification System recognizes 15 basic soil groups. The groups are shown on each of the accompanying charts: Chart I – Soil Identification Procedure, and Chart II – Soil Use. The following description of the various tests is an amplification of the information on soil identification procedures outlined in Chart I.

2. All samples should be representative of the soil or aggregate being tested. All material larger than 75 mm (3 in.) should be removed from the sample.

3. Organic materials are identified by their dark colours, their odour of decay, and their general composition of plant and animal remains. Peat is a brown fibrous material consisting largely or entirely of decaying vegetable matter. Usually many individual fibres are easily visible. Soils in which this vegetable matter has decayed has become dispersed amongst the silt and clay are termed loams.

4. In Chart I, dark highly organic loams are termed 'muck'. When this distinction is made, 'loam' applies to a soil consisting of a friable, i.e., easily crumbled or pulverized, mixture of varying portions of completely decomposed organic particles and inorganic materials of the clay, silt and sand sizes. When loams have an organic content of 30% or less, they are considered as good agricultural soils. Chart I classifies the black or dark silts containing about 10% or more organic fines, which are usually surface soils found in basin areas or around swamps, as organic silt. Organic clays are composed predominantly of clayey soils, but as little as 10% of black organic fines, i.e. black carbonaceous particles of silt and clay sizes, can change the colour and behaviour of

Chart 1. Soil Identification Procedure [72]

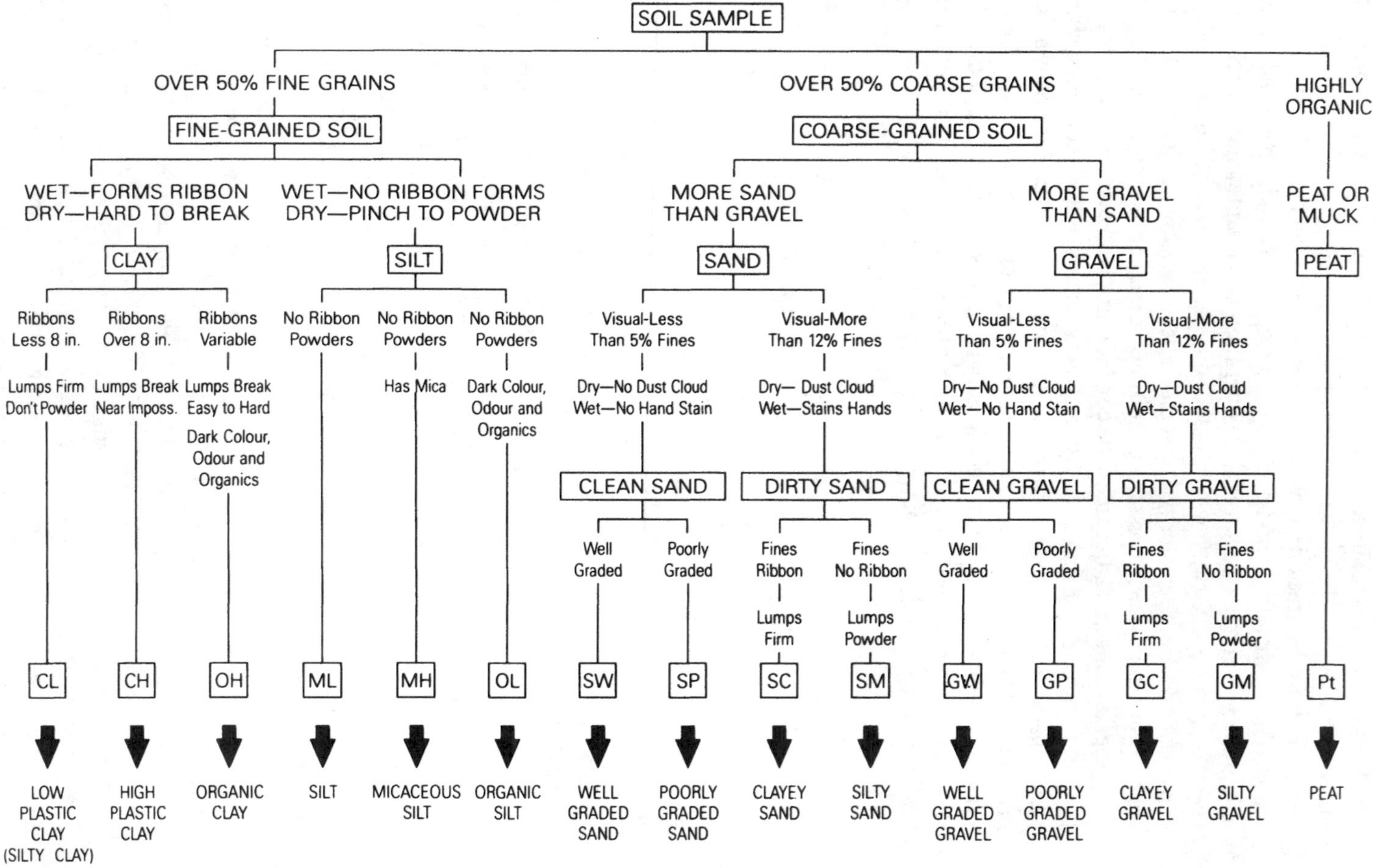

SOIL SAMPLE

- OVER 50% FINE GRAINS → FINE-GRAINED SOIL
  - WET—FORMS RIBBON / DRY—HARD TO BREAK → CLAY
    - Ribbons Less 8 in. — Lumps Firm Don't Powder → CL → LOW PLASTIC CLAY (SILTY CLAY)
    - Ribbons Over 8 in. — Lumps Break Near Imposs. → CH → HIGH PLASTIC CLAY
    - Ribbons Variable — Lumps Break Easy to Hard; Dark Colour, Odour and Organics → OH → ORGANIC CLAY
  - WET—NO RIBBON FORMS / DRY—PINCH TO POWDER → SILT
    - No Ribbon Powders → ML → SILT
    - No Ribbon Powders — Has Mica → MH → MICACEOUS SILT
    - No Ribbon Powders — Dark Colour, Odour and Organics → OL → ORGANIC SILT
- OVER 50% COARSE GRAINS → COARSE-GRAINED SOIL
  - MORE SAND THAN GRAVEL → SAND
    - Visual-Less Than 5% Fines — Dry—No Dust Cloud, Wet—No Hand Stain → CLEAN SAND
      - Well Graded → SW → WELL GRADED SAND
      - Poorly Graded → SP → POORLY GRADED SAND
    - Visual-More Than 12% Fines — Dry— Dust Cloud, Wet—Stains Hands → DIRTY SAND
      - Fines Ribbon — Lumps Firm → SC → CLAYEY SAND
      - Fines No Ribbon — Lumps Powder → SM → SILTY SAND
  - MORE GRAVEL THAN SAND → GRAVEL
    - Visual-Less Than 5% Fines — Dry—No Dust Cloud, Wet—No Hand Stain → CLEAN GRAVEL
      - Well Graded → GW → WELL GRADED GRAVEL
      - Poorly Graded → GP → POORLY GRADED GRAVEL
    - Visual-More Than 12% Fines — Dry—Dust Cloud, Wet—Stains Hands → DIRTY GRAVEL
      - Fines Ribbon — Lumps Firm → GC → CLAYEY GRAVEL
      - Fines No Ribbon — Lumps Powder → GM → SILTY GRAVEL
- HIGHLY ORGANIC → PEAT OR MUCK → PEAT → Pt → PEAT

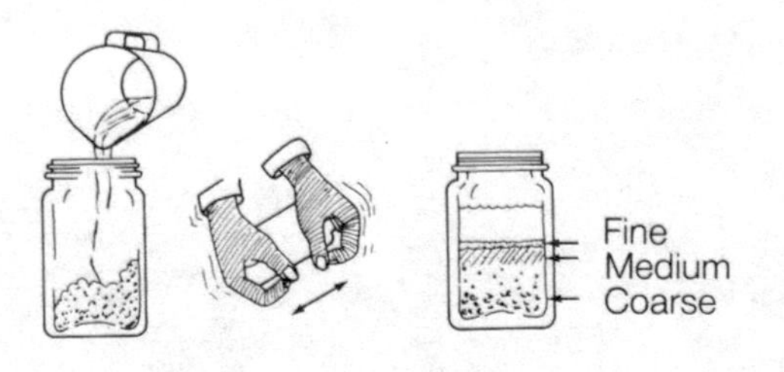

Fig. 58 *Settlement test*

silty because clay-size particles migrate downward with the passage of time.
10. The amount of fines in a sample, i.e., that material which passes a U.S. No. 200 sieve (0.075 mm), of possible base or surfacing material can be estimated by separating the gravel and sand using a U.S. No. 4 sieve (5 mm) or a piece of 6–7 mm (¼ in.) metal mesh cloth. A visual estimate of the gravel portion (greater than 5 mm) is made. The 'sand plus fines' portion (less than 5 mm) is poured into a straight-sided jar containing about 13 cm (5 inches) of water (Figure 58). The mixture is agitated or stirred until the soil is completely broken up into individual particles. The jar is then shaken vigorously, end to end, to get all the soil particles into suspension. The jar is quickly placed on a flat surface and the top level of settlement after 25 seconds is marked. This settlement represents the sand portion of the sample. Silt sizes settle in about 60 seconds and clay sizes settle in about 1½ hours. When the water clears or 1½ hours has elapsed, whichever comes first, measure to the top of the settled material. The total height is equal to the percentage of the 'sand plus fines' portion estimated above. The 25-second height represents the sand percentage of material in the bottle. The percentage of sand in the whole sample is found by multiplying the sand percentage in the bottle by the estimated percentage of the 'sand plus fines' material. While these percentages are only approximate, they can be used to determine if a course-grained soil is classified as sand or gravel in Chart I. The total percentage of both sand and gravel, as found above, when subtracted from 100%, results in the percentage of fines in the sample. This percentage of fines value can be used to confirm visual observations called for in Chart I to determine if a course grained soil is clean or dirty.
11. Another sample of the same granular material must be checked for the plasticity of the fine material. This sample should be separated by using a U.S. No. 40 (0.4 mm) sieve or a piece of window screen. Place three to four heaped tablespoons of the minus 0.4 mm material in a small bowl and add water, a few drops at a time, until a golf-ball size portion of the thoroughly mixed and kneaded sample can be formed which is not excessively wet. Form the test specimen into a 20 mm (¾ in.) roll, adding a little more water if necessary. With the palm of the hand down, using light finger pressure, roll the specimen on a flat waterproof surface (Figure 59). If the surface becomes wet or muddy, the soil is too wet and must be reworked to reduce the moisture content. Continue to roll the specimen, reducing its diameter. The diameter at which the specimen

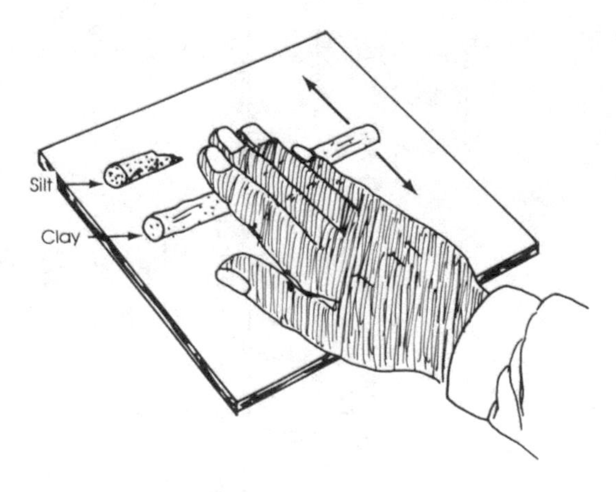

Fig. 59 *Moulding test*

Guide Values for Plasticity Test [72]

Diameter at which Roll Crumbles	*Suitability and Use*
3 mm (1/8 in.) or less	Unsuitable for gravel base or wearing surface aggregate
Over 3 mm to less than 6 mm	Suitable for gravel wearing surface aggregate
6 mm (1/4 in.) or larger	Suitable for gravel base aggregate

breaks or crumbles is a guide to its plasticity.

12. Fines, i.e., material passing a U.S. No. 200 sieve (0.075 mm opening), affect granular soils in the following two ways:

(i) the presence of fines retains moisture in granular materials. Therefore a small portion of fines, i.e., 5% minimum, retains moisture in a granular surface, increasing the cohesive characteristics of the surface during dry periods and consequently reducing material losses due to ravelling. However, when gravel is covered with a waterproof membrane, such as a bituminous seal coat or pavement, this retained moisture cannot escape. Consequently, the moisture content of granular base courses under a waterproof surface increases with each additional intrusion of water, eventually saturating the fines in the granular base and introducing failure to the overlying pavement due to the reduced structural strength capability of the base course. The reduction of fines in the granular base material to less than 5% prevents this phenomenon; and

(ii) the presence of too large a percentage of fines in a granular soil surface reduces the effect of gradiation in the course materials, and increases the effects normally associated with the fine soils. Gradiation provides

CHART II. Soil Use [72]

SOILS	SYMBOL	PROPERTIES[1]			COMPACTION	USES	
		Permeability	Load carrying ability	Frost suscep-tibility	EQUIPMENT[2]	Base course[3]	Wearing course[4]
WELL-GRADED GRAVEL	GW	Pervious	Excellent	None to very slight	Vibratory[5] Rubber tyre Steel wheel[6]	Excellent[7] to good	Fair
POORLY-GRADED GRAVEL	GP	Very pervious	Good	None to very slight		Fair	—
SILTY GRAVEL[8]	GM	Semi-pervious to impervious	Good to fair	Slight to medium	Vibratory Rubber tyre Sheepsfoot Steel wheel	Excellent to fair	Good to fair
CLAYEY GRAVEL[8]	GC	Impervious	Good to fair	Slight to medium		Good to poor	Excellent to good
WELL-GRADED SAND	SW	Pervious	Excellent	None to very slight	Vibratory Rubber tyre Steel wheel	Poor	Fair
POORLY-GRADED SAND	SP	Pervious	Good	None to very slight		Poor	—
SILTY SAND[8]	SM	Semi-pervious to impervious	Good to fair	Slight to hight	Vibratory Rubber tyre Sheepsfoot Steel wheel	Fair to poor	—
CLAYEY SAND[8]	SC	Impervious	Good to fair	Slight to high		Fair to poor	Good to fair

CHART II. **Soil Use** [cont.]

		PROPERTIES[1]			COMPACTION EQUIPMENT[2]	USES	
SILT	ML	Semi-pervious to impervious	Fair	Medium to very high	Rubber tyre Segmented wheel Steel wheel		
MICACEOUS SILT	MH		Fair to poor				
ORGANIC SILT	OL		Poor				
SILTY CLAY	CL	Impervious	Fair	Medium to high			
HIGH PLASTIC CLAY	CH		Poor	Medium	Rubber tyre Sheepsfoot Steel wheel		
ORGANIC CLAY	OH		Very poor	Medium			
PEAT AND	Pt	Remove from subgrade					

[1] Qualitative values listed below are for properly compacted soils.[2] Equipment listings are in order of efficiency – first is best.
[3] Qualitative values listed are for bases on high traffic roads.
[4] Gravel road wearing surfaces for roads with less than 100 vehicles per day.
[5] Crawler tractors can be used as vibratory equipment.
[6] Steel-wheel rollers are best used as grade finishers.
[7] Well-graded gravels are usually very difficult to compact.
[8] These materials cover a considerable quality range – from a low percentage of fines (5%–12%) and well graded to a high percentage of fines (over 12% to about 20%) and poorly related.

mechanical stabilization, i.e., the physical interlocking of the various sand and gravel size particles. As the percentage of the fines increase, the presence of clay which increases the cohesive characteristics of a granular soil surface during dry weather, i.e. binds the soil grains together through cohesion, makes the surface slippery and prone to soften and to rut under traffic [51]. The maximum percentage of fines allowable in a granular soil surface material varies in accordance with environmental conditions and other soil characteristics. Different sources indicate different allowable maximum percentages of from 10–12% [72] to 15–25% [51].

13. The tests described above lead to a 'most likely' classification of the soil through use of Chart I, Soil Identification Procedure. The properties of the soils so identified are noted in Chart II, Soil Use. Some additional considerations concerning the use of the identified soils are listed below.

(i) The drainage properties (drainability) of a soil are noted in Chart II. The groupings include [72]: (a) very pervious – instant drainage, (b) pervious – fast drainage, (c) semi-pervious – slow to very slow drainage, and (d) impervious – essentially no drainage. When the subgrade or natural soil under a surfacing is less permeable than the surfacing material, the subgrade must be sloped to drain and compacted in that shape or it will trap water at the bottom of the surface material causing a substantial weakening of the surface during the rainy season. Furthermore, the adoption of 'box construction', whereby a granular soil surface is contained within shoulders of a less permeable material, ensures the entrapment of water within the surface even if the subgrade is properly shaped and compacted. Granular soil surfacing must be 'daylighted', i.e., extended to the edge of the structure, or contained by a more permeable soil to retain any degree of strength during periods of precipitation.

(ii) Clay fines in granular soil slow drainage more than silt fines.

(iii) Fine grained soils attract moisture from subsurface water through capillary or wicking action. Therefore, fine grained subgrades can be softened by a high water table or by water standing in roadside ditches. Silts and fine sands are the most susceptible to rapid capillary action; clays can draw subsurface water to greater heights, up to ten metres, but the action is considerably slower. Considerable improvement in the structural capabilities of road surfaces built on or of materials containing fine sands and/or silts can be achieved, especially in flat terrain or poorly drained basins, by elevating the road surface approximately 0.5 metres above the existing terrain to avoid the capillary rise in those materials. Capillary action in clay embankments can be prevented by building the embankment on a layer of sand.

(iv) Clays derive their strength from cohesion (stickiness). As the water content in clay decreases through the drying process, the capillary and surface tension of the moisture film cause the clay particles to move closer together; continued drying forms a strong material in lump form, indicated by surface cracking. Conversely, increased moisture causes clay to expand and lose strength. Wet clay can lose as much as 75% of its compacted strength [72]. Clay surfaces become slippery as soon as the surface is coated by moisture from rainfall, mist or fog.

(v) Compaction improves the physical properties of all soils. Compaction is a mechanical process which increases the soil's density by squeezing the air out of it, that is, by reducing the air voids. In general, compaction makes a soil tighter, denser, and stronger, and keeps it drier, minimizes settlement, and reduces volume changes from swelling and shrinking through wetting and drying cycles. For each soil, there is a certain moisture content at which a relatively high density can be achieved with a minimum compactive effort. This is called the optimum moisture content and occurs when there is just enough moisture to lubricate the soil particles and to fill most of the voids after compaction. A rough approximate of the proper moisture content for fine grained soils and dirty gravels and sands is that state in which the soil is thoroughly damp, but not wet and sticky. In this condition, squeezing a handful of soil firmly will form a ball that retains its shape, without sticking to the hand. A further check for the correct misture content is to drop that ball of soil about 50 cm onto a hard surface. The ball should only crack apart but not shatter on impact.

(vi) The erodibility of a soil varies according to its classification. Fine sand is the most erodible soil under the action of flowing water, followed by (in a compacted state) some silts, fine gravels and clays, and coarse gravel. Shallow sidecuts in silty material are much less susceptible to water erosion if excavated side slopes are vertical rather than sloped. Silts are the most likely to suffer wind erosion, followed by fine sands.

14. Lateritic gravel is a particular granular material occurring in tropical areas. Laterite soils are usually red in colour because of the presence of iron oxides. The term lateritic gravel, as used here, refers to a lateritic soil in which the iron oxides occur as spheroid concentrations, i.e., little balls, in a soil matrix. It is sometimes called pea laterite because the spheroid particles are usually the same size as small peas. Lateric gravel represents one stage of the laterization or weathering process. It is gap graded, that is, some gradation sizes are missing, and is frequently located in lenses or seams from half to one and half metres in depth, which are overlaid by up to two metres of topsoil. The specific gravity and water absorption characteristics of lateritic gravel appear to offer some indication of the laterites' performance as a road surfacing material. A study of Ghanaian laterite [79] resulted in the following classifications:

Performance	*Specific Gravity*	*Water Absorption (24 hrs)*
Excellent	>2.85	<4%
Good	2.85–2.75	4–6%
Fair	2.75–2.58	6–8%
Poor	<2.58	>8%

The lateritic nodules can vary considerably in hardness, corresponding to the degree of laterization, i.e., to the content of iron oxides, and hence to the specific gravity. Those with high specific gravities are usually very strong and are comparable to hard rock. Those with low specific gravities may be quite soft, the nodules breaking down under traffic into a dust which blows away. Consequently, the regravelling cycle for some lateritic gravels under a given traffic volume may be shorter than with other gravels. To minimize this degratiton, compaction should be light and shaping kept to a minimum to avoid high shear stresses. For nodular laterite, steel-wheeled or pneumatic-tyred rollers in the 4500 to 7300 kg range coupled with good moisture control are often the most effective and least damaging compaction equipment [79].

Technical Note 4.03
Size and Type of Surface Material

1. The size of granular material used for road surfacing should not exceed 20 mm [51, 83] to provide a good riding surface which can be easily bladed or broomed to restore an even profile. The presence of coarser material both promotes corrugation formation during the dry season and increases the difficulty in removing those corrugations.
2. Lateritic gravels usually provide excellent road surfaces as long as care is taken not to include the clayey soils which may be inter-layered with the lateritic gravel and commonly underlie the natural lateritic gravel deposits. Quartzitic gravels from land deposits are often equally effective. Both should contain a portion of fines (Technical Note 4.02, Para 12) to provide cohesion in dry weather. River gravels are less suitable: usually lacking proper gradiation, being rounded from the action of flowing water, and lacking suffficient fines. Most gravels lacking sufficient fines for surfacing can be improved by proper blending with more clayey soil. In mountainous areas, suitable gravels can often be found in glacial moraines. However these will frequently contain stones larger than 20 mm which should be removed before the gravel is used as a road surface. Some areas, notably river deltas, are devoid of naturally occurring road surfacing gravels. In such areas sand-clay surfaced roads can be made which are suitable for light traffic. Artifical aggregates, such as broken brick, can also be used for building road surfaces, as is practised in India and Bangladesh [51].

Technical Note 4.04
Surfacing Material Thickness

1. This text is based on the premise that access improvements is in many ways a trial and error exercise. Perhaps the best example of this process is the application of a granular surface to motorable rural access infrastructure. Proper pavement design involves assessing the strength of the soil under the road and of the various layers of the pavement in relation to the traffic loads

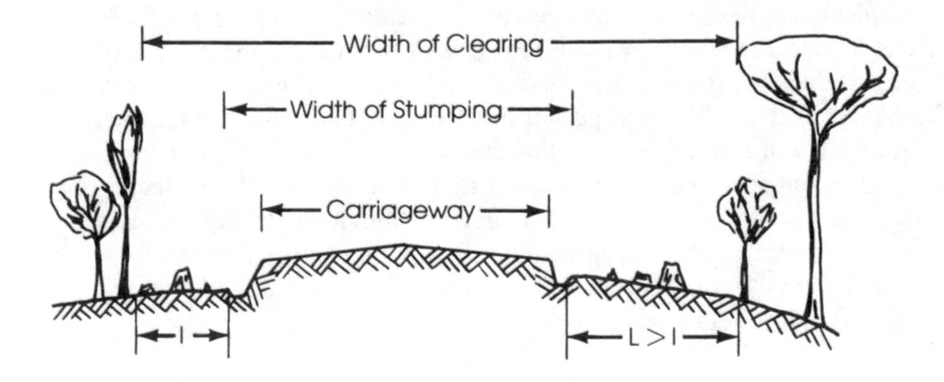

Openings for roadways in the forest. Left, a narrow opening suitable for sandy soils, near small trees and with early exposure to the sun; right, a wide opening for clayey soils, near tall trees and with late exposure to the sun. Roads running North-South need a wider clearing than East-West roads to have the same duration of daily sunlight on the carriageway. Savannah areas require clearing only in selected segments, if at all.

Fig. 60 *Clearing to encourage rapid drying*

expected on the road. It is a refined process used to design roads which are to be surfaced with asphaltic materials or portland cement concrete. The relatively high cost and 'permanent' nature of these surfaces justify the investment in proper design. But on gravel surfaced roads such refined design is neither possible nor necessary. Judgements can be made based on local experience of the thickness of gravel needed, and it is a comparatively simple matter to add more gravel where the gravel pavement strength proves to be deficient.

2. Several of the countries studied for this project have adopted standard layer thicknesses of granular surfacing. These vary from 10 to 20 cm in all but a few cases. However, given the lack of quality control and attention to proper spreading techniques, 10 cm of specific uncompacted thickness often leads to minimal or no actual compacted surface thickness in some locations. Since any application of granular surfacing is an expensive operation which depletes unrenewable resources, and the compaction of granular surfacing is known to reduce the frequency of both regrading (reprofiling) and resurfacing requirements, this handbook recommends 20 cm of uncompacted surface thickness as an initial application of surfacing material; followed by a conscientious compaction effort, including the application of water when necessary. This will reduce the thickness to a compacted layer of approximately 15 cm.

3. In practice, application of granular surface material to improve the structural strength lacking in saturated subgrade material will only succeed if the subgrade material is properly shaped and compacted to shed the water that penetrates through the granular material. Indeed, proper shaping of the subgrade material, installing suitable drainage facilities to remove standing

surface water next to the travelled way, and compacting the subgrade material to render it less permeable, may eliminate the need for added strength from a granular surface. Such steps will also ensure the maximum strength of any standard thickness of granular material required. Cutting back the trees on both sides of the roadway to permit sun and wind to dry out the road surface (Figure 60) will reduce the time that the roadway is in a saturated condition [58]. If granular surface is required, low traffic volume permits low technology usage such as the application of a practical minimum thickness, usage of low-grade material, i.e., local granular soil, and a low factor of pavement safety [37]. Additional layers of properly graded gravel can be added later if required as additional steps in a stage construction process.

Technical Note 4.05
Rainfall and Soil Formation

1. Soils are produced from parent rock by physical and chemical processes of weathering. In dry areas the weathering is predominantly physical, i.e., caused by temperature changes, in some places aided by frost. Thus in areas of low rainfall (considered as less than 750 mm/year in central Africa) [79], massive rock formations weather to discrete interlocking particles or aggregates, with the petrological characteristics of the parent rock greatly influencing the nature of the soil.
2. In wetter areas chemical decomposition of minerals in the rocks intensifies the weathering process. Thus higher rainfall (more than 750 mm/year) produces more intensive disintegration which in turn produces soils rich in the more insoluble mineral (quartz and mica) and increasing proportions of clay formed from the decomposition products of the rock. In areas of highest rainfall (more than 1000 mm/year), considerable quantities of dissolved compounds of iron, aluminium, and calcium move in the ground-water to lower strata (Figure 61). Where drainage is impeded, these compounds accumulate as a result of evaporation, forming deposits of nodular lateric and calcaerous gravels. Clay also accumulates in areas of poor drainage, which together with evaporation is an additional soil-forming factor [79].

Technical Note 4.06
Field Evaluation for Drainage Structures

1. Watershed areas, delineated by a line showing the topographical high points which define the limits from which surface water can flow to the stream or water body being studied, may be drawn on topographical maps if such mapping is available. Such data are useful for drainage structure location and, if rainfall intensity data is available, for drainage structure sizing. Standard runoff coefficients for various watershed surfaces applied in design calculations such as the commonly used rational formula for small drainage areas, i.e.,

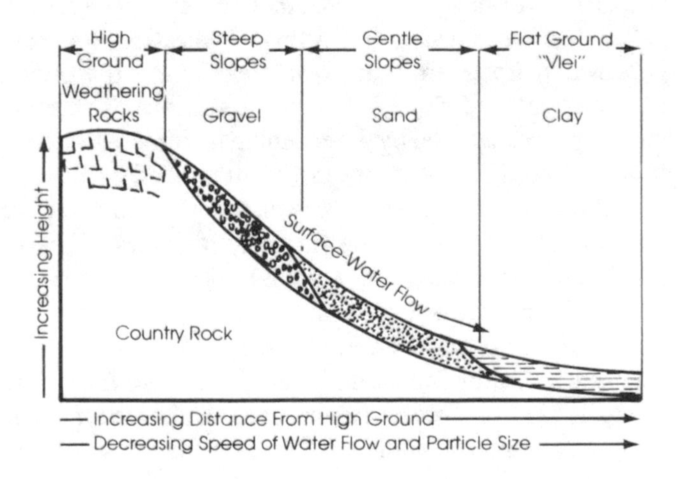

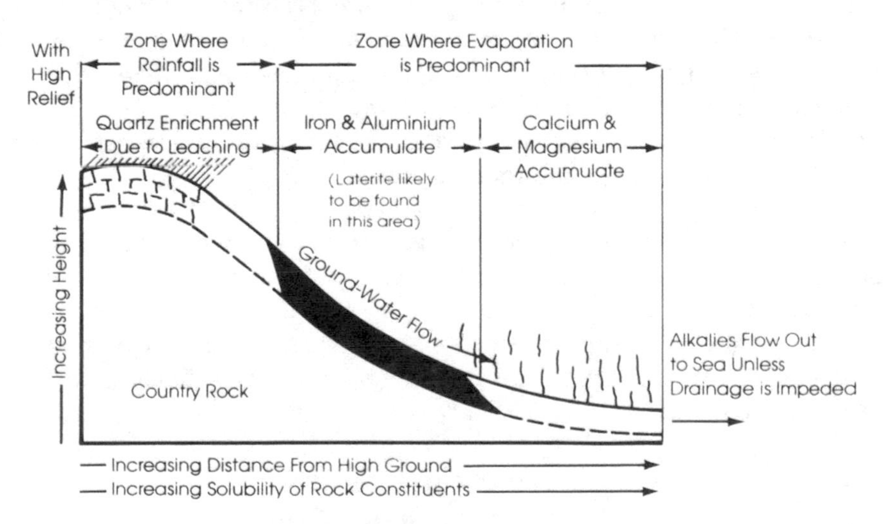

Fig. 61 *Effect of hydrological factors on soil location*

under 0.8 km², roughness coefficients for various channel materials, and maximum permissible velocities for erodible channel materials can be found in most drainage-oriented texts [80, 81].

2. Drainage catchment areas may also be determined in the field by estimation, by pacing, or by a stadia-compass traverse. The data needed to determine the time of concentration and runoff coefficient should also be noted in any such preliminary survey. However, when the rainfall intensity is unknown, the

application of theoretical design procedures often results in an erroneous output masked by unwarranted mathematical computations.

3. The required capacity of drainage structures may sometimes be estimated at their proposed location when water is flowing. Typical sections of the waterway can be measured upstream and downstream from the proposed crossing along with the streambed slope between the sections. The approximate stream velocity may be calculated by determining the time it takes a light object to float the measured distance between the streambed sections. The cross-sectional area of the present flow as calculated from the measured sections and the flow velocity will determine the quantity of water presently flowing in the stream if the reach between sections is typical, i.e., if the stream has the same slope and shape between sections.

4. Debris caught on nearby vegetation can frequently identify the flooded depth of the stream. This indicator should be independently verified by local residents' observations if possible. The computations of the water quantity flowing at the flood stage require additional input from an experienced drainage engineer.

5. If no measurements are possible, the observer must rely on experience in similar watersheds to size the drainage crossing. This method should be used cautiously. Considerable care must be taken to compare like areas, like topography, and like vegetation, as well as similarity of watershed shape. The possibility of existing crossings of the same stream nearby should not be overlooked, nor should restricted areas of the streambed which may offer the same waterway width as the proposed crossing.

6. The existing stream location should also be investigated for signs of erosion or unstable channel alignment. Since both erosion and siltation are the result of changes in stream velocity, such occurrences often foretell future problems downstream of a new drainage structure if the structure accelerates the flow, or at the inlet if the structure retards the flow. These occurrences can also be symptomatic of possible problems in drainage channels and ditches built in the same soil nearby.

Technical Note 4.07
Design Storm Definition

The expected frequency of occurrence of the design discharge for a drainage area is called the design storm. If the rainfall intensity which produces the design discharge has the probability of being equalled or exceeded once a year, it is a one-year design storm [81]. Drainage facilities designed to carry a one-year design storm have a low first cost, but maintenance costs are high because channels will be damaged by storm runoff almost every year. Proper design for drainage structures for roads in the through traffic network calls for the selection of design storms which produce the lowest annual cost, i.e., the total expenditure for construction and maintenance, and anticipated repair or replacement costs, divided by the facility's design life, rather than an arbitrary design storm frequency. However, selection of a proper design storm does not preclude the possibility of a larger storm destroying the drainage structure immediately after it is built, since the selection is based on statistical probabilities.

Technical Note 4.08
Controlling Travelway Erosion

1. Studies [79] have shown that earth and gravel surfaces with a cross-slope of less than 5% tend to retain water in surface irregularities. These puddles are caused by poor quality control during construction or irregularities introduced by passing traffic. Cross-slopes of greater than 5% increase the risk of erosion and, in the case of motor vehicle traffic, may affect safety and convenience in vehicle operation. Therefore, travel surfaces should have a 5% cross slope if the gradient is less than 5% [88]. On a side hill location this slope is usually an outslope that slopes down away from the hillside except on curves around promontories where, for safety reasons, surfaces carrying wheeled vehicles must slope downward towards the hillside, and cross drainage must be provided. Surfaces with a full width cross slope are more prone to erosive damage than are crowned surfaces with the same slopes and total width.
2. When the gradient on a path, trail, or track exceeds 5%, the cross-slope is not longer effective. Flowing water is instead diverted from the travelled surface to prevent erosion [88]. Footpath diversions can include waterbars (Figure 62) which are logs anchored across the path at a 45° angle, grade dips consisting of short sections of the path constructed with a grade of at least 10% in a direction opposite the prevailing path grade, terraced steps (Figure 63), or water curves made by turning the alignment to follow a contour for a short distance while allowing the surface water to continue down the slope. Trail diversions include grade dips and waters curves. Track diversions consist primarily of grade dips.
3. Earth and granular surfaced roads normally have drainage ditches on one or both sides of the roadway. Therefore the cross-slope is never completely removed from road surfaces no matter how steep the longitudinal grade becomes. Frequently road surfaces are crowned, with the high point in the centre and a 5% downslope, after compaction, towards both sides [16, 34, 35, 49, 58, 79], although they may have a 5% outslope or inslope, or be superelevated on curved sections. When no compaction equipment is available, the cross-slope if often increased to 7% to ensure a 5% cross-slope after traffic-induced compaction. Road surface erosion on steep longitudinal grades can ber controlled with dipouts (Technical Note 4.10) or with narrow concrete channels covered with a steel grating (i.e. a cattle grid) running diagonally across the road.

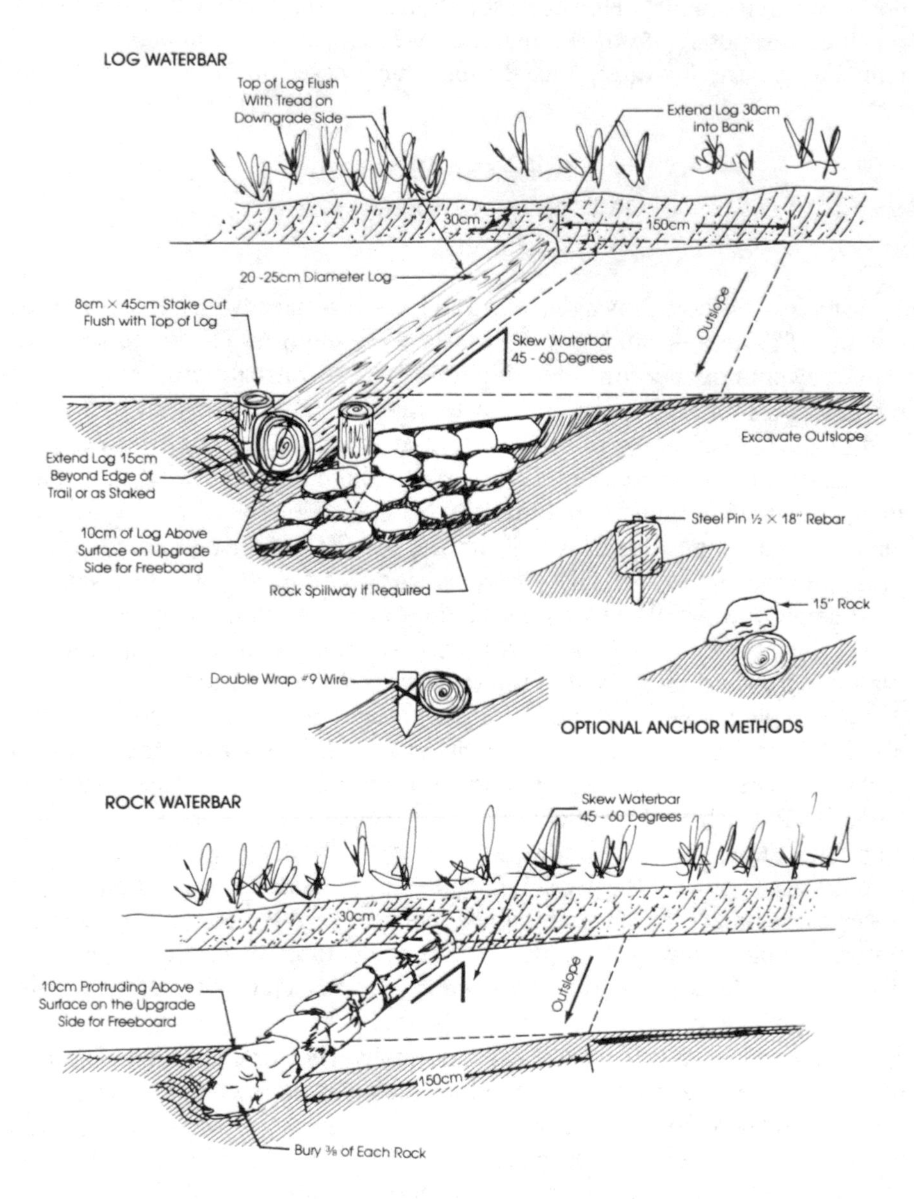

Fig. 62 *Water bars for use on paths*

Technical Note 4.09
Surface Water Diversion Spacing

1. Surface water running down paths, trails and tracks must be diverted from their alignments before it begins to erode their surfaces. Surface water collected in roadside ditches must be diverted from those ditches into the

Table 15. Surface Water Diversion Spacing [88]

Soil Type	*Gradient of Path, Trail, Track, or Trapezoidal Road Side Ditch*						
	2%	*4%*	*6%*	*8%*	*10%*	*12%*	15%
Loam	110 m	45 m	30 m	23 m	15 m		
Clay sand	150 m	110 m	60 m	45 m	30 m	15 m	
Clay or clay-gravel		150 m	90 m	60 m	45 m	23 m	
Gravel (rounded)			230 m	150 m	110 m	75 m	45 m
Shale or angular rock			250 m	180 m	120 m	90 m	75 m

surrounding countryside before it begins to erode the ditch bottoms. Mitre drains or turnouts are used to relieve the flow of water in roadside ditches and may be necessary to disperse water diverted from paths, trails and tracks by waterbars or grade dips. Water bars, grade dips and mitre drains must be spaced at appropriate intervals if erosion is to be prevented. This spacing is a function of the longitudinal slope and the soil type of the water-carrying infrastructure i.e., the path, trail, track, or side drain. Table 15 is a matrix which presents a suggested spacing as a function of soil and slope. V drains require closer spacing because their narrow inverts concentrate the erosive flow.

2. Where possible, such surface water diversions and mitre drains should coincide with stream courses and other lines of natural drainage. Care should always be taken to prevent the flow through a mitre drain from generating soil erosion at its outlet by fanning out the discharge end of the drain to reduce the velocity of the flow.

Technical Note 4.10
Drainage Relief Systems

1. The cross-road drainage relief for earth roads usually consists of a dip-out which is a skewed depression constructed across the road with some sort of armour or waterproofing. Granular surfaced roads frequently have a ditch relief culvert under the roadway.

2. A second form of ditch relief consists of a turn-out, i.e., a mitre, contour or diversion drain, to convey water from the downhill side ditch along a contour line (Figure 64). The water is disposed by soakage or by spreading over the natural surface where its velocity is small and the likelihood of erosion damage is reduced [80]. When a road runs perpendicular to the terrain slope, mitre drains are employed from both side ditches and the dirt from the mitre drains is disposed of on the uphill side of the drains to divert the sidehill drainage from the mitre drains.

3. In areas where sheet flow occurs, i.e., the rain water runoff moves in a sheet along the ground, and the infrastructure improvements will block or receive that flow, cutoff ditches or catch drains should be placed between the sheet flow area and the infrastructure (Figure 65). Cutoff ditches prevent scour

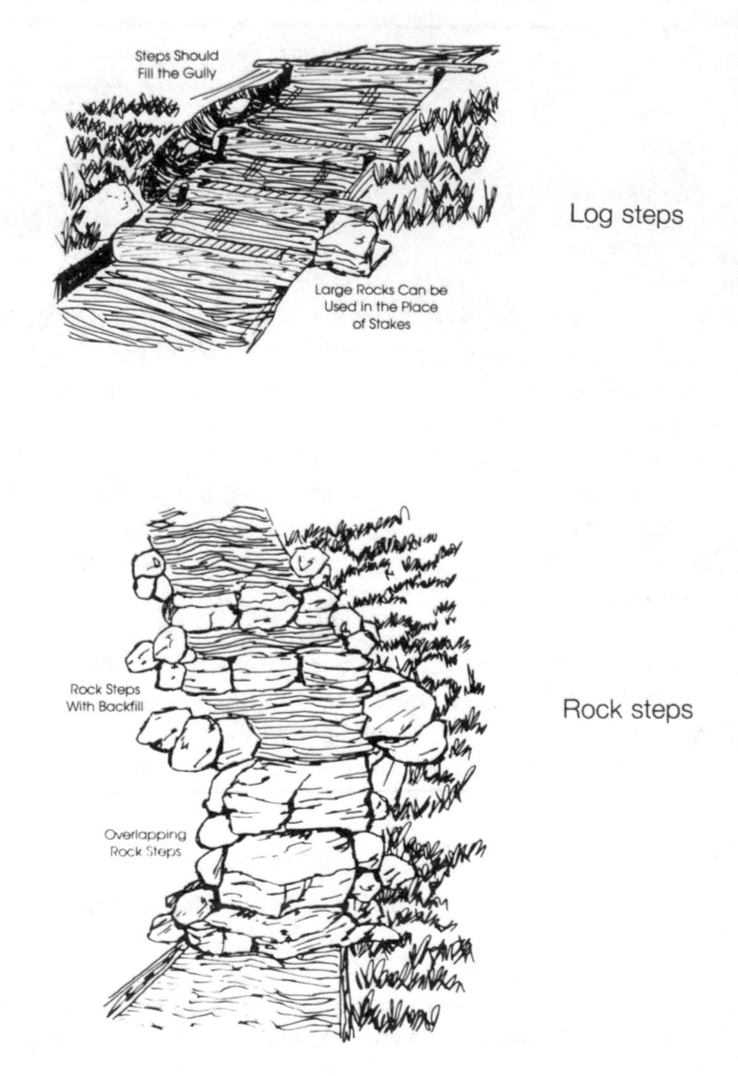

Fig. 63 *Methods of constructing terraced steps*

from water flowing down any cut face (batter) or along the toe (bottom) of any embankment. They are open channels with spoil (dirt from the ditch excavation) placed on the lower side of the channel to form a bank [80]. On slopes which are potentially unstable, such cutoff ditches can be a source of weakness by providing a path through which water can penetrate and weaken the subsoil. This weakening can cause a cut face to slide down onto the roadway below or an embankment to slide down the hillside. When cutoff drains are planned for soils susceptible to such weakening, the cutoff ditches should be lined with concrete.

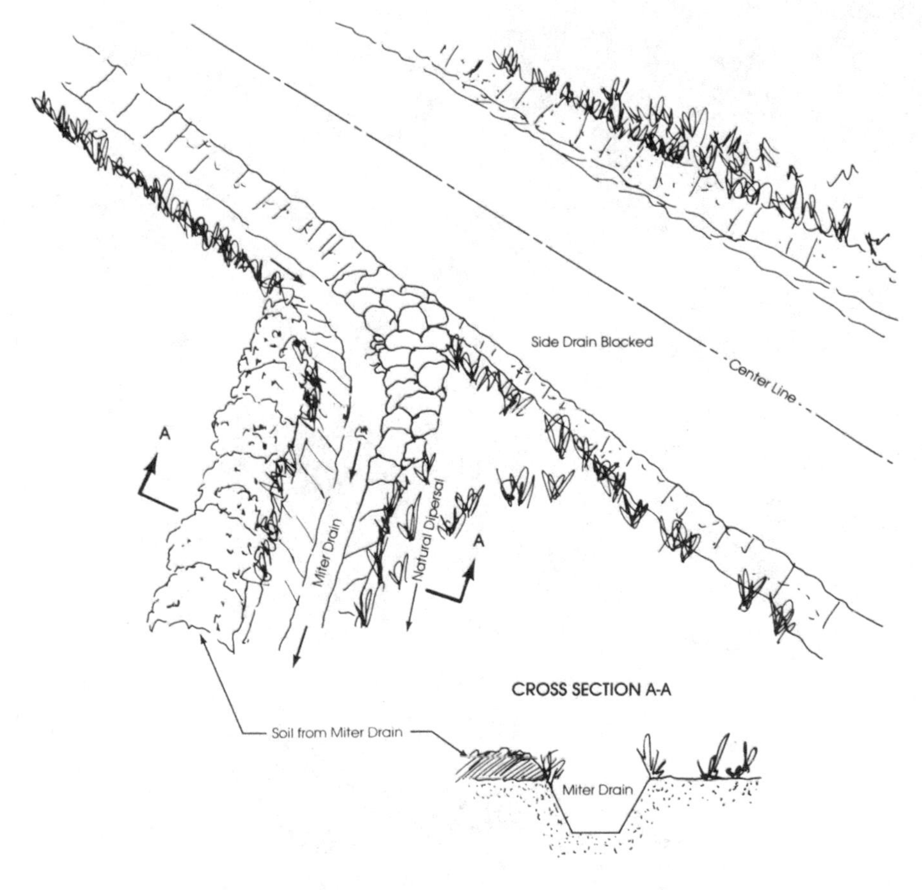

Fig. 64 *Mitre drain construction*

Technical Note 4.11
Stage Construction of Water Crossings

1. Stage construction of water crossings is necessarily site-specific. Paths can cross streams on stepping stones (Figure 66), fords, foot logs (Figure 67) or wood or rope bridges. Although all these solutions are not practical for all streams, stepping stones or fords may provide access during periods of low water; while foot logs or bridges provide all-weather passage, which may not be required in the first stage of development.
2. Trails can initially use fords as crossings, but all-weather access in some cases will require wooden bridges if wood is available. Areas of swift flowing water (0.9 m/sec) of 60 cm depth or more and locations with very soft stream bottoms are not suitable for fords for trails or paths [88].
3. Fords (Figures 68, 69) may be suitable for tracks, earth, granular or gravel surfaced, and paved roads and, under the stage construction concept, should

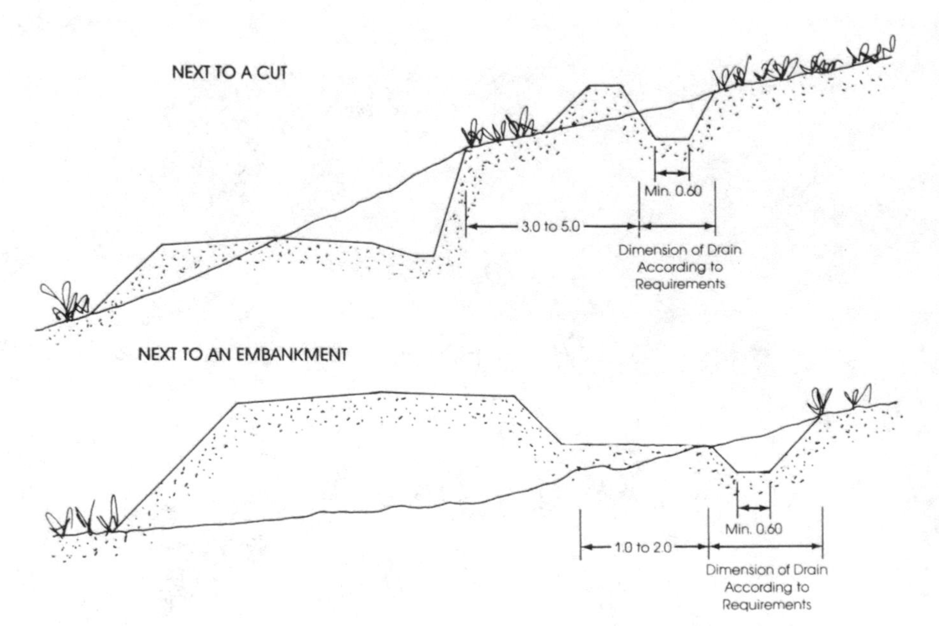

Fig. 65 *Cutoff ditch construction*

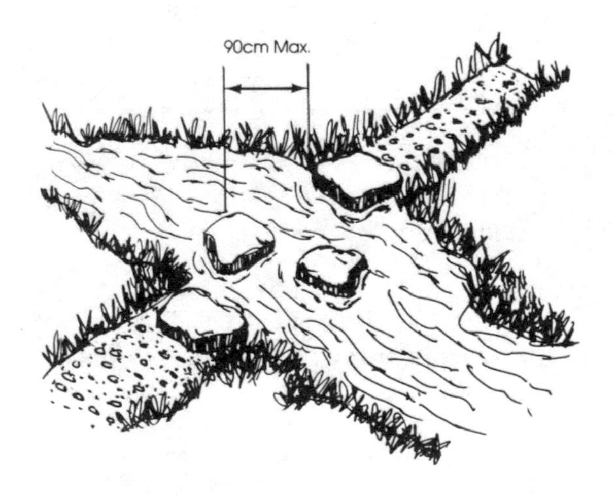

Fig. 66 *Stepping stones*

always be evaluated before choosing a more expensive form of water crossing. Where low flows occur often, a vented ford, culverted ford or Irish bridge, which permits low flows to pass through culverts underneath the ford's surface, may provide a crossing in all but severe storms. Vehicular fords, however, require relatively long approaches, so narrow streams may sometimes be crossed by culverts at less cost than constructing a ford [81].

4. Traditionally, natural fords have been used as stream crossing sites. Fords

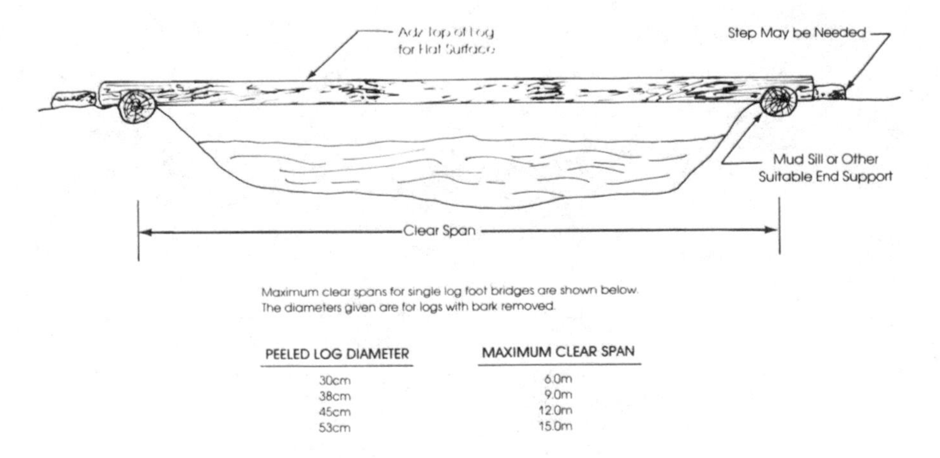

PEELED LOG DIAMETER	MAXIMUM CLEAR SPAN
30cm	6.0m
38cm	9.0m
45cm	12.0m
53cm	15.0m

Fig. 67 *Foot log details*

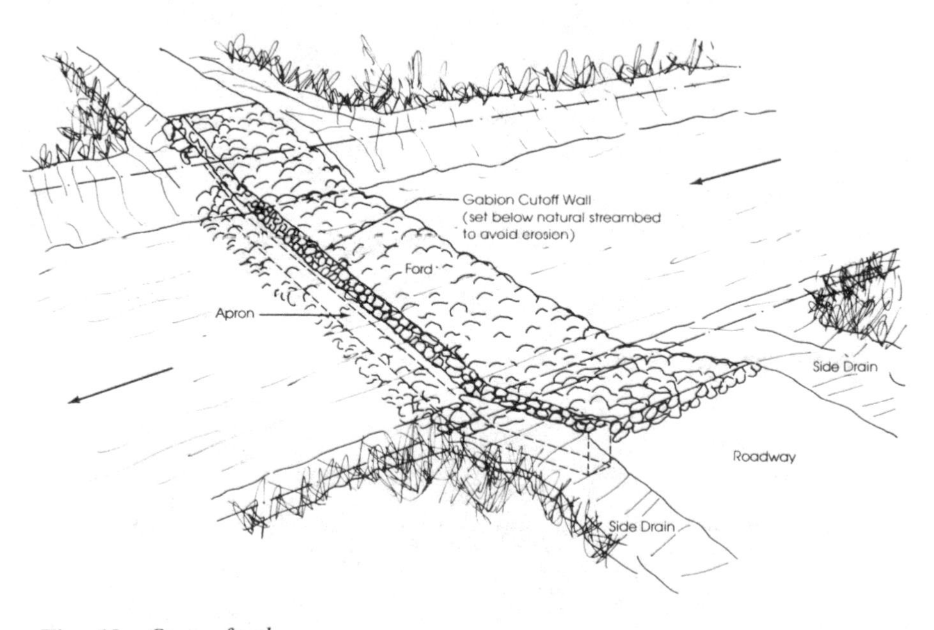

Fig. 68 *Stone ford*

are best suited for infrastructure in drier regions where stream flooding is infrequent and of short duration; where other infrastructure provides bypasses during periods of unfordable depths, and where the crossings are short and water depth seldom exceeds fordable limits. Where stream or river flow cannot be forded, the construction of submersible bridges, which are those designed to be overtopped by flood waters, can be considered. Submersible bridges are relatively uneconomical, and are usually only considered for lengths of over 60 metres [76].

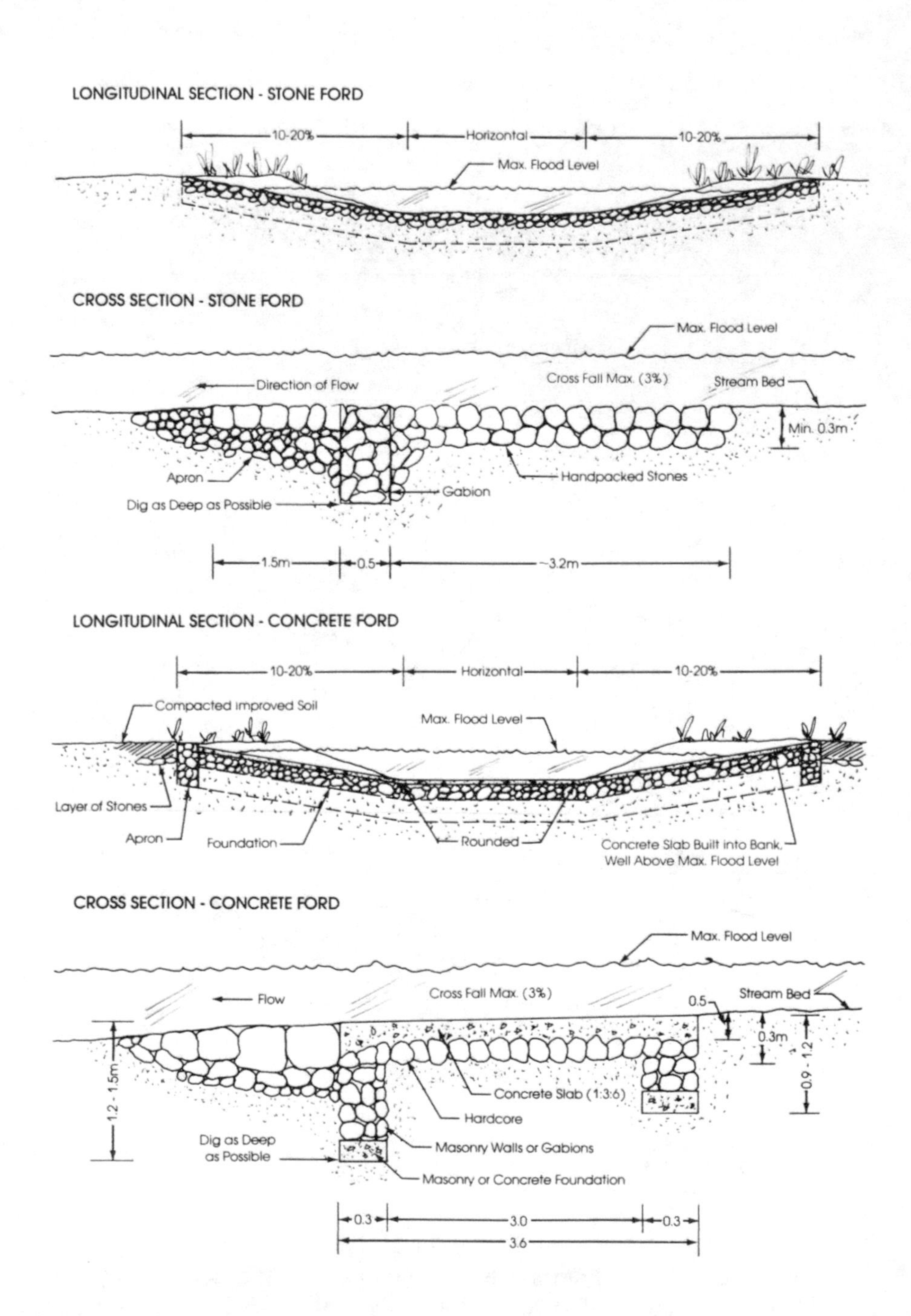

Fig. 69 *Longitudinal & cross sections of stone and concrete fords*

5. Some of the structures described above are considered low water stream crossings (LWSC) because their design includes provisions for structural submersion. The following tables [24] suggest criteria for their use with motorable infrastructure.

Table 16. Low Water Stream Crossing Selection Criteria for Motorable Infrastructure

Criteria	*LWSC most favourable when*	*LWSC least favourable when*
Average daily traffic (ADT)	less than five vehicles/day	more than 200 vehicles/day
Average annual flooding	less than two times/year	more than ten times/year
Average duration of traffic interruption per occurrence	less than 24 hours	more than three days
Extra travel time for detour	less than one hour	more than two hours

Table 17. Factors to Consider When Evaluating LWSC

A. GEOMORPHIC AND LAND USE	
1. Drainage area shape	Long and narrow (length three to four times width)
2. Stream and basin slope	Steep
3. Channel and overbank	Low valley storage upstream. Located in a stable stream reach
B. STRUCTURAL	
General	
1. Vertical curves at structure	Gentle
2. Orientation of structure across stream bed	Straight; skew should be avoided as much as possible
3. Approach length	Long, to provide sufficient sight distance for warning signs
4. Height of pavement above stream bed	Less than 1.2 m
Fords	
1. Normal daily flow depth	Less than 10 to 15 cm
2. Pavement material	May vary from riverbed gravel to concrete
3. Erosion	End walls and gabion protection may be desirable
Vented Fords	
1. Pavement and fill materials	Should be dense packed; heavy to withstand erosion and washout. May be encased in concrete for narrow crossings
2. Vents	Pipes of various materials can be used. Should be anchored in the ground; both ends bevelled to allow easy passage of debris; more than one vent should be used, but fewer lines of large pipes desirable
3. Erosion protection	Cut-off walls and splash aprons may be needed. Rip-rap protection of slope may be considered

Table 17. (*contd.*)

Submersible Bridges	
1. Pavement	Heavy and solid. Light and loose pavements such as bituminous or gravel pavements are not desirable
2. Bridge Deck	Must be heavy to withstand drag, uplift and lateral forces due to overflow of upstream water Must have one-way camber downstream Upstream and downstream edges should be rounded
3. Erosion protection	Cut-off walls and impervious aprons may be desirable

6. When LWSC are not practical, conventional bridges (Figure 70) can be constructed with wood abutments, piers, stringers and decks; or with concrete abutments and piers, and wooden stringers and decks; or with concrete abutment and piers, steel stringers and wooden decks; or with concrete abutments, piers and decks over steel stringers; or all of concrete. The economics of each method are area-specific; but if wood is available and the site does not suffer insects that rapidly attack wooden structures, wood bridges are a viable stage construction activity. Timber bridges are not only inexpensive, but they also use local resources, e.g., materials, skills, and utilize labour-based construction methods. However, wooden decks, like gravel surfaces, suffer traffic deterioration and usually require frequent repair and replacement.

7. Fixed bridges, which are the conventional bridges described above, are those designed for a specific site. The final design must consider the approaches; the water flow characteristics; the bottom conditions; the bridge length; the availability of local materials; and navigation requirements. Bridge design must be undertaken by competent structural engineers using specific site data for foundation design to reduce costly overdesign or bridge failure. A rule-of-thumb for minimum cost design is that the cost of the substructure should equal the cost of the superstructure (less the decking cost) [76]. One-lane bridges meet the traffic volume requirements of low-volume roads being considered in this handbook; however, bridges or culverts with a span of less than 6 metres are sometimes built to accommodate two-lanes if the rest of the road is two-lanes wide. Parapets should always be provided on narrow crossings and they should always be prominently marked for safety reasons.

8. When water crossing site conditions make the cost of fixed bridges prohibitive, floating bridges and ferries must be considered. Both are inconvenient and often costly alternatives. Ferries have very limited capacity and their costs for short and medium crossings of less than 100 metres in most cases far exceed bridge costs, even if operational labour is inexpensive [76]. Ferries may vary from one or more wooden prams, i.e., flat-bottomed boats with squared-off ends, with timber decking to shipyard-constructed steel vessels. They may be pulled or pushed by hand or be powered by stream flow using a cableguide, but where crossings are long or the river is navigable, they must be self-powered. Ferry landings often require special engineering input.

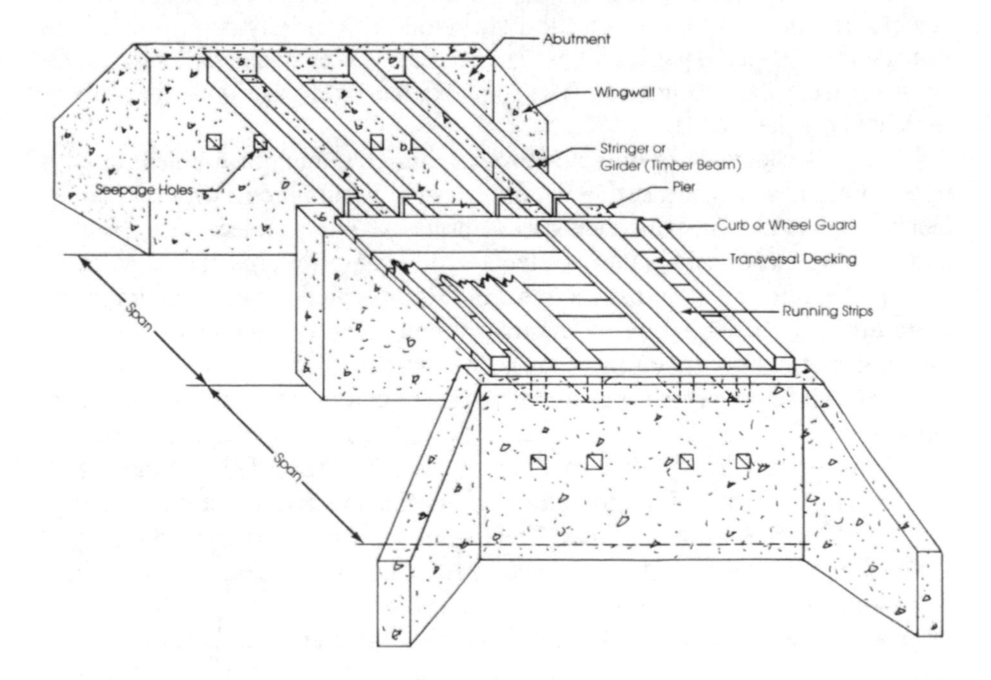

Fig. 70 *Bridge component nomenclature*

Technical Note 4.12
Recording Road Physical Inventory Notes

1. Inventory notes can be recorded in a standard survey notebook. They should be recorded from the bottom of the page upward so that features which appear to the left of the observer will be recorded on the left side of the right page. The left page can be used to record odometer readings and notes. The line up the middle of the right page is commonly used to represent the road centreline. Features being inventoried which appear on the sketch on the right page and in the notes on the left page should be coded to save space and speed up the inventory process. However, such coding should be agreed to before the inventory is begun so that all inventories will be consistent and readable to other people who may used the data from the inventory books. Usually the coding is indicated in the first few pages of each book.
2. Coding must be devised for such road features as changes in horizontal and vertical alignment and treadway and should widths since the centerline in the sketch on the right hand page appears as a straight line. Other features commonly coded are the relationship of the roadway surface to the terrain, i.e., elevated, depressed, or side hill; drainage flows; ditch types; embankment or cut face heights; land usage and vegetation types on each side of the road; habitations and specific buildings such as churches, clinics, and administrative

offices; burial locations; and road, track, trail and path intersections. Drainage structures are usually coded on the right hand page, briefly described in the notes on the left hand page, and, where necessary, sketched and dimensioned on a separate page. Signs can be coded on the right page with their texts indicated on the left page.

3. The exact location of the inventoried features is a major problem in any inventory. Physically measuring the entire route in rural areas with surveying instruments is too time-consuming to be practical and is unnecessary in most cases. However, selected measurements with a compass, tape and hand level should be made of ditch cross-sections, the slope of steep grades, all drainage structures and the water course features at their locations, and in areas where high embankments or deep cuts occur. The location of the individual features should be close enough so they can be located on a map and so that another person can physically relocate the features on the ground at a later date.

4. Consequently most rural road inventories are recorded by odometer readings to the tenth of a kilometre. Odometers are calibrated by equating their readings to known distances. The longer the known distance, the better the calibration will be. Therfore, the most common method of calibrating odometers is to record their readings on roads that have kilometre posts. The odometer reading is recorded at the beginning kilometre marker, at several intermediate markers, and at the end. Then all distances between the various markers are computed to identify any selected markers that may have been moved or improperly placed. Once the general relationship of odometer to kilometre post distnce is established, the longest valid distance is used to determine the odometer correction factor. Odometer calibration should be rechecked as frequently as possible.

5. The distances recorded in the field book are normally actual odometer readings. If possible, the corrected readings should also be recorded in the field book to prevent later errors in applying the correction factors. Frequently a third distance is also recorded in the same location in the field book when long routes are inventoried. This is the corrected distance from the beginning of the route to the inventoried feature, i.e., the cumulative distance corrected not only for the odometer factor but also for gaps in the odometer readings caused by side trips, detours, or an inventory of several days duration. This last value is the inventory location of record. When other parties wish to locate a specific inventoried feature, e.g., a culvert, in the field, they reverse their odometer correction factor and apply it to the distance of record. When they reach the beginning of the inventoried route, they add their calculated odometer distance to the current odometer reading to determine their odometer reading at the correct location.

Technical Note 4.13
Recording Conditions and Needs Roads Inventories

1. Existing conditions on each link must be evaluated in reference to (i) the type of transport activity, (ii) the proposed types of transport aids, and (iii) the desired accessibility. Most experienced engineers have no difficulty in determining accessibility constraints for motor vehicles. Technical Note 2.3, Suggested Infrastructure Requirements for Transport Aids, will help in determining generic constraints but local investigations are necessary to relate accessibility to the specific transport aids used in-country.

2. If a particular transport aid is currently used in the project area, the engineer should ask the users to show him or her examples of infrastructure that constrain the transport aid's use. The local population is a good source of general intelligence about transport constraints on all the local infrastructure. This may include information on how often streams flood, how high the water rises, where trucks have bogged down in the past, how they were freed, etc. The local people should not be expected to know access constraints in other areas or to known if trucks do not come for economic reasons rather than because of access constraints. Therefore, they should not be asked to prioritize improvements or to analyze the technical aspects of the constraints, although any reason they offer for constraint causes should be investigated since the local farmers work constantly with the local soils. Neither should they be asked to transfer individual transport aid problems to other unrelated transport aids. Local inhabitants can also be very helpful in finding material sources. One useful method of encouraging local people to look for suitable aggregate is to leave several bottles of aggregate at a village and offer a small reward to any person who locates similar material that is subsequently used on the roadway [34].

3. The following form represents one method of recording specific transport constraints. It contains a space for a sketch of the rural access infrastructure link being evaluated, and a list of specific constraints for evaluation. Other information such as population, locations of crops being cultivated, and storage, are including for the project planner's use. The form also provides space for listing the features which limit accessibility for specific transport aids and for recording suggested solutions to these limitations. Annexe 5 includes several Technical Notes which may be useful for determining the causes and possible solutions to access problems that are correctable by spot improvements. A brief description of some of the items listed on the form is included after the form.

Level 1 RAI Conditions and Needs Inventory Form

RAI from ....

Sketch of RAI showing approximate alignment, node connections to other RAI and roads, and main physical features. KM

RAI width									
Cleared width									
Sharp curve									
Steep gradient									
RAI surface type									
Drainage structures									
Condition:									
– RAI surface									
– roadside drainage									
Terrain									
Natural soils									
Gravel pits									
Village/settlement									
(population)									
Agricultural crops									
Storage facilities									
Features limiting access									
Suggested solutions to attain desired accessibility									

Inventoried by: Date:

Suggested solution by: Date:

EXPLANATION OF TERMS ON LEVEL 1 RAI CONDITION AND NEEDS INVENTORY FORM

RAI Rural Access Infractructure.
RAI Width Record effective width which can be used as a travelway to a tenth of a metre, and location of any width changes.
Cleared Width Record width of clearing available for passing where widths may be a constraint.
Sharp Curve Indicate only horizontal curves which present obstacles to specific transport aids being considered for use on the link.
Steep Gradient Indicate only gradients which hamper access by proposed transport aids. Record gradient length and per cent slope. Also record whether gradients are unsuitable only during rainy seasons or throughout the entire year.
RAI Surface Type Indicate each treadway surface type, such as earth, gravel, paved, etc., and its limits, i.e., beginning and ending locations.
Drainage Structure Note main dimensions of bridges, culverts, and fords. Include supplementary detailed sketches of each drainage structure and condition of each item such as bridge decking, beams, piers, abutments, etc.
Condition
– *RAI Surface* Record weather at time of survey and try to estimate and record likely surface condition when different weather conditions prevail, e.g., dry weather-good, wet weather-poor.
– *Roadside Drainage* List apparent condition and adequacy of ditches and ditch relief system. Include ditch lengths and side of the RAI on which the ditch occurs. Indicate condition and adequacy of mitre drains.
Terrain List terrain category such as flat, rolling, or mountainous, and location of terrain changes. List terrain being traversed, not surrounding terrain.
Natural Soils List soils types as described in Technical Note 4.02.
Gravel Pits Indicate locations and types of granular soil deposits.
Village/Settlement Include name and population.
Agriculatural Crops Indicate locations of cultivated land and crops grown thereon. Also indicate fallow land and its cropping cycle. Since this inventory is limited to the areas adjacent to the RAI, agricultural data for the entire zone of influence must be gathered separately.
Storage Facilities Indicate locations and include supplemental list of types, sizes, and condition of storage facilities near the infrastrcuture.

Technical Note 4.14
Level 2 RAI Inventory Form

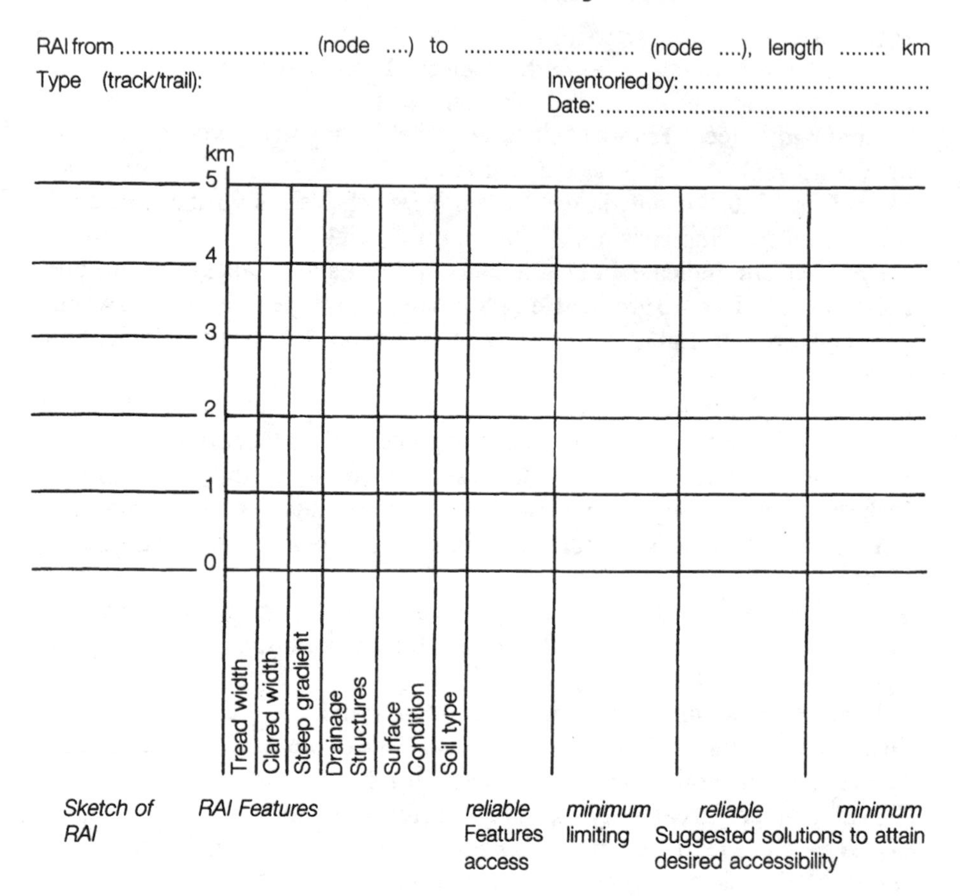

RAI from .... (node ....) to .... (node ....), length .... km

Type (track/trail):

Inventoried by: ....

Date: ....

km

5

4

3

2

1

0

Tread width | Clared width | Steep gradient | Drainage Structures | Surface Condition | Soil type

Sketch of RAI | *RAI Features* | *reliable* Features access | *minimum* limiting | *reliable* Suggested solutions to attain desired accessibility *minimum*

Technical Note 4.15
Field Engineering for Realignment

For most low-volume feeder road construction, it is possible to set the alignment by eye; or to set the tangents (straights) by eye and the curves using stakes, a tape, and string lines. A simple method of laying out curves in open country, based only on the determination of a suitable tangent length, is shown in Figure 71 [16].

Stakes and string lines are suitable to delineate road edges and ditch alignments. Abney levels can provide sufficient vertical control of profiles, cross slopes, and ditch gradients. Camber boards (Figure 15) can be used to check both the camber and smoothness of the finished road surface.

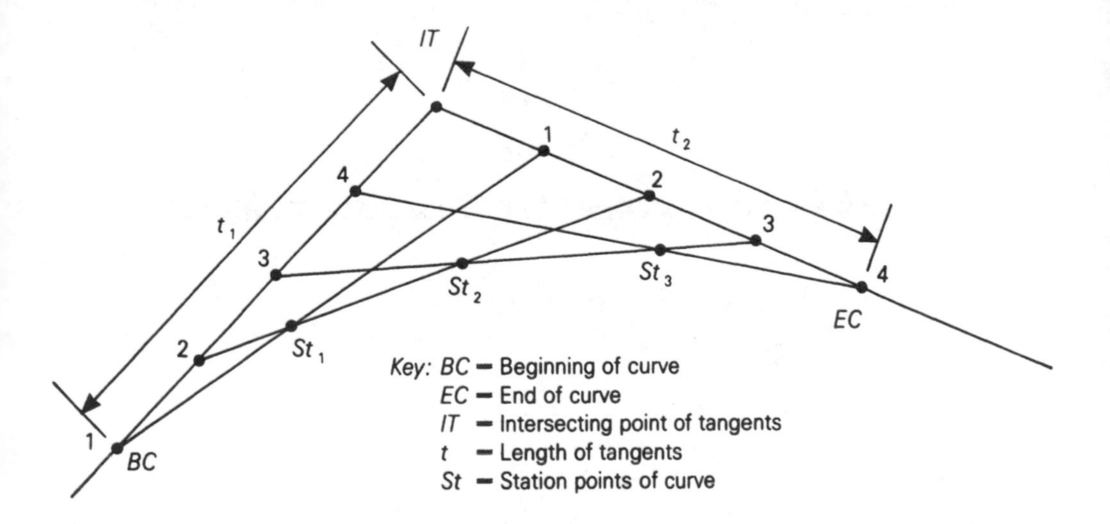

Note: Where site conditions permit, the curve should be set out from equal tangents, so that the steeper portion is not pushed nearer to one tangent than to the other. When necessary, as, for example, when an obstruction has to be avoided, one tangent can be shortened so as to suit the ground conditions.

The method of setting out is as follows:

1. The intersecting point of the tangents is located by ranging. The most suitable lengths of tangent are determined and measured.
2. The tangent points *BC* and *EC* are then fixed on the straights.
3. Each tangent is divided into an equal number of parts and numbered as shown in the figure.
4. Points on the curve are fixed on the ground by lining in with string lines.
   Station 1 is at the intersection of 1-1 with 2-2.
   Station 2 is at the intersection of 2-2 with 3-3.
   Station 3 is at the intersection of 3-3 with 4-4.

Fig. 71 *Setting out curves*

ANNEXE 5

Stage Development and Spot Improvements

This annexe contains ten technical notes:
T.N. 5.01, *Cost Evaluation of Spot Improvements*, is referenced in Paras 4.15, 6.03, 8.26 and 13.05.
T.N. 5.02, *Stage Construction Features*, is referenced in Para 13.11.
T.N. 5.03, *Road Surface Potholes*;
T.N. 5.04, *Travel Surface Longitudinal Erosion*;
T.N. 5.05, *Muddy Surface, Rutting Under Traffic*;
T.N. 5.06, *Slope Erosion*;
T.N. 5.07, *Ditch Erosion*;
T.N. 5.08, *Waterway Crosses Roadway*;
T.N. 5.09, *Bridge Deficiencies*; and
T.N. 5.10, *Pipe Culvert Deficiencies*; are all referenced in Paras 6.03, 13.16, 13.17, and 13.19.

Technical Note 5.01
Cost Evaluation of Spot Improvements

1. Stage construction involving 'spot improvements' as they prove to be necessary, is manifestly the appropriate concept in planning rural transport infrastructure (RAI). Besides keeping the original investment within bounds it provides elasticity to accommodate the changes in pace and direction which are inevitable in any rural development scheme. It thus provides a means to secure at minimum total cost a transport network which will meet the transport needs as the scheme develops. One possible objection is the likely difficulty in securing particular items such as bridges later when the need arises and when the initial construction facilities have been dispersed. But with careful planning this difficulty can be avoided.
2. To illustrate the economic value of this approach two scenarios are investigated to determine the cost effectiveness of spot-improving a feeder road as the first phase of the stage construction process after preliminary investigations have shown that improving the feeder road is viable. The evaluation method used in this technical note is based on the 'present worth' concept of two alternative choices. This involves a concept of the time value of money, i.e., that values of future expenditures and revenues must be discounted by an appropriate factor to give their present day values.
3. The following example illustrates the need to recognize the time value of money. If you were given a choice between receiving $100 today and receiving

the same amount a year from now, which would you prefer? Even when the possible risk of not being able to obtain the $100 a year from now or devaluation of the currency is ignored, you would prefer to receive the money now, since you can put it in a savings account and receive interest. Thus, the value of a dollar today is not the same as the value of a dollar tomorrow. In other words, 'time is money' [4].

4. In short, the scenario which defers an equal expenditure to a future time should cost less in today's capital. Therefore, the first scenario in this evaluation is the spot improvement of infrastructure to accommodate short-term access and transport needs until the development process indicates further improvements are necessary. This alternative not only preserves options and resource investments at the time when uncertainty exists, but also permits access improvements to remain in step with the development progress. The second scenario is to build the infrastructure to satisfy the perceived end needs for access and transport immediately. This alternative becomes the benchmark against which the cost effectiveness of the first scenario is tested.

5. Any investigation must begin by developing a timetable of construction phases. The evaluation is based on that timetable. Once the proposed phasing of the first stage has been determined, the cost streams for the two scenarios are calculated and their present values compared. The differences in user savings are ignored in these evaluations because, at the traffic levels considered in this handbook, they are significant between types of transport aids but do not offer large aggregate savings to the users of specific transport aids. In this technical note it is assumed that stage one consists of spot improvements while stage two construction involves improving the infrastructure to its final form. The latter improvements may include upgrading the route previously spot-improved or abandoning that location for a new alignment. In practice, several stages may be undertaken, either as a result of the initial evaluations or as a result of the re-evaluation of access needs that should be made before the second stage is undertaken.

6. In order to compare deferred construction costs to present construction costs, the first scenario must include the following:

(i) The cost of the spot improvements now; and
(ii) the cost of routine maintenance of the spot-improved infrastructure during the time that the new construction is deferred.

Implicit in this evaluation is the lack of periodic maintenance costs on the spot-improved route. When, or if, periodic maintenance is required, the access situation should be re-evaluated to detemine if another stage in the construction is required to keep the accessibility in tune with the actual requirements of the development process instead of proceeding directly to the construction of the route in its final form.

7. The second scernario, immediate full construction, must include:

(i) the cost of full construction at this time;
(ii) the cost of routine maintenance of the newly constructed route during the

time-frame selected for evaluation, i.e., the time construction is deferred in the first scenario; and

(iii) the prorated costs of periodic maintenance in that same time-frame. This periodic maintenance cost may be significant in cases where regravelling costs constitute a sizeable portion of original construction costs.

Item (iii) is necessary to equate the ultimate infrastructure costs so that the ultimate costs of both scenarios will represent the finished product in the same condition at the end of the evaluation period, i.e., the time the ultimate construction was deferred.

8. The development of cost streams for the two scenarios expresses the costs of the spot improvements as a percentage of the costs of immediate construction of the ultimately required infrastructure. In real terms, construction costs are assumed to remain the same over time. The following terms have been used to develop these two cost streams:

A = the percentage (as a decimal) of the new construction costs to be spent for spot improvements if new construction is to be deferred,

C = new construction costs,

$P_1, P_2 \ldots P_n$ = present worth factor applied to year 1, 2, ...n. These factors are included in Table 18, Discount Factors, for 1% to 50% over a time-span of one through ten years,

n = the expected lifetime of the spot improvements or the time when the deferred construction is to take place,

M = the percentage of new construction costs assigned yearly to provide routine maintenance for the newly constructed route,

m = the percentage (as a decimal) of the maintenance cost M assigned yearly to provide routine maintenance on the spot-improved route during the period when new construction is deferred. 'm' may be smaller or greater than M. To facilitate the evaluation, it is expressed as a percentage of M, although it would be calculated independently,

B = the percentage (as a decimal) of the new construction costs allocated for periodic maintenance of the new construction during the time span of n years, i.e., the cost of wear and tear on the road during the evaluation period if the new road is constructed at the beginning rather than the end of the period under consideration. For presentation purposes, the value of periodic maintenance costs over n years is represented as one value in year n.

9. Using the above definitions, the cost stream for deferring new construction becomes:

$$AC + P_1\, m\, MC + P_2\, m\, MC + \ldots\ldots + P_{n-1}\, m\, MC + P_n C \qquad \ldots. (1)$$

and the cost stream for new construction over the same period n becomes:

$$C + P_1\, MC + P_2\, MC + \ldots\ldots + P_{n-1}\, MC + P_n\, BC \qquad \ldots\ldots. (2)$$

The present value of the cost of the spot improvement solution is equal to or less than the present value of the cost of building the new construction now, when expression (1) is equal to or less than expression (2). To determine the present worth factor for which this relationship is valid in any given time-frame, involves solving the following relationship:

$$P_{n1} \leqslant \frac{1 - A + M(1 - m)(P_1 + P_2 + \ldots. + P_{n-1})}{1-B} \qquad \ldots\ldots (3)$$

A simplified solution may be obtained by ignoring the routine maintenance costs:

$$P_n \leqslant \frac{1 - A}{1 - B} \qquad \ldots\ldots (4)$$

10. To illustrate the use of Table 18, consider a road with a present construction cost of $50,000 per km which can be spot-improved for $25,000 per km to provide reliable access for seven years (A = 0.5). Periodic maintenance costs per km required to return the newly constructed road to like-new standards after seven years, total $12,500 ($B$ = 0.25). Then:

$$P_7 = (1.0 - 0.5)/(1.0 - 0.25) = 0.667$$

Table 18 indicates that a discount factor of 0.665 represents a 6% value. Therefore, an opportunity cost of capital of 6% or more will result in the spot-improvement scenario costing less in today's dollars than immediate full construction.

11. Using simplified expression (4), Tables 19 through 22 have been developed. They compare, for specific time-spans of three, five, seven and ten years, and for various relationships of spot improvements' costs, as a proportion of new construction costs (A), and for various relationships of periodic maintenance costs to new construction costs (B), the critical interest rates for the opportunity cost of capital, or the rates for which expression (4) becomes an equality. The critical interest rates have been determined by using the discount factors of Table 18, and are rounded upwards. For instance, if 13.2% will result in equality, 14% is shown in the tables.

12. Based on the assumption that the opportunity cost of capital in most developing countries lies between 10% and 15%, those values are enclosed in Tables 19 through 22. Parameters, i.e., values of A and B, which result in answers below or to the left of the enclosed figures are suitable for the spot improvement scenario for that particular timespan. Values above or to the right indicate selection of the full construction scenario. Values within the enclosed area need to be evaluated further.

13. The full evaluation includes routine maintenance costs. It calls for the determination of the value of the expression on the right side of inequality (3). For example, if $A = 0.6$, $B = 0.15$, $M = 2.5\%$ (Technical Note 7.2), $m = 110\%$ (i.e., the costs of maintaining the spot-improved road are 10% greater than the costs of maintaining a newly constructed road), where $P_{1-6} = 0.893 + 0.797 + 0.712 + 0.636 + 0.567 + 0.507$. Solving the expression gives $P_7 = 0.459$. Again, looking at Table 18, one finds that the P_7 = 0.459 value indicates an interest rate between 11% and 12%. Therefore, at an opportunity cost of capital of 12% or more, spot improvements are the preferred scenario.

14. It is noted that when the maintenance costs (m) of the spot-improved road are less than the maintenance costs of the new road (M), the numerator of the expression on the right side of the inequality sign of expression (3) is always larger than the numerator of the expression on the right side of expression (4). Under these conditions, the enclosed values of Tables 19–22 need not be further evaluated as described in Para 13, if prevailing opportunity costs of capital are greater than or equal to the relevant percentage of one of these tables, since the spot improvement scenario will be the preferred one.

15. A sensitivity analysis of the values in the enclosed areas of Tables 19 through 22, indicates that in the three and five years' cases, a value of m of 125% or less will increase the critical interest rate by no more than one per cent, while in the seven years' case it will increase the critical interest rate by no more than 2%. In the ten years' case, an m value of 120% or less will increase the critical interest rate by no more than 2%. In other words, the margin of error resulting from using expression (4) in Tables 19–22 rather than expression (3) is relatively small if $m < 125\%$.

16. An evaluation of the data in Tables 19 through 22 leads to the following conclusions:

(i) when spot improvements can postpone new construction for three years, they are preferable to new construction if they cost less than 40% of new construction costs. Above 40%, their feasibility should be tested using the solution method of Para 13;

(ii) when spot improvements can pospone new construction for five years, they are preferable to new construction if they cost less than 50% of new construction costs. Above 50%, their feasibility should be tested;

(iii) when spot improvements can postpone new construction for seven years, they are preferable to new construction, if they cost less than 60% of new construction costs. Above 60%, their feasibility should be tested; and

(iv) when spot improvements can postpone new construction for ten years, they are preferable if they cost less than 70% of new construction costs. Above 70%, their feasibility should be tested.

Table 18. Discount Factors

%	1	2	3	4	5	6	7	8	9	10
					No. of years					
1	.9901	.9803	.9706	.9610	.9515	.9421	.9327	.9235	.9143	.9053
2	.9804	.9612	.9423	.9239	.9057	.8880	.8706	.8535	.8368	.8404
3	.9709	.9426	.9151	.8885	.8626	.8375	.8131	.7894	.7664	.7441
4	.9615	.9246	.9151	.8885	.8626	.8375	.8131	.7896	.7664	.7441
5	.9524	.9070	.8638	.8227	.7835	.7462	.7107	.6768	.6446	.6139
6	.9434	.8900	.8396	.7921	.7473	.7050	.6651	.6274	.5919	.5584
7	.9346	.8734	.8163	.7629	.7130	.6663	.6228	.5820	.5439	.5083
8	.9259	.8573	.7938	.7350	.6806	.6302	.5835	.5403	.5003	.4632
9	.9174	.8417	.7722	.7084	.6499	.5963	.5470	.5019	.4604	.4224
10	.9091	.8265	.7513	.6830	.6209	.5645	.5132	.4665	.4241	.3855
11	.9009	.8116	.7312	.6587	.5935	.5346	.4817	.4339	.3909	.3522
12	.8929	.7972	.7118	.6355	.5674	.5066	.4524	.4039	.3606	.3220
13	.8850	.7832	.6931	.6133	.5428	.4803	.4251	.3762	.3329	.2946
14	.8772	.7695	.6750	.5921	.5194	.4556	.3996	.3506	.3075	.2697
15	.8696	.7561	.6575	.5718	.4972	.4323	.3759	.3269	.2843	.2472
16	.8621	.7432	.6407	.5523	.4761	.4104	.3538	.3050	.2630	.2267
17	.8547	.7305	.6244	.5337	.4561	.3898	.3332	.2848	.2434	.2080
18	.8475	.7182	.6086	.5158	.4371	.3704	.3139	.2660	.2255	.1911
19	.8403	.7062	.5934	.4987	.4191	.3521	.2959	.2487	.2090	.1756
20	.8333	.6944	.5787	.4823	.4019	.3349	.2791	.2326	.1938	.1615
21	.8264	.6830	.5645	.4665	.3855	.3186	.2633	.2176	.1799	.1486
22	.8197	.6719	.5507	.4514	.3700	.3033	.2486	.2038	.1670	.1369
23	.8130	.6610	.5374	.4369	.3552	.2889	.2348	.1909	.1552	.1262
24	.8065	.6504	.5245	.4230	.3411	.2751	.2218	.1789	.1443	.1164
25	.8000	.6400	.5120	.4096	.3277	.2621	.2097	.1678	.1342	.1074
26	.7937	.6299	.4999	.3968	.3149	.2499	.1983	.1574	.1249	.0992
27	.7874	.6200	.4882	.3844	.3027	.2383	.1877	.1478	.1164	.0916
28	.7813	.6104	.4768	.3725	.2910	.2274	.1776	.1388	.1084	.0847
29	.7752	.6009	.4658	.3611	.2799	.2170	.1682	.1304	.1011	.0784
30	.7692	.5917	.4552	.3501	.2693	.2072	.1594	.1226	.0943	.0725
31	.7634	.5827	.4448	.3396	.2592	.1979	.1510	.1153	.0880	.0672
32	.7576	.5739	.4348	.3294	.2495	.1890	.1432	.1085	.0822	.0623
33	.7519	.5653	.4251	.3196	.2403	.1807	.1358	.1021	.0768	.0578
34	.7463	.5569	.4156	.3102	.2315	.1727	.1289	.0962	.0718	.0536
35	.7407	.5487	.4064	.3011	.2230	.1652	.1224	.0906	.0671	.0497
36	.7353	.5407	.3975	.2923	.2149	.1580	.1162	.0855	.0628	.0462
37	.7299	.5328	.3889	.2839	.2072	.1512	.1104	.0806	.0589	.0429
38	.7246	.5251	.3805	.2757	.1998	.1448	.1049	.0760	.0551	.0399
39	.7194	.5176	.3724	.2679	.1927	.1387	.0998	.0718	.0516	.0371
40	.7143	.5102	.3644	.2603	.1859	.1328	.0949	.0678	.0484	.0346
41	.7092	.5030	.3568	.2530	.1794	.1273	.0903	.0640	.0454	.0322
42	.7042	.4959	.3493	.2460	.1732	.1220	.0859	.0605	.0426	.0300
43	.6993	.4890	.3420	.2391	.1672	.1170	.0818	.0572	.0400	.0280
44	.6944	.4823	.3349	.2326	.1616	.1122	.0779	.0541	.0376	.0261
45	.6897	.4756	.3280	.2262	.1560	.1076	.0742	.0512	.0353	.0243
46	.6849	.4691	.3213	.2201	.1507	.1032	.0707	.0484	.0332	.0228
47	.6803	.4628	.3148	.2142	.1457	.0991	.0674	.0459	.0312	.0212
48	.6757	.4565	.3085	.2084	.1408	.0952	.0643	.0434	.0294	.0198
49	.6711	.4504	.3023	.2029	.1362	.0914	.0613	.0412	.0276	.0185
50	.6667	.4444	.2963	.1975	.1317	.0878	.0585	.0390	.0260	.0173

Table 19:[1] **Three-Year Lifetime**
Cost of Capital – Critical Interest Rate

B = Periodic Maintenance Proportion	*A = Spot Improvements as Proportion of New Construction Costs*							
	.20	.30	.40	.50	.60	.70	.80	.90
.10	4%	9%	15%	22%	32%	45%	>50	>50
.15	3%	7%	12%	20%	29%	42%	>50	>50
.20		5%	11%	17%	26%	39%	>50	>50
.25		3%	8%	15%	24%	36%	>50	>50
.30			6%	12%	21%	33%	>50	>50
.35			3%	10%	18%	30%	49%	>50
.40				7%	15%	26%	45%	>50
.45				4%	12%	23%	41%	>50
.50					8%	19%	36%	>50

[1] In this and the following tables, the critical cost of capital is that cost for which, given the parameters A and B, the present value of both scenarios are approximately equal. For any cost of capital equal to or greater than the critical value, the spot improvement scenario will be preferred. Enclosed values indicate a more detailed evaluation should be undertaken to include routine maintenance costs.

Table 20. Five-Year Lifetime
Cost of Capital – Critical Interest Rate

B = Periodic Maintenance Proportion	*A = Spot Improvements as Proportion of New Construction Costs*							
	.20	*.30*	*.40*	*.50*	*.60*	*.70*	*.80*	*.90*
.10	3%	6%	9%	13%	18%	25%	36%	>50
.15	2%	4%	8%	12%	17%	24%	34%	>50
.20		3%	6%	10%	15%	22%	32%	>50
.25		2%	5%	9%	14%	21%	31%	50%
.30			4%	7%	12%	19%	29%	48%
.35			2%	6%	11%	17%	27%	46%
.40				4%	9%	15%	25%	44%
.45				2%	7%	13%	23%	41%
.50					5%	11%	21%	38%

Table 21. Seven-Year Lifetime
Cost of Captial – Critical Interest Rate

B = Periodic Maintenance Proportion	A = Spot Improvements as Proportion of New Construction Costs							
	.20	.30	.40	.50	.60	.70	.80	.90
.10	2%	4%	6%	9%	13%	17%	24%	37%
.15	1%	3%	6%	8%	12%	16%	23%	36%
.20		2%	5%	7%	11%	15%	22%	35%
.25		1%	4%	6%	10%	14%	21%	34%
.30			3%	5%	9%	13%	20%	33%
.35			2%	4%	8%	12%	19%	31%
.40				3%	6%	11%	17%	30%
.45				2%	5%	10%	16%	28%
.50					4%	8%	14%	26%

Table 22. Ten-Year Lifetime
Cost of Capital – Critical Interest Rate

B = Periodic Maintenance Proportion	*A = Spot Improvements as Proportion of New Construction Costs*							
	.20	.30	.40	.50	.60	.70	.80	.90
.10	2%	3%	5%	7%	9%	12%	17%	25%
.15	1%	2%	4%	6%	8%	11%	16%	24%
.20		2%	3%	5%	8%	11%	15%	24%
.25		1%	3%	5%	7%	10%	15%	23%
.30			2%	4%	6%	9%	14%	22%
.35			1%	3%	5%	9%	13%	21%
.40				2%	5%	8%	12%	20%
.45				1%	4%	7%	11%	19%
.50					3%	6%	10%	18%

Technical Note 5.02
Stage Construction Features

Some of the technical and economic features which pertain specifically to stage construction activities include [82]:
(1) The main criteria for building new rural access infrastructure planned for later improvement by staged construction is to adopt the alignment to the topography and river crossings so as to effect minimum changes in the existing ecological regime.
(2) Fills on existing alignments should be widened only on one side. The existing slope should be benched to key in the widened section. The new side-slope should be built in compacted lifts, not by dumping material over the side (Figure 72).
(3) Culverts should be extended on the same gradient if possible. If not, the downstream section should be the steeper one, to prevent siltation at the joint. The new culvert construction should be of the same material as the existing section.
(4) If future wheel loads will cause problems with existing subgrade material like black cotton soil, other expansive clays or fine materials in areas with high groundwater, a sand layer should be placed below any fill (Figure 73), even if the fill is to be widened later.
(5) In areas where the subsoil has a high moisture content, the clearing activity should precede any further activity by a long enough time period to allow the sun and wind to dry the subsoil.
(6) Before any surfacing material is placed, the existing surface, which will become the new subgrade, should be reshaped and compacted to allow proper compaction of the new surface material. A sand blanket over the subgrade may be required to prevent subgrade intrusion into the surface material, and to encourage drainage of water that penetrates the new surface. This sand blanket must always be installed above normal flood level.

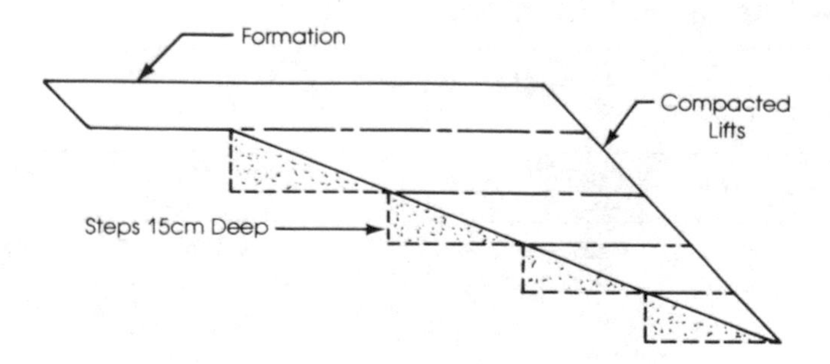

Fig. 72 *Benching*

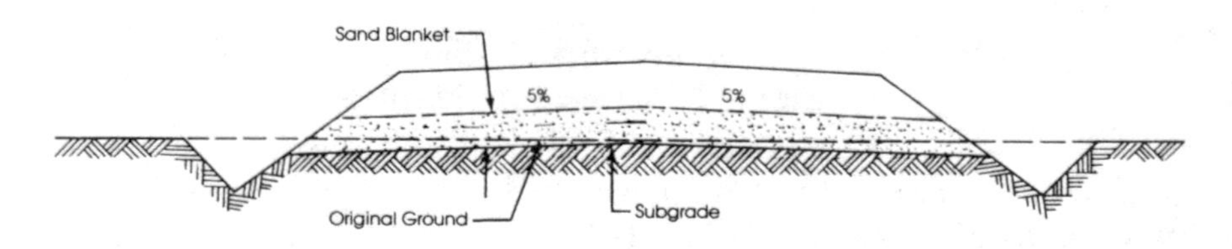

Fig. 73 *Sand blanket placement in areas with high groundwater*

(7) Any location that requires rock cut should be trimmed to the final shape in the first stage. Good use can usually be made of excavated rock in adjacent drainage structures.
(8) Excavating side slopes or batters is often likely to expose loose soil and rock which will be subject to erosion and danger of slippage. Slopes should be trimmed to the natural stable slope evident in surrounding ground. Where this is not possible, the roadway should be widened to accommodate slips when they occur and arrangements made for rapid clearing of slides or slipped material after heavy rains.
(9) On feeder roads and tracks, massive structural construction of retaining walls or headwalls should be undertaken only when events indicate its necessity. Such construction should be properly located when originally built to accommodate any anticipated future upgrading.
(10) Stage construction should not result in additional operational safety hazards, nor should it introduce such safety hazards in any future stages.
(11) While the purpose of stage construction is to retain as many access-related options as possible, it is not economically prudent to retain options for circumstances that are unlikely to occur or for which other access routes are suitable. For instance, if a developing area will never produce enough transportation demand for a two-lane road, that phase of stage construction should not influence any design considerations.

Technical Note 5.03
Road Surface Potholes

This technical note outlines the causes of potholes and suggests possible actions to eliminate their formation in a variety of route locations.

Cause: Rainwater remains on travelled surface
Solution: Remove water faster
Actions possible: 1–11, 14

Cause: Rainwater from surrounding area flows onto or just below road surface
Solution: Deflect surface flow
Actions possible: 2–12, 14

Cause: Groundwater or springs seep onto or just below surface
Solutions: Intercept flow
Actions possible: 5, 6, 11, 13, 14

ACTIONS (Terminology shown in Figure 74)	Depressed section in flat terrain	Cut section in non-flat terrain	Road level with flat terrain	Road even with non-flat terrain	Side hill section	Low fill (<0.5 metres)	Fill (> 0.5 metres)
1. Reshape to proper camber and compact	X	X	X	X	X	X	X
2. Add uphill side-ditch					X	X	
3. Deepen uphill side-ditch					X	X	
4. Widen uphill side-ditch					X	X	
5. Add side-ditches		X	X	X	X	X	
6. Deepen side-ditches		X		X	X	X	
7. Widen side-ditches		X	X	X	X	X	
8. Construct or add dipouts (Figure 75)				X	X	X	
9. Construct or add relief culverts (Fig. 76)				X	X	X	X
10. Construct or add diversion drains				X	X	X	
11. Raise road surface	X	X	X	X	X	X	
12. Add interceptor drains (Figure 77)		X			X		
13. Place longitudinal subsurface drains		X			X		
14. Relocate road	X						

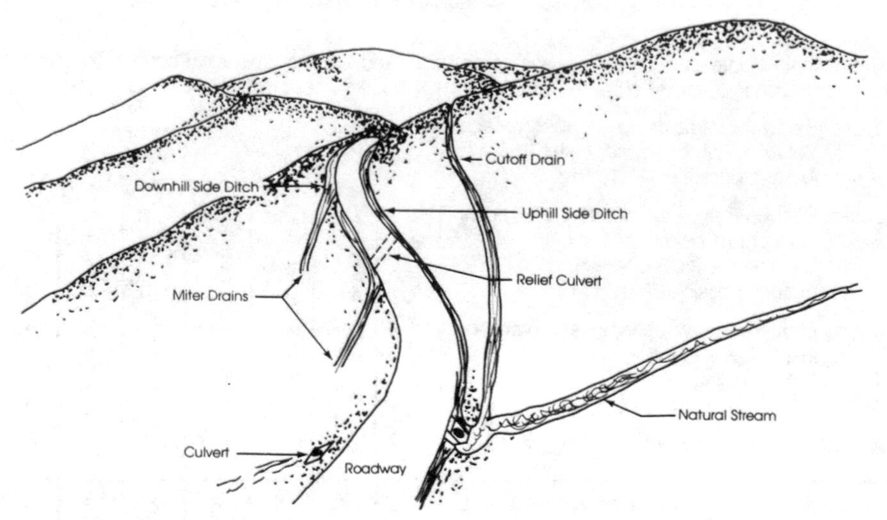

Fig. 74 *Drainage structure nomenclature*

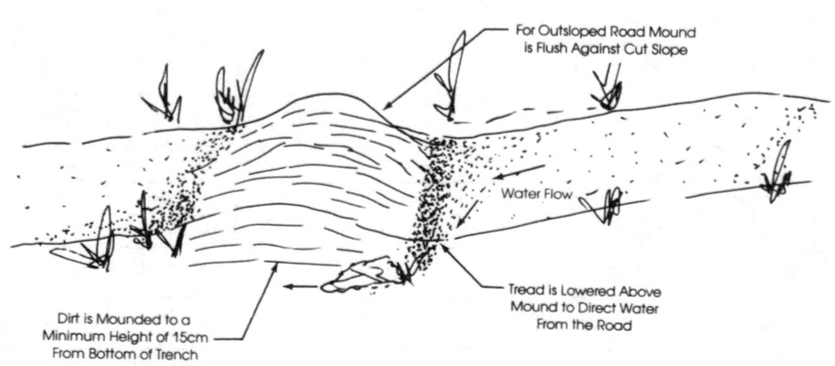

Fig. 75 *Simple dipout*

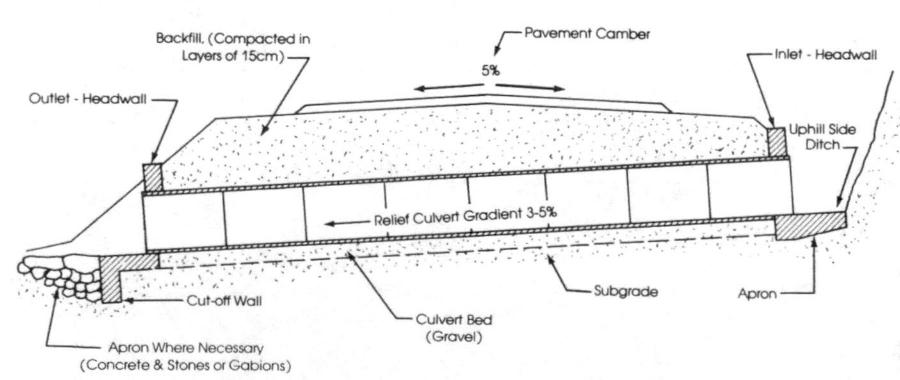

Fig. 76 *Typical relief culvert*

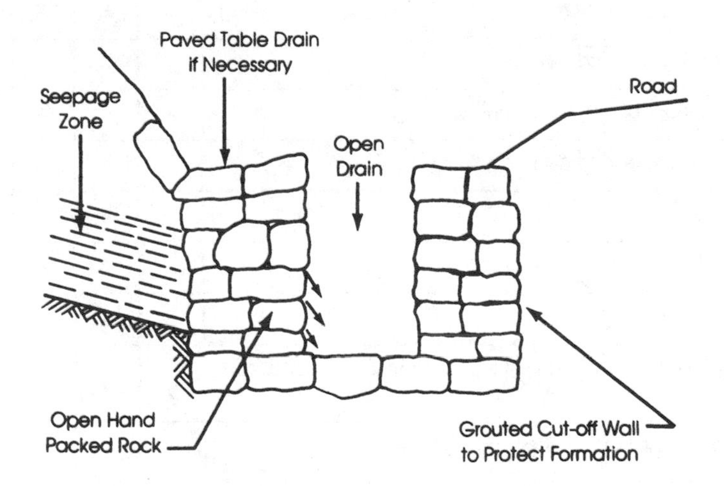

Fig. 77 *Open intercepting drains*

Technical Note 5.04
Travel Surface Longitudinal Erosion

This technical note suggests possible actions to eliminate longitudinal erosion in a variety of route locations.

Cause: Surface is of possibly uncompacted or under-compacted erodable material
Solution: Densify by compaction
Actions possible: 1–14
Solution: Add compacted granular material
Actions possible: 1–15
Solution: Add asphaltic surface seal coat on steep hills where granular surface erodes
Actions possible: 1–16

ACTIONS	Cut section, non-flat terrain	Road even with non-flat terrain	Side hill section	Low fill (<0.5 metres)	Fill (>0.5 metres)
1. Reshape to proper camber	X	X	X	X	X
2. Densify by compaction	X	X	X	X	X
3. Add uphill side-ditch			X	X	
4. Deepen uphill side-ditch			X	X	
5. Widen uphill side-ditch			X	X	
6. Add side-ditches	X	X	X	X	

Technical Note 5.04 (*condt.*)

Action	Col. 1	Col. 2	Col. 3	Col. 4	Col. 5
7. Deepen side-ditches	X	X	X	X	
8. Widen side-ditches	X	X	X	X	
9. Construct or add dipouts	X	X	X	X	X
10. Construct or add relief culverts	X	X	X	X	
11. Construct or add diversion drains		X	X	X	
12. Raise road surface	X	X	X	X	
13. Add interceptor drains	X		X		
14. Construct cattle grids	X	X	X	X	X
15. Add and compact granular surface material after above steps	X	X	X	X	X
16. Add asphaltic seal after above steps	X	X	X	X	X

Technical Note 5.05
Muddy Surface, Rutting Under Traffic

This technical note outlines the causes of muddy surfaces and surfaces that rut in the presence of traffic and water. It suggests possible actions to eliminate these problems in a variety of route locations.

Cause: Rainwater remains on roadway surface
Solution: Remove water faster
Actions possible: 1–11, 16–18

Cause: Rainwater from surrounding area flows onto or just below surface
Solution: Deflect surface flow
Actions possible: 2–12, 16–18

Cause: Groundwater or springs seep onto or just below surface
Solution: Intercept flow
Actions possible: 5–6, 11, 13, 16–18

Cause: Surface is of clay or certain silty materials
Solution: Provide better surface
Actions possible: 1–11, 16–18

Cause: Roadway is crossing bog or swamp, surface is organic material
Solution: Elevate roadway
Actions possible: 14–18

ACTIONS	Depressed section in flat terrain	Cut section in non-flat terrain	Road level with flat terrain	Road even with non-flat terrain	Side hill section	Low fill (<0.5 metres)	Fill (>0.5 metres)
1. Reshape to proper camber and compact	X	X	X	X	X	X	X
2. Add uphill side-ditch					X	X	

Technical Note 5.05 (*contd.*)

3. Deepen uphill side-ditch					X	X	
4. Widen uphill side-ditch					X	X	
5. Add side-ditches		X	X	X	X	X	
6. Deepen side-ditches			X		X	X	X
7. Widen side-ditches		X	X	X	X	X	
8. Construct or add dipouts					X	X	X
9. Construct or add relief culverts				X	X	X	X
10. Construct or add diversion drains				X	X	X	
11. Raise road surface	X	X	X	X	X	X	
12. Add interceptor drains		X			X		
13. Place longitudinal subsurface drains		X			X		
14. Remove organic material if possible			X	X		X	
15. Construct causeway embankment				X			X
16. Improve surface by mechanical stabilization	X	X		X	X		X
17. Add and compact granular surface material as required after taking above steps	X	X	X	X	X	X	X
18. Relocate road	X						

Technical Note 5.06
Slope Erosion

This technical note suggests possible actions to eliminate slope erosion on fill or embankment slopes and cut or batter slopes.

Cause: Embankment face too high to stand without erosion control measures
Solution: Take erosion control measures
Actions possible: 1–3, 5, 6, 8, 10

Cause: Cut face eroding from flowing water
Solution: Take erosion control measures
Actions possible: 1–4, 6, 7, 9, 11

ACTIONS	Embankment slopes	Excavated slopes
1. Flatten slope	X	X
2. Round top and bottom of slopes	X	X

Technical Note 5.06 (*condt.*)

3. Plant vegetal cover, and where applicable mulch slopes before cover grows	X	X
4. Construct cutoff dike or ditch		X
5. Construct berm and gutter	X	
6. Construct outlet drains down slope	X	X
7. Construct subdrain to intercept water veins		X
8. Divert surface water from abutments and headwalls	X	
9. Bench Slopes		X
10. Place material in compacted lifts	X	
11. Make nearly vertical faces in silty materials		X

Technical Note 5.07
Ditch Erosion

This technical note outlines erosion prevention measures for various drainage ditches.

Cause: Ditch slope too steep or too much flow in ditch
Solution: Reduce hydraulic grade and/or amount of water carried or increase erosion resitance of channel
Actions possible: All actions but 1, 11, 12 and 15 for side ditches along non-paved roads

ACTIONS	Roadside ditches	Other drainage ditches
1. Plant grass in ditch		X
2. Construct or add diversion ditches	X	X
3. Change shape from V to trapezoidal (Figure 78)	X	X
4. Widen trapezoidal ditch	X	X
5. Place ditch checks, sticks driven into ditch bottom	X	X
6. Place ditch checks, stone rubble (Figure 79)	X	X
7. Place ditch checks, logs or timber planks	X	X
8. Place ditch checks, rubble masonry	X	X
9. Place ditch checks, concrete with downstream aprons	X	X
10. Line ditch with stones or logs (Figure 80)	X	X

Technical Note 5.07 (*contd.*)

11. Line ditch with asphaltic concrete		X
12. Line ditch with portland cement concrete		X
13. Outlet with paved sluce	X	X
14. Provide energy dissipator at end of outlet	X	X
15. Relocate to flatter alignment		X

Technical Note 5.08
Waterway Crosses Roadway

This technical note suggests appropriate water crossings for various types of rural access infrastructure.

Cause: Alignment runs at an angle to natural drainage flow
Solution: Provide method of crossing
Actions possible: See right columns for infrastructure type

ACTIONS	Path	Trail	Track	Earth road	All-weather road
1. Place stepping stones (Figure 66)	X				
2. Place footlogs (Figure 67)	X				
3. Build ford with gabions (Figure 81)	X	X	X	X	
4. Build paved ford (Figure 82)			X	X	
5. Build paved, vented ford (Figure 83)				X	
6. Place log culvert (Figure 84)		X	X	X	X
7. Place timper culvert		X	X	X	X
8. Place pipe culvert	X	X	X	X	X
9. Place masonry culvert (Figure 85)	X	X	X	X	X
10. Place concrete box culvert				X	X
11. Build timber bridge	X	X	X	X	X
12. Build timber bridge with masonry substructure	X	X	X	X	X
13. Build timber bridge with concrete substructure (Figure 70)			X	X	X
14. Build bridge with concrete substructure and steel girders				X	X
15. Build concrete bridge				X	X
16. Install ferry crossing	X	X	X	X	X

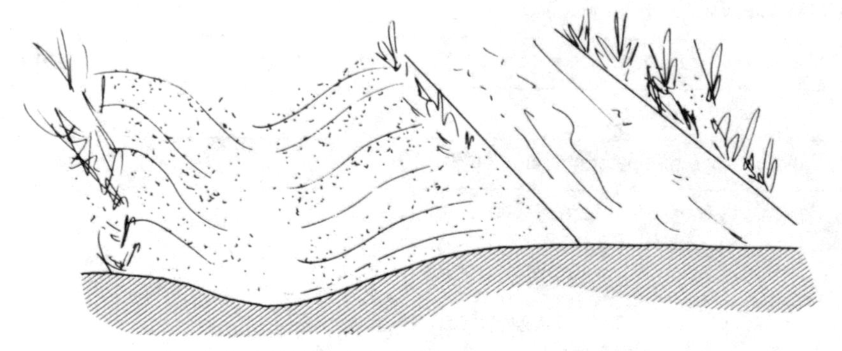

a. A parabolic ditch is the most efficient ditch section, and the least subject to erosion.

b. A compromise between hydraulic efficiency and ease of construction is achieved wih this trapezoidal ditch. If the corners are well rounded, erosion is not severe. It is also mush safer for traffic than triangular ditches.

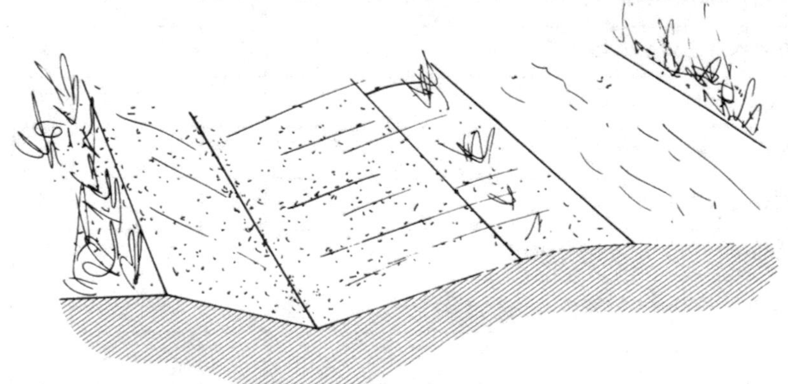

c. The triangular ditch section, illustrated here, is the easiest ditch to construct. Howerver, it is hydraulically inefficient and subject to erosion.

Fig. 78 *Ditch shapes*

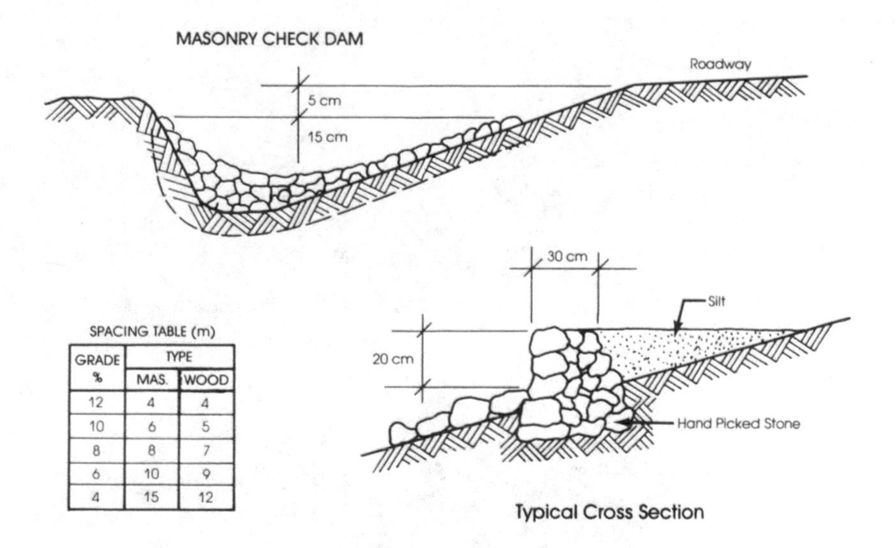

SPACING TABLE (m)

GRADE %	TYPE	
	MAS.	WOOD
12	4	4
10	6	5
8	8	7
6	10	9
4	15	12

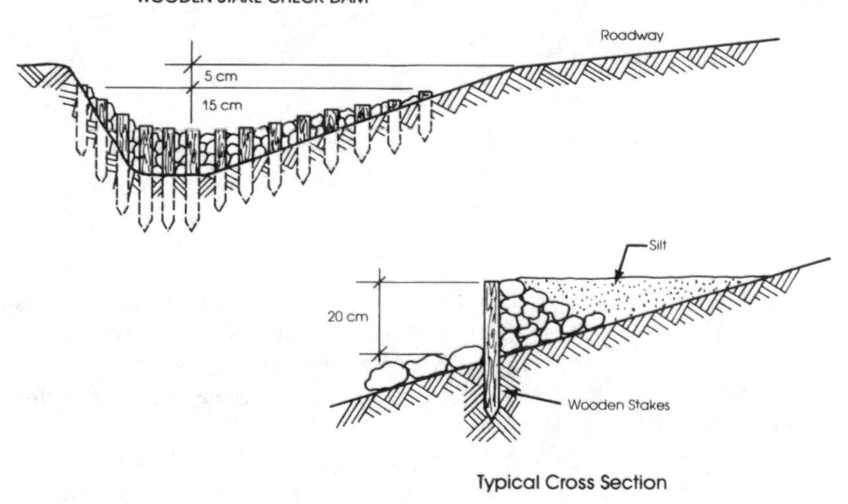

Fig. 79 *Ditch checks*

Technical Note 5.09
Bridge Deficiencies

APPROACH PROBLEMS:

(1) Potholing in surface at ends of bridge causes undue impact on bridge and vehicle.

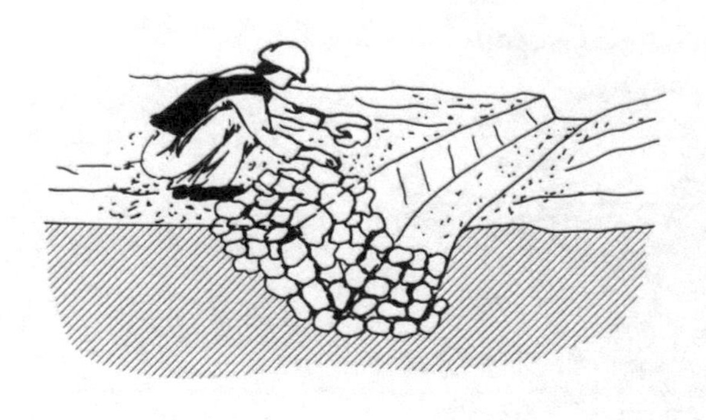

Fig. 80 *Reinforcing ditches with stones or logs*

Solution: Fill holes with well-mixed, moist, clayey sand and gravel, and compact thoroughly (Figure 27). If routine grading appears inadequate for keeping road level with deck, apply short bituminous strip just in front of deck to move potholing problem from deck edge to joint between surface material and asphalt where it is easier to repair.

(2) Surface material spreads onto deck, increasing dead weight load, interfering with deck drainage, and encouraging deck deterioration or rotting.

Solution: Remove all surface material from deck by dragging off with back of grader blade or by shovelling.

(3) Depressions at end of deck due to fill consolidation.

Solution: Patch to maintain level transition.

(4) Extensive settlement or cracking of approach fill, which is a sign of sliding or serious erosion.

Solution: Protect approach slopes with riprap.

(5) Sideslope eorsion from surface water running down the road and off the bridge.

Solution: Flatten fill slopes, introduce vegetation on slopes, install paved drains, or install berms on approaches.

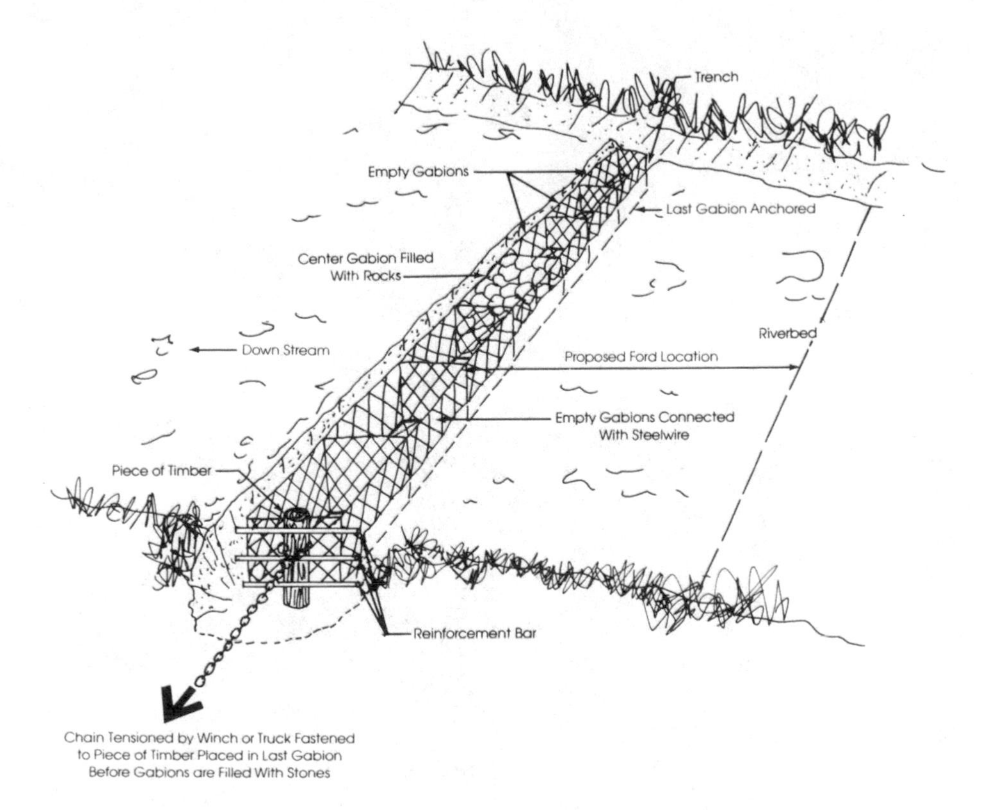

Fig 81. *Placing gabions for ford construction*

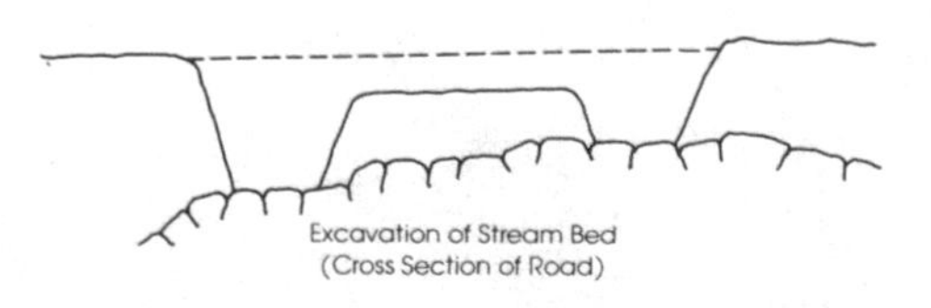

The Stones are Placed in the Excavation, Laid on Beds of Mortar, the Largest Ones Lowest

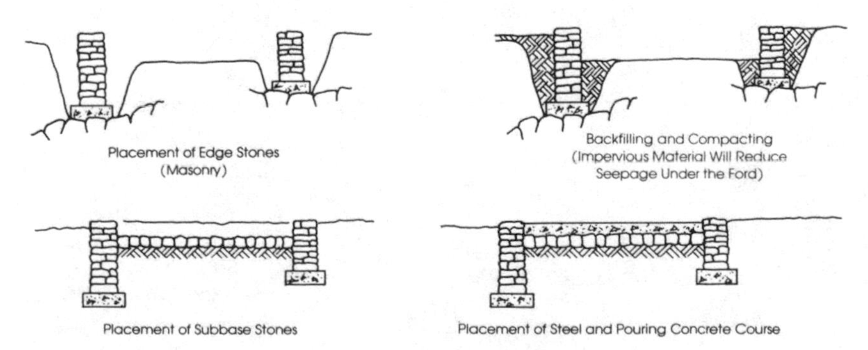

Fig. 82 *Construction method for paved ford*

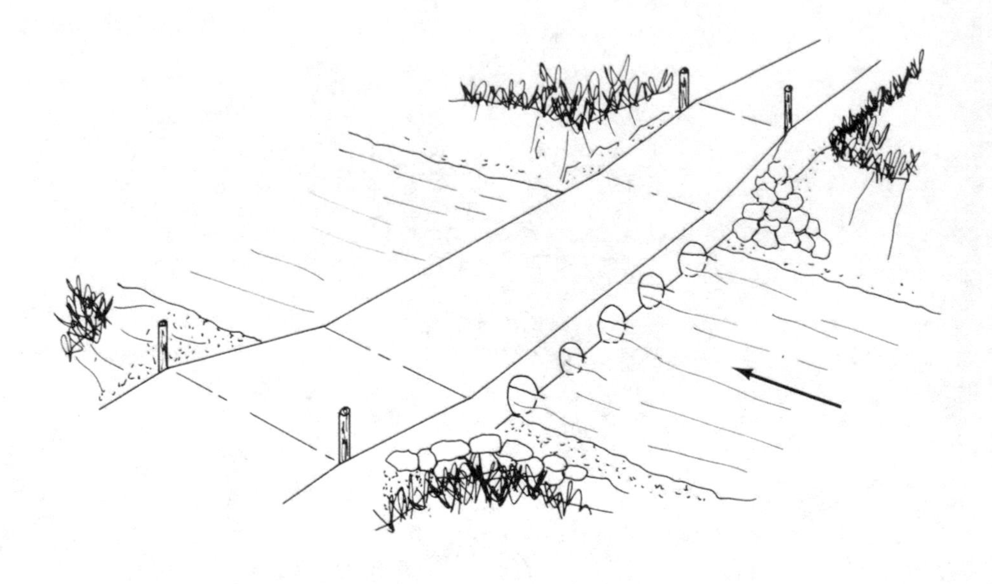

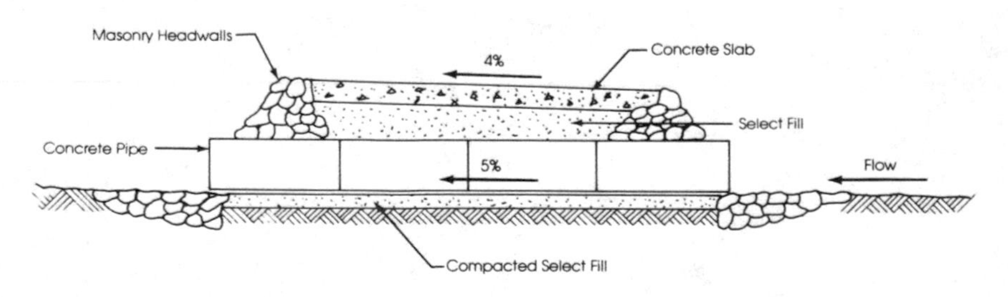

Fig. 83 *Vented ford details*

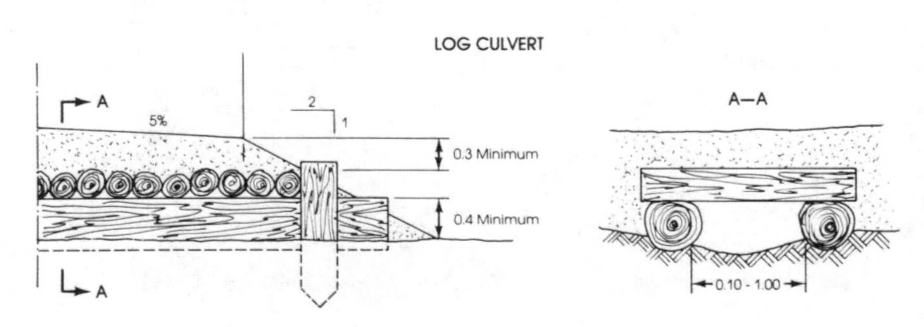

Fig. 84 *Log culvert*

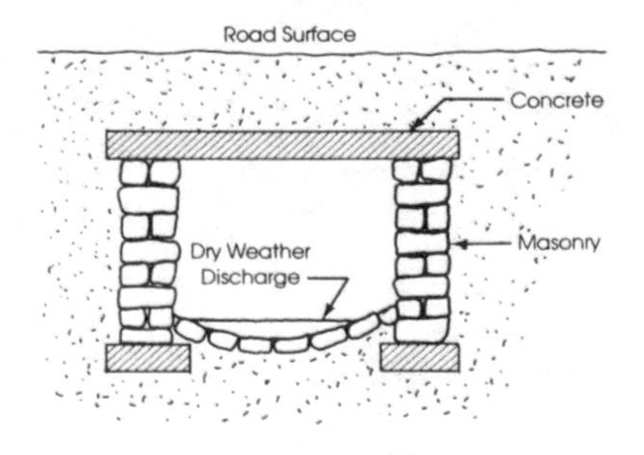

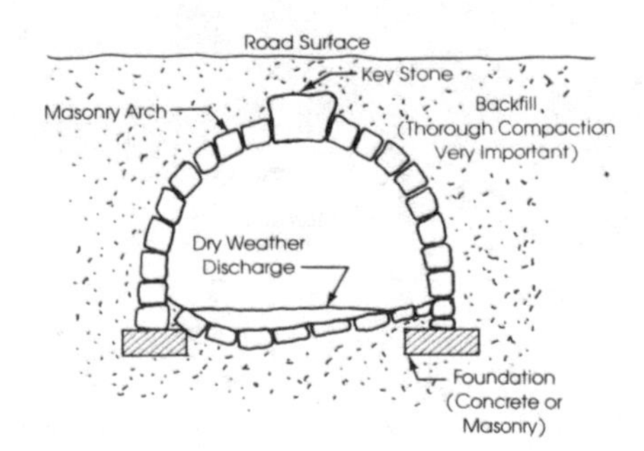

Fig. 85 *Masonry culverts*

WATERWAY PROBLEMS:

(1) Fences across waterways, fallen trees, and other debris, which slow stream flow, resulting in silting that can deflect the main channel toward wingwalls, abutments, and approach slopes, causing subsequent erosion and undercutting.

Solution: Remove obstructions from stream.

(2) Natural stream silting which causes serious stream deflection problems.

Solution: Rechannel and realign stream flow.

SUBSTRUCTURE PROBLEMS:

(1) Evidence of significant scour or undercutting (Figure 86).

Solution: Conduct engineering evaluation of crossing to determine cause and correction.

(2) Cracks and spalls in concrete and masonry structure below bearing points on abutments and piers.

Solution: Clean to fresh concrete, chip crack so that depth is twice width, and patch with 1 to 2 cement mortar.

(3) Wingwalls cracked away from main abutments.

Solution: If wingwall is not tilting or moving, is adequately serving its purpose as a retaining wall, and if it appears that the crack may be relieving water pressure behind the abutment, leave the crack alone. Tilting wingwalls, causing sliding or erosion of the approach fill, should be braced or replaced.

(4) Moisture accumulations on face of abutment or wingwall, or water seeping through cracks and joints during dry weather.

Solution: Provide outlet for water accumulated behind structure by providing a drain in back of the abutment or wingwall, or by drilling through the structure to relieve the water pressure.

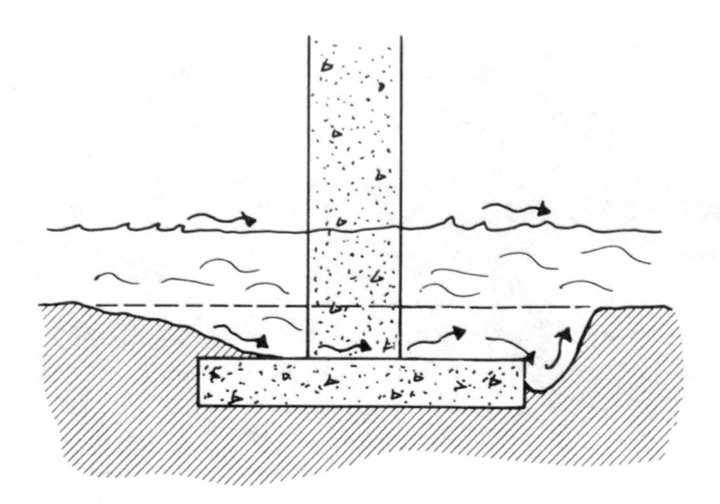

Fig. 86 *Scouring*

(5) Movement in timber pile or bent abutments.

Solution: Install deadman behind abutment and connect to additional wales with steel cable (Figure 87).

(6) Decay, or marine life attack on timber pile bents.

Solution: Add new piles or replace existing piles. If possible use treated piles.

(7) Unusual movement of pile caps under heavy loading; for instance, when trucks pass over the bridge.

Solution: Provide additional bracing.

SUPERSTRUCTURE PROBLEMS:

(1) Portions of stringers, floor beams, and pile caps no longer bearing due to either settlement, or crushing of wood.

Solution: Repair may be possible by use of heavy jacks and the addition of metal or hardwood shims (wedges).

(2) Structural steel or timber partially encased in substructure concrete deteriorating at face of exposure.

Solution: Remove dirt and debris periodically to prevent slow release of retained water which produces a prolonged rusting or deterioration period. Replace member if rusting is in an advanced state or if member is cracked.

(3) Nailing planks on top of concrete or steel stringers have slipped out of alignment.

Solution: Realign and renail nailing planks. Replace if rotted.

(4) Stringers canted, twisted, or out of alignment due to deck not being firmly anchored, or to bridge seat deterioration or damage.

Solution: Add lateral bracing, repair cause.

PILE ABUTMENT

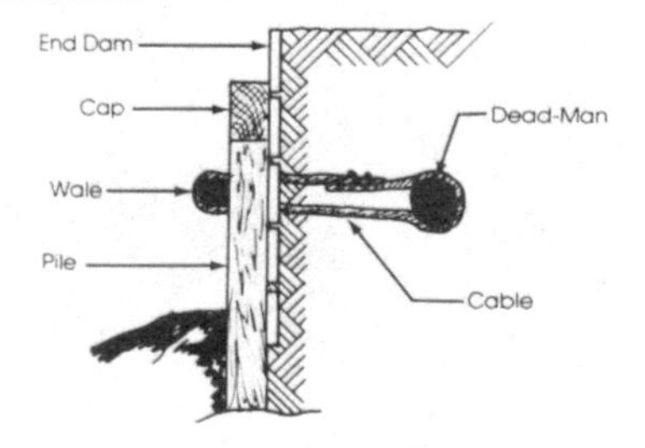

TIMBER BENT ABUTMENT

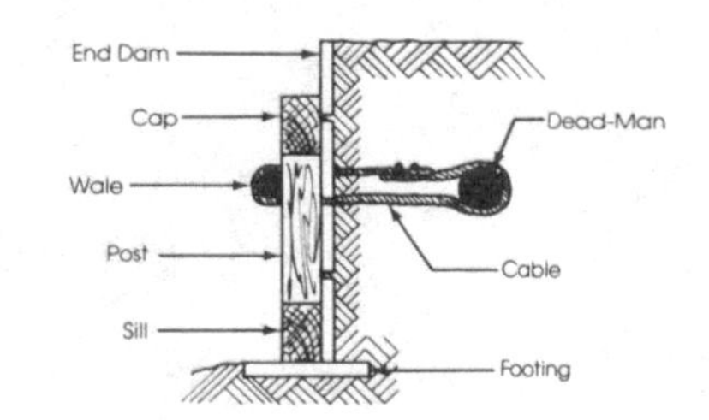

TIMBER SILL ABUTMENT

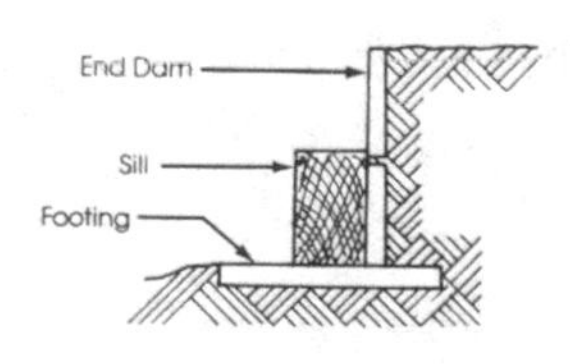

TIMBER PILE BENT

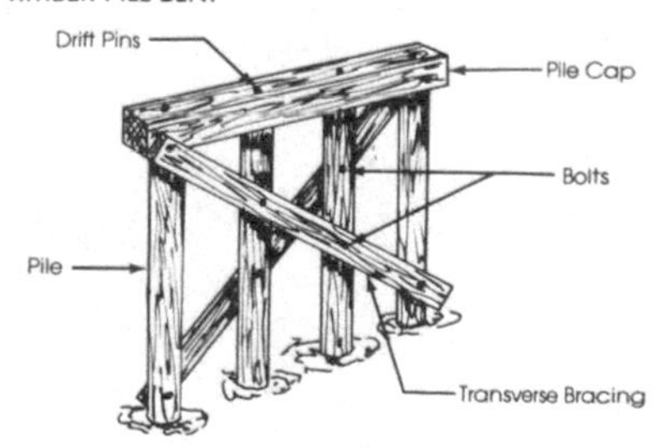

Fig. 87 *Timber abutments and piers*

Fig. 88 *Proper nails for fastening decking*

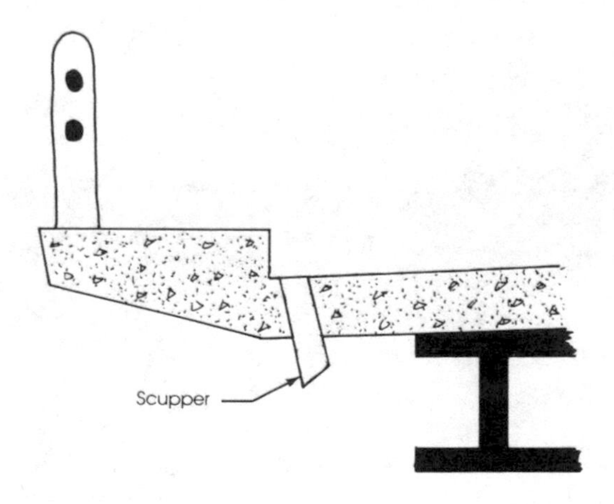

Fig. 89 *Drainage scuppers for concrete bridges*

(5) Deck planking warping upward during the dry season.

Solution: Pull spikes and drive new ones at other points where the wood is sound (Figure 88).

(6) Deck planks rotted or broken.

Solution: Replace with new planks of same thickness.

(7) Deck planks under attack by insects and dry rot.

Solution: Replace with treated planks.

Note: Treated planks do not significantly increase resistance to vehicle wear.

(8) Longitudinal timber treads deteriorate.

Solution: Same as deck planks. Tread plank ends must be firmly anchored.

(9) Timber wheel guards out of line.

Solution: Rebolt firmly in place.

(10) Water collecting on concrete deck.

Solution: Check to see if the approaches need to be cut or graded down. Drill drain holes (Figure 89) or build up low spots.

Technical Note 5.10
Pipe Culvert Deficiencies

1. Pipe culverts may function improperly because of poor design or poor construction techniques. Some culverts are both poorly designed and poorly built. The symptoms of both design and construction problems can be very similar. Therefore the actual cause of the symptoms must be determined before any corrective action is attempted.

2. This Technical Note contains three separate lists (Paras 4, 5 and 6). The

first list describes some of the more common symptoms of culvert distress. The second list indicates some of the more common causes of the various symptoms. The third list offers solutions for the various causes. The lists are followed by two matrices with a common axis. The symptoms are identified by corresponding capital letters on the left-hand side of the first matrix. The possible causes are identified across the top of the matrix by corresponding lower case letters. The field observer can therefore read across the list to the right from the observed symptom to locate the possible causes and then determine the type of investigation to be made.

3. The left-hand side of the second matrix contains a list of numerals keyed to possible solutions. Once the observer has identified the possible cause or causes, reading down and across the list to the left will determine the options which may solve the symptoms originally observed. For example, if the culvert shows signs of siltation only at its entrance, the upper matrix would be entered on line A. Possible causes are identified by dots in columns a., Blockage in culvert, and b., Invert too flat, i.e., the culvert was designed and/or built with too little slope. If no blockage is observed, column b. is followed down to the second matrix where two possible solutions for flat inverts are suggested: line 2. Increase maintenance activity, i.e. dig out the material deposit after every storm, and line 3. Excavate, realign, and recompact, i.e. dig the culvert up and relay it at the proper slope using proper construction methods. Each solution has its cost, the maintenance cost is a recurring cost while the reconstruction cost is an immediate cost. However the choice between solutions must not only be weighted by cost considerations, but also by evaluating the added risk of failure during a long or intense storm when proper maintenance may be impossible.

***4.* List of Symptoms**

A. Siltation at culvert entrance.
B. Siltation at culvert exit.
C. Siltation at both culvert ends.
D. Ponding at entrance.
E. Ponding at exit.
F. Headwalls tipped.
G. Erosion at entrance.
H. Scour at exit (Figure 90).
I. Channel protection failure.
J. Embankment washout.
K. Erosion on upstream embankment face.
L. Erosion on downstream embankment face.
M. Cracks on road surface over culvert.
N. Deep holes on surface over culvert.
O. Road surface depressed over culvert.

***5.* List of Causes**

a. Blockage in culvert.

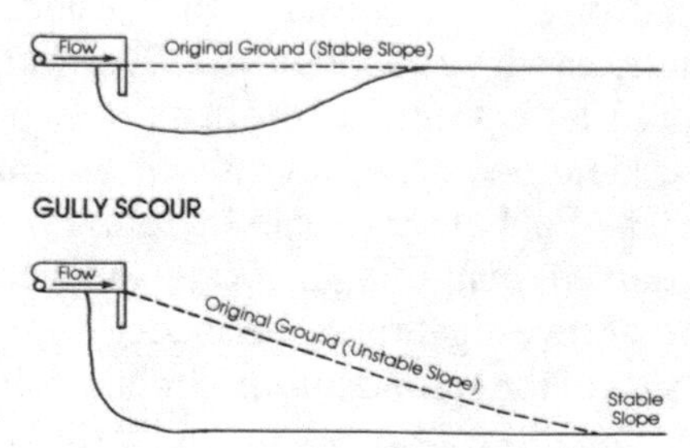

Fig. 90 *Types of scour at culvert outlets*

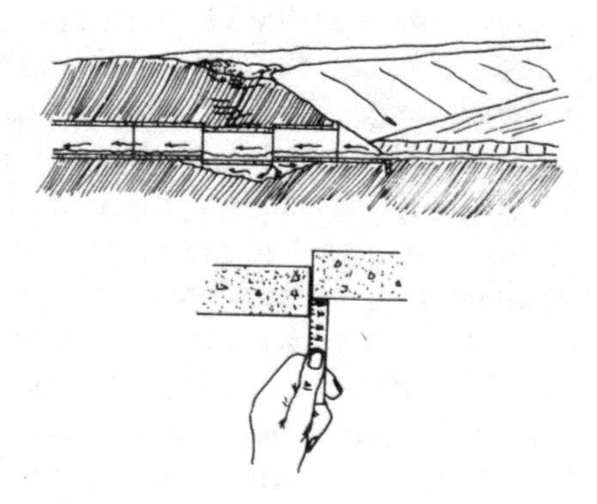

Fig. 91 *Pipe settlement causes surface defects*

b. Invert too flat.
c. Invert too steep.
d. Invert too low.
e. Invert too high.
f. Culvert too small.
g. No apron.
h. Headwall foundation too shallow.
i. Change in stream direction.
j. No headwall.
k. No wingwall.
l. Existing channel protection overtopped.
m. No cut off wall at end of apron.
n. No sand blanket under riprap.
o. Improper compaction of bedding or backfill (Figure 91).
p. Embankment being overtopped.
q. Water running off at low point in road profile.
r. Culvert broken, not enough cover.
s. Seepage due to culvert displacement.

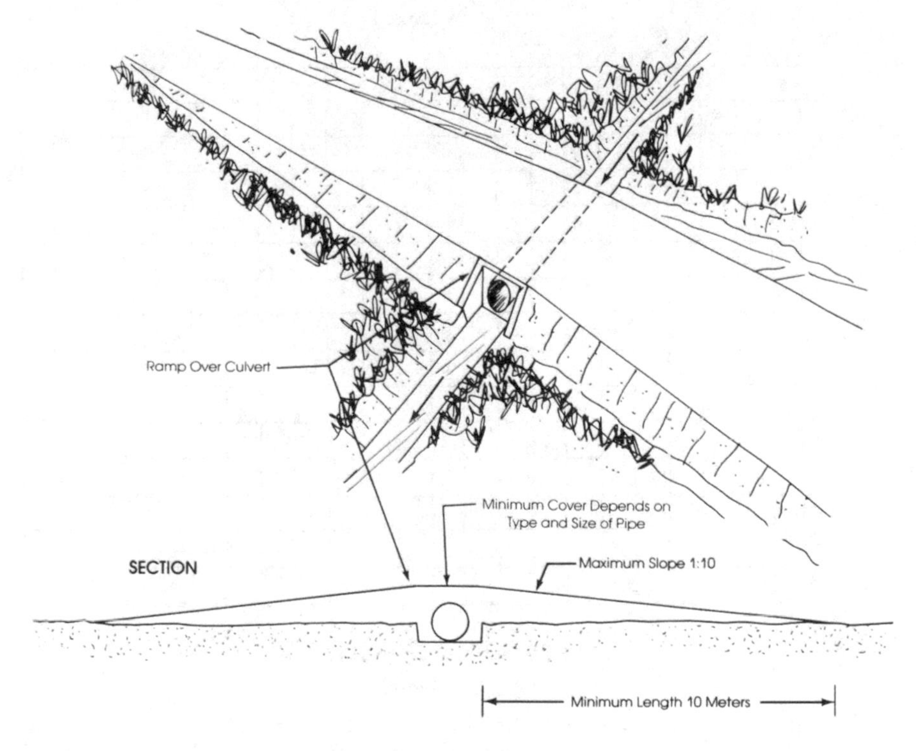

Fig. 92 *Adding cover where pipe culverts cannot be lowered*

t. Seepage due to unsealed pipe joints.
u. Culvert pipe bells pointing downstream.
v. Culvert broken, embankment too high.
w. Improper crown, water retained in wheel tracks.

***6.* List of Possible Solutions**

1. Clear obstruction.
2. Increase maintenance activity.
3. Excavate, realign, and recompact.
4. Add exit apron.
5. Add and compact material.
6. Place additional culvert.
7. Remove and replace culvert.
8. Replace with deeper foundation.
9. Install cutoff wall.
10. Install wingwall/s.
11. Provide channel protection.
12. Install headwall.
13. Remove riprap and reset on sand blanket.
14. Excavate and recompact.

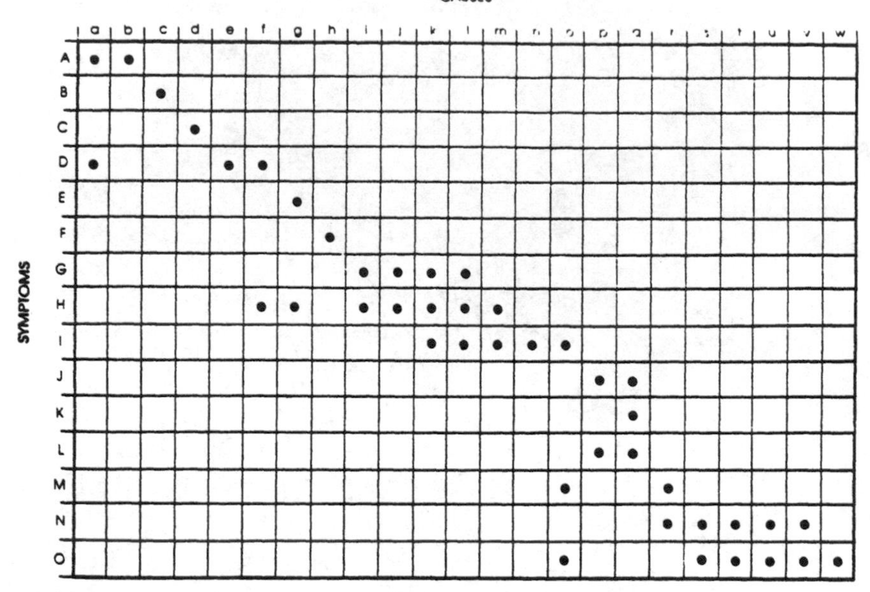

PIPE CULVERT DEFICIENCIES MATRIX

CAUSES

SYMPTOMS	a	b	c	d	e	f	g	h	i	j	k	l	m	n	o	p	q	r	s	t	u	v	w
A	●	●																					
B			●																				
C				●																			
D	●				●	●																	
E							●																
F								●															
G									●	●	●	●											
H						●	●		●	●	●	●	●										
I											●	●	●	●	●								
J																●	●						
K																	●						
L																●	●						
M															●			●					
N																		●	●	●	●	●	
O															●				●	●	●	●	●

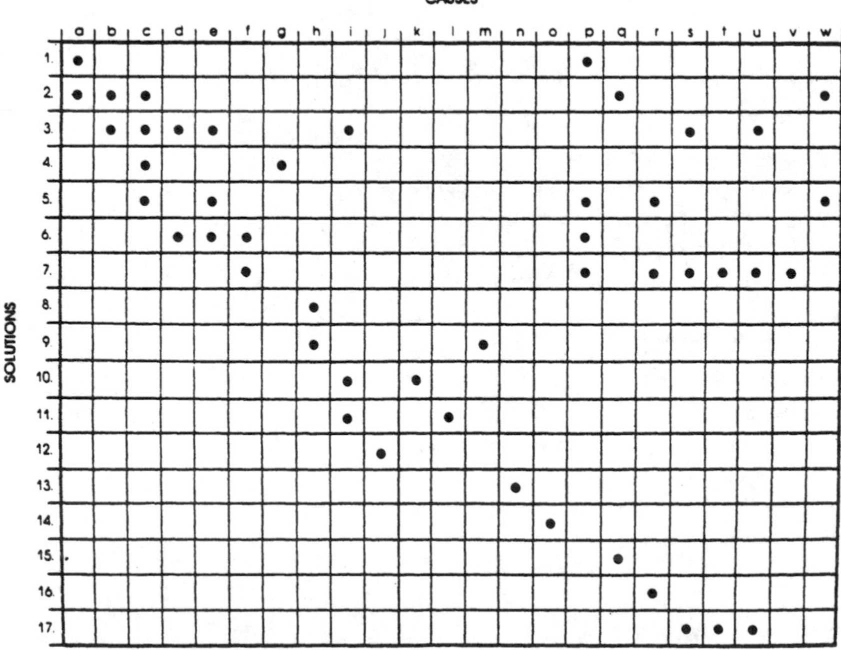

PIPE CULVERT DEFICIENCIES MATRIX

CAUSES

SOLUTIONS	a	b	c	d	e	f	g	h	i	j	k	l	m	n	o	p	q	r	s	t	u	v	w
1.	●															●							
2.	●	●	●														●						●
3.		●	●	●	●				●										●		●		
4.			●				●																
5.			●		●											●		●					●
6.				●	●	●										●							
7.						●										●		●	●	●	●	●	
8.								●															
9.								●					●										
10.									●		●												
11.									●			●											
12.										●													
13.														●									
14.															●								
15.																	●						
16.																		●					
17.																			●	●	●		

15. Rechannel surface water to sluiceway.
16. Replace broken sections and increase cover (Figure 92).
17. Reseal joints.

ANNEXE 6

Construction of Rural Access Infrastructure

This annexe contains seven technical notes:
T.N. 6.1, *Pros and Cons of Labour-Based Construction*, is referenced in Para 14.05.
T.N. 6.2, *Earthwork Considerations for Labour-Based Construction*, is referenced in Para 14.12 and Technical Note 6.5.
T.N. 6.3, *Trade-Offs between Using Force Account or Contracting*, is referenced in Para 14.26.
T.N. 6.4, *Construction Site Supervision*, is referenced in Para 14.30.
T.N. 6.5, *Indirect Compaction*, is referenced in Para 14.34.
T.N. 6.6, *Productivity Data for Compaction by Selected Equipment Units*, is referenced in Para 14.35 and Technical Note 6.2.
T.N. 6.7, *Gravel Surfacing*, is referenced in Para 14.36.

Technical Note 6.1
Pros and Cons of Labour-Based Construction

1. The advantages of labour-based construction include [44]:

(i) amount of employment created per unit of investment is potentially large;
(ii) economic conditions in rural areas are improved due to the construction workers' wages;
(iii) new individual skills are created;
(iv) pressure on foreign exchange is reduced due to lower equipment, spare parts and fuel requirements;
(v) impact of delays due to unreliable supply of spare parts and fuel are reduced; and
(vi) underutilization of equipment with associated high user costs is reduced.

2. The disadvantages of labour-based construction include:

(i) increased initial costs until labour-based experience is developed;
(ii) necessity of retraining supervisory engineers, training para-professionals or sub-professionals in low technology methods, and changing institutional attitudes towards low or intermediate technology;
(iii) increased overhead costs due to a different type of management;
(iv) increased need for lower-echelon field supervisors due to large labour

force – a gang leader is required for 20–25 men, and an overseer from headquarters is needed for 3–4 gangs [16]; and

(v) certain outputs, particularly compaction and surface smoothness, compare unfavourably with standards achieved with equipment.

Technical Note 6.2
Earthwork Considerations for Labour-Based Construction

1. Material movement breakpoints [16] for labour-based feeder road projects are considered as:

(i) shovels, 5 metres;
(ii) headloading and shoulder yokes on flat terrain, 20 to 30 metres; these haul methods are also well suited to shorter hauls with steep grades;
(iii) wheelbarrows, 150 metres, if the rise is limited to 2 metres;
(iv) animal drawn scrapers, 100 metres;
(v) animal carts, 1 km with pneumatic tyres and flat terrain; and
(vi) flat bed trucks and tractor-trailers, over 500 metres.

The rule of thumb for haul routes, that one metre of rise is equivalent to 10 metres on the level, can be used to compare the merits of various locations for borrow pits for labour-based construction.

2. Further suggestions for feeder road earthwork construction include [16]:

(i) when embankments are constructed, the initial camber for naturally compacted roadways should be at a 7% cross-slope which will become 5% after natural compaction is complete;
(ii) preferably, mechanical or towed compactors should be used (Figure 93 and Technical Note 6.6); and
(iii) the need for water for compaction can be minimized by doing earthworks soon after it rains if working conditions permit, by excavating to a vertical face to prevent excessive exposure to drying conditions, by compacting the soil as soon as possible after spreading, and by moistening the material in the pit for more even distribution and to ease the excavation of naturally compacted and dry materials.

3. The Transport and Road Research Laboratory (TRRL) reports [49] that in Kenya's labour-based road construction programme, the proper camber is formed using material excavated from standard size side ditches which is spread over a predetermined width of the already levelled road. Therefore proper camber is dependent on correct ditch cross-section and road width. If either is changed, the camber will change. TRRL states that proper camber is probably the single most important parameter of these roads in ensuring their satisfactory performance.

4. Tracks, as defined in this handbook, do not require a specific compactive

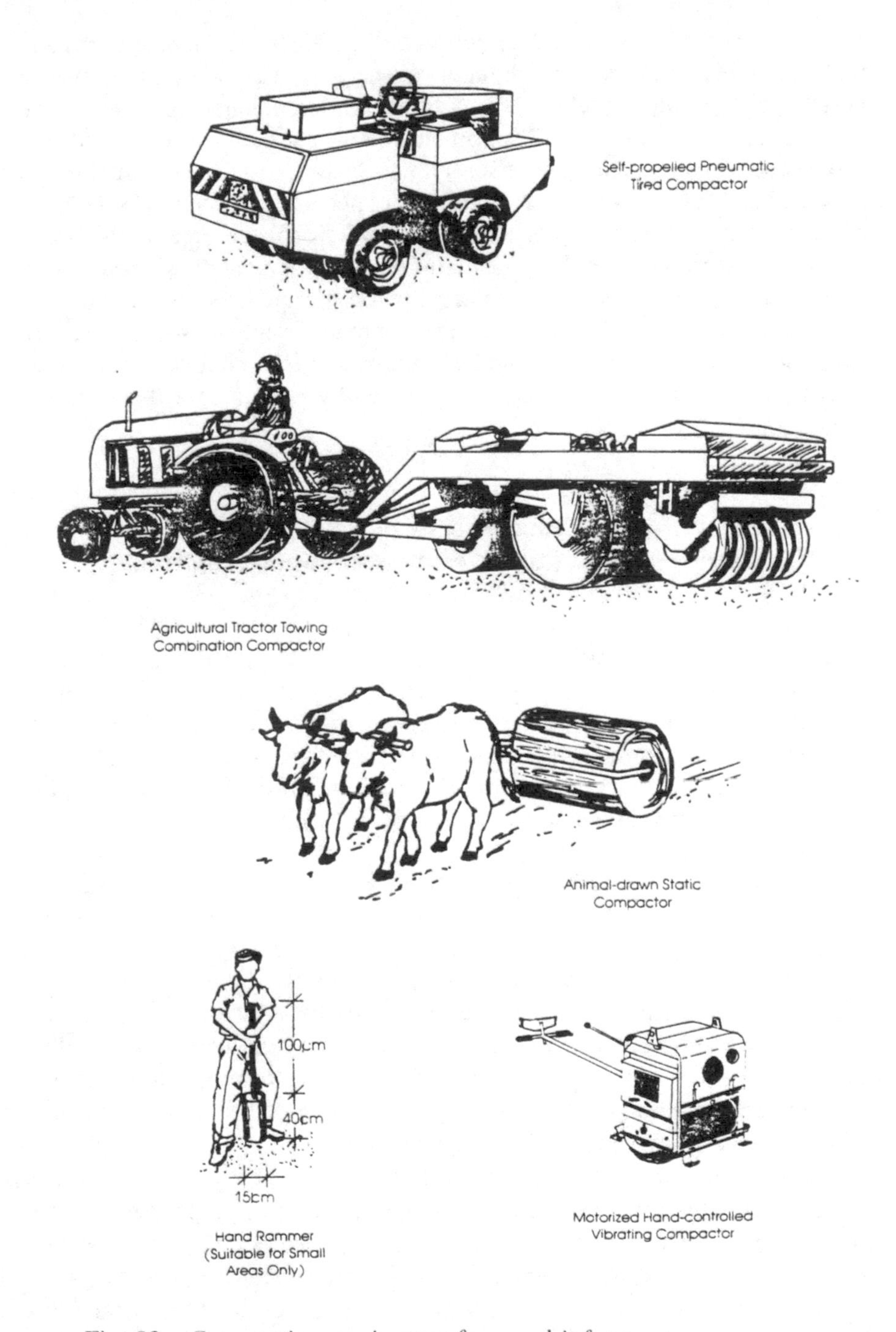

Fig. 93 *Compaction equipment for rural infrastructure*

effort, i.e. they are compacted by nature and traffic. Often simple roads and tracks are much easier to build along watersheds than in valleys. However in the type of hilly terrain where this is true, the settlements are often in the valleys near sources of water. Thus an alternative solution [41] to providing a feeder road connection linking the settlements by the most direct route through the valley is to build tracks on the ridges (a concept said to be responsible for the term highway) with connections to the individual settlements. Not only would the earthworks be reduced, a major benefit when building labour-based infrastructure, but also the drainage structure requirements would be considerably less and the costs of hauling water and compacting embankments would be eliminated. Such a ridge route alignment may provide suitable access to areas of low productivity or initial access in the early phases of a development programme.

Technical Note 6.3
Trade-Offs Between Using Force Account or Contracting

1. Force account operations can work to the *ad hoc* design standards appropriate to rural access infrastructure spot improvement construction; contracted work often requires detailed specifications and bills of quantities.
2. Departmental construction crews can adapt their work to given budget constraints; sudden budget constraints on contracted work may involve lengthy negotiations, modified specifications, changes in payment conditions or claims for compensation.
3. Departmental crews are adaptable to varying technologies and materials; most established contractors have developed expertise in equipment usage and are unlikely to opt for labour-based construction methods or ill-defined construction activities carried out on an *ad hoc* basis.
4. Capable contractors assume management responsibility, relieving government supervisors for other duties, and in principle have the necessary flexibility and incentives for carrying out construction projects efficiently; departmental forces are not financially self-supporting so they do not have to complete their work and have few incentives for cost efficiency.
5. Suitably capitalized contractors relieve the government of the necessity of investing capital funds in plant and equipment, and of the expertise required for the management, maintenance, and repair of that equipment. However, departmental forces can be organized to begin a project as soon as funds are approved, and can undertake small projects that are uneconomical for a contractor [44].
6. Contracted construction should be more expensive than force account construction by the amount of the contractor's profit. This presumes that the two organizational types are equally efficient and neglects the spur to efficiency provided by competitive tendering. In practice, prices bid by well organized, motivated and experienced contractors are usually below the true total cost of

force account construction in a competitive contracting environment. Sometimes attempts are made to bring force account organizations into the competitive field by having them submit tenders in parallel with contractors. When this is done great care must be taken to be sure that the prices submitted by force account organizations truly reflect their real costs, including for instance all overheads and sufficient funding to buy new equipment at realistic prices. In a non-competitive situation, contract costs may exceed force account costs by substantial amounts due either to poor management or excessive profit-taking. Force account costs may be excessive if management and motivation are poor, if the permanent staff and equipment resources are allowed to increase beyond those required to accomplish the projected work load, or if equipment is not properly used, maintained, and repaired due to a lack of experience, supplies, or supervision. The real cost of force account construction is not obvious and is often not known.

Technical Note 6.4
Construction Site Supervision

The site supervisor's responsibilities [16] can include:

(i) planning – setting production targets, estimating budgets, and selecting construction methods:

(ii) design – choosing alignment, selecting specific construction criteria within the agency's policy limits, determining location and types of drainage structures, selecting quarries and testing local materials; and

(iii) supervision – setting out work, instruction staff supervisors and foremen, determining labour tasks and balancing labour activities, conducting on-the-job training, maintaining quality control, monitoring output and costs, and paying the workers. The ultimate success of the project depends on the site supervisor's ability to plan resource requirements in advance and to maintain adequate stocks of funds, tools, materials, fuel, oil and spare parts for timely use at the site; an especially difficult task on small rural works projects in developing countries.

Technical Note 6.5
Indirect Compaction

1. Reference [35] contends that if a road bed is exposed to weather and traffic for a considerable time between the excavation-to-level and ditching activities described in Technical Note 6.2, Para 3, indirect compaction will occur before the final shaping with material from the ditches is undertaken.
2. Reference [35] further states:
'One very important reason why the indirect compaction method can work

lies in the organization of the labour-based maintenance system. The main strength of this system lies in the fact that there always is a man on the spot who can attend to the faults before they deteriorate further. It is evident that the condition of the rural roads after construction is much more dependent on the proper execution of the maintenance than on the standards of initial construction'.

3. The main disadvantage of indirect compaction is recognized in Reference [35] as the level of erosion during the construction period. Extra man-days are therefore required to reshape the road one or more times before the required compaction standards have been achieved.

Technical Note 6.6
Productivity Data for Compaction by Selected Equipment Units

Average output of various compaction units before output coefficients are applied. See next page.

Technical Note 6.7
Gravel Surfacing

1. Gravel surfacing requires special consideration on single-lane rural roads. Turning bays or cleared areas are needed for gravel trucks to turn around. These should be spaced at a maximum distance of 100 m intervals [34]. Key turning bay locations should coincide with planned passing bay locations to reduce passing bay construction costs. (Technical Note 4.01)

2. During construction, gravel surfacing mateial should be dumped from moving trucks rather than in piles, to reduce the number of labourers required for spreading in labour-based construction, to reduce the spreading time in equipment-based work, and to avoid segregation of the granular materials.

3. Gravelling should begin at the most distant points on each side of the gravel pit and proceed towards the pit to keep the trucks from interfering with the spreading, watering and compaction activities, except when loaded truck compaction is used to supplement other compaction efforts as frequently occurs in labour-based construction. In such cases the trucks or tractors hauling the gravel must be driven about 30 cm to the left or right of the vehicle ahead. Such techniques are very difficult to enforce since the operator's natural inclination is to follow the existing ruts [69].

4. When gravelling roads are to be maintained by labour, 3 cubic metre piles of gravel should be left at 100 m intervals for repairing potholes and ruts [34].

Technical Note 6.6
Productivity Data for Compaction by Selected Equipment Units [16]

Average output of plant before output coefficients are applied[1]

Type of plant	*Width compacted by plant (m)*	*Speed of rolling (m/min)*	*No. of passes req'd*	*Area compacted per hour (m²)*	*Depth of compacted layer (mm)*	*Output of compacted soil per hour m³)*	*Soil types and conditions for which the plant is suitable*
2.8 t smooth-wheeled roller	1.30	50	8	488	130	63	Suitable for most soil types under proper moisture conditions
8 t smooth-wheeled roller[2]	1.78	50	4	1320	150	198	Suitable for all soil types except soft wet clay and uniformly graded sand
13.5 t grid roller with 80 hp track-laying tractor	1.60	125	7	1715	200	343	Suitable for all types of soil over a wide range of moisture conditions by adjusment of ballast
13.5 t grid roller with 150 hp wheeled-tractor	1.60	250	8	3000	200	600	See remarks above but with the qualification that this combination is not suitable for uniformly graded sand or soils in very wet conditions
12 t pneumatic-tyred roller[2]	2.08	50	4	1560	130	203	Suitable for most soils under proper moisture conditions and particularly wet cohesive materials
200 kg vibrating roller hand-propelled	0.61	10	8	46	80	3.7	Suitable for granular soils only
1000 kg tandem vibrating roller	0.81	20	4	243	150	36	Suitable for granular soils only
1.7 t double vibrating roller	0.84	15	4	189	110	21	Suitable for granular soils only

Technical Note 6.6 (cont.)

Type of plant	*Width compacted by plant (m)*	*Speed of rolling (m/min)*	*No. of passes req'd*	*Area compacted per hour (m^2)*	*Depth of compacted layer (mm)*	*Output of compacted soil per hour m^3*	*Soil types and conditions for which the plant is suitable*
3.8 t towed vibrating roller	1.83	40	6	730	250	180	Suitable for all soil types except uniformly graded soils
7.7 t self-propelled vibrating roller	1.83	80	6	1460	150	220	Suitable for all soil types except uniformly graded sand
12 t towed vibrating roller	2.08	40	3	1660	300	498	Suitable for all soil types. (Vibrating device must be turned off on wet clay)
200 kg vibrating plate compactor	0.38	10	3	76	150	11.4	Suitable for granular soils only
660 kg vibrating plate compactor	0.61	15	4	137	200	27	Suitable for most soil types
2.0 t vibrating-plate compactor	0.86	10	2	258	300	77	Suitable for most soil types except heavy clay
100 kg power rammer	Area 0.05 m^2	60 blows/min	6 blows	30	150	4.5	Suitable for the reinstatement of trenches and compaction in confined areas of all types of soils
610 kg frog rammer	Area 0.43 m^2	50 blows/min	12 blows	108	300	32	Suitable for the reinstatement of large trenches of all soil types
600 kg dropping weight compactor	Area 0.093 m^2	25 blows/min	2 blows	70	600	42	Suitable for trench reinstatement with most types of soils and wide range of moisture conditions
55 kg vibro-tamper	0.28	5	3	28	100	2.8	Suitable for the reinstatement of trenches and compaction in confined areas of all soil types
100 kg vibro-tamper	0.40	8	3	64	200	13	Suitable for the reinstatement of large trenches and compaction in confined areas of all soil types

[1] See Chapter 17 for discussion of output coefficients.
[2] Towed combination rollers with steel wheels and pneumatic tyres are available.

ANNEXE 7

Maintenance

This annexe contains seven technical notes:
T.N. 7.1, *Maintenance Technology Considerations*, is referenced in Paras 7.19 and 15.09.
T.N. 7.2, *Lengthman Maintenance Contract System*, is referenced in Paras 7.08 and 15.23 and Technical Notes 7.4, 7.6 and 8.1.
T.N. 7.3, *Effect of Corrugations on Soil Surface Maintenance*, is referenced in Paras 7.08, 10.08 and 15.28.
T.N. 7.4, *Lengthman Output According to Various Sources*, is referenced in Para 15.28 and Technical Note 8.1.
T.N. 7.5, *Hand Tools for Road Maintenance*, is referenced in Para 15.28.
T.N. 7.6, *Maintenance Programme Checklist*, is referenced in Paras 7.07 and 15.33.
T.N. 7.7, *Reducing Maintenance Constraints*, is referenced in Paras 7.11 and 15.34.

Technical Note 7.1
Maintenance Technology Considerations

Technical considerations affecting maintenance technology include:

(1) ditch configuration – hand-dug side ditches with steep side slopes and narrow (one shovel width) flat inverts are impossible to clean and maintain with common road maintenance machinery, as are grassed or paved ditches or ditches containing ditch checks. The typical 'V' ditch is easily cleaned and reshaped by motor graders while the shoulders are being maintained; but the labour-based maintenance of 'V' ditches requires considerable labour input because of the surface area involved [75];

(ii) surface width – within limits, narrowness minimizes the amount of maintenance required on all rural transport infrastructure. However, maintenance of a 5 metre wide surface requires the same number of motor grader passes as a 6 metre width, while labour-based maintenance of a 6 metre wide surface requires at least 20% more labour than 5 metre wide surface. A correct continuous camber is also more difficult for labourers to maintain on wider roads. If the gravelled portion of the road surface is narrow (4 metres or less), a motor grader may contaminate the gravel with inferior material from the shoulders [34];

(iii) surface profile – granular surfaced roads constructed by labour-based methods are often uneven; motor grader maintenance will move the granular surfacing into the low areas causing uneven thickness and perhaps even exposing the subgrade material [34];

(iv) surface smoothness – labour-based maintenance cannot maintain the same degree of smoothness as equipment in locations where a smooth suface is of prime importance [34]; and

(v) type of infrastructure – paths, trails and tracks that are constructed by labour are frequently too narrow to be maintained by equipment [88].

Technical Note 7.2
Lengthman Maintenance Contract System

1. A review of the lengthman maintenance contracting system used in Kenya [22] indicates the following advantages:

(i) it is inexpensive, the annual cost of routine maintenance in Kenya is on the order of 2.5% of the construction cost;

(ii) the skills developed in the construction of the roads built using labour-based construction methodology can be utilized in their maintenance, thus limiting the need for training;

(iii) no mechanized equipment is involved, therefore spare parts and skilled labour are not problems; timely payment and close supervision become the critical factors in ensuring proper maintenance;

(iv) each contractor has his own section of road which means that he takes pride in it, and the local people know on whom to exert pressure if the road is not being maintained;

(v) administration is simplified by having a contract which specifies both the output expected and the payment for services. Payment is based on 12 days of effective work per month regardless of the actual time the lengthman works. If the maintenance is inadequate, payment can be withheld; and

(vi) the part-time involvement of the lengthman adds income to farming areas where there is under-employment in farming activities.

2. The Kenyan system is well suited for implementation by local authorities lacking the sophisticated support services required of equipment-based maintenance. The contract document, reproduced here, is equally simple and clear.

Road Maintenance Contract Form [16]

Department of Public Works

Name ....
ID Certificate No: ....
Dear Sir,
ROAD MAINTENANCE CONTRACT

1. I am directed by the Chief Engineer (Roads) to offer you a contract to provide services for the maintenance of .... from km ....

2. The terms and conditions of the contract are as follows:
   (1) You will be paid for providing services at the rate of .... per day for .... working days per month.
   (2) You are expected to carry out the following road maintenance works:
      (a) Fill and compact potholes and ruts. Ensure surface is kept in good order.
      (b) Return gravel from shoulders and slopes back to the carriageway.
      (c) Keep drains and culverts clean and free running.
      (d) Cut grass and bush on shoulders, ditch, verges and slopes.
      (e) Report major damage to the Maintenance Inspector at ....
   (3) You will be provided with the tools required for the work. You are personally responsible for their safe keeping and should any loss occur you will be financially responsible for their replacement.
   (4) The works will be inspected monthly and should it be apparent that the work is not being carried out in proper manner then payment shall be withheld until such time as the Maintenance Inspector is satisfied that the work has been properly carried out.
   (5) The contract is automatically cancelled should either party fail to carry out their obligations under the contract.

3. If you wish to accept the contract on the terms and conditions set out above you should sign the acceptance below.

Yours faithfully,

Maintenance Inspector
for: *CHIEF ENGINEER (ROADS)*

Acceptance:
I accept the contract on the terms and conditions set out in the above letter.

Date .... Signature ....

Technical Note 7.3
Effect of Corrugations on Soil Surface Maintenance

1. At very low volumes, less than ten vehicles per day (vpd), environmental conditions such as erosive rainfall and wheel rutting exert as much influence on the road surface as corrugation formation. Therefore, labour-based routine maintenance automatically deals with corrugations at the same time as it fills potholes, retrieves surface material, and smooths the cross-slope.

2. At ten vpd or more, supplemental dragging (Figure 94) and periodic passes with a grader provide positive economic returns. To increase the effectiveness of grading, it should be done immediately following the rainy season when there is still moisture in the upper layers [71]. The International Labour Organization (ILO) [34], suggests that a road should be regraded after approximately 1500 vehicle passes. Dragging frequencies depend on the type of drag (Figure 95), traffic density, and soil type. ILO suggests the following rule of thumb for tractor-towed drags (Figure 96):

(i) heavy drags (steel joists – two times per week at 100 vpd (350 vehicles passes); every 10 days at 50 vpd (500 vehicle passes); and
(ii) very light drags (discarded tyres) only for roads carrying up to 20 vpd.

3. Regular dragging reduces the need for regarding [33, 71, 93]. Drags are most effective on loose friable surfaces where they can fill in ruts and smooth out surface irregularities. On rocky soils and hard crusted surfaces they are less effective since they cannot remove corrugations once they are formed [77, 85]. Dragging is most effective when there is a small amount of moisture in the soil [77]. Consequently, it is least effective in arid regions or during rainy seasons. It is commonly assumed [85] that drags will not restore camber or lost material. Tractors must be driven at 5 to 10 km per hour, depending on the material, or they will ride over irregularities without correcting them and raise large quantities of dust [85].

4. An evaluation [71] of the economics of maintenance in Kenya, suggests that, given a standard level of routine maintenance and 30 rubber tyre draggings per year, the motor grading frequency for lateritic surfaces should be in the range of once for approximately 7000 vehicle passes; or once every 140 days – approximately five months – for a 50 vpd road; and once per year for 20 vpd. The vehicle passes per grading for sandy clay surfaces are assumed to be approximately 5000, while earth roads are assumed to require regrading after approximately 3000 passes. These are only order of magnitide values, and vary from the ILO recommendations in Para 2. Regrading is more effective and long lasting if the regraded surface is recompacted, with the addition of water if necessary, at the same time.

5. Using the assumptions in paragraph 4 [71] and allowing for economic losses for road closures, the approximate break-even economic criteria in Kenya on the roads studied, at the costs in effect at the time of the study (1975), indicated that lateritic gravel could economically be added on 5 to 6 metre

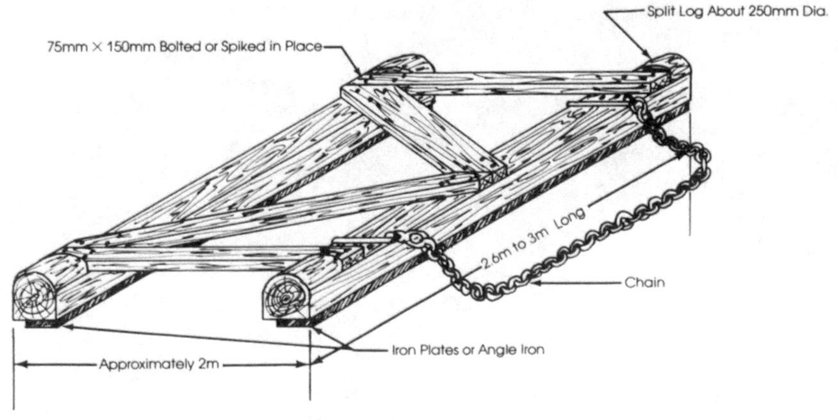

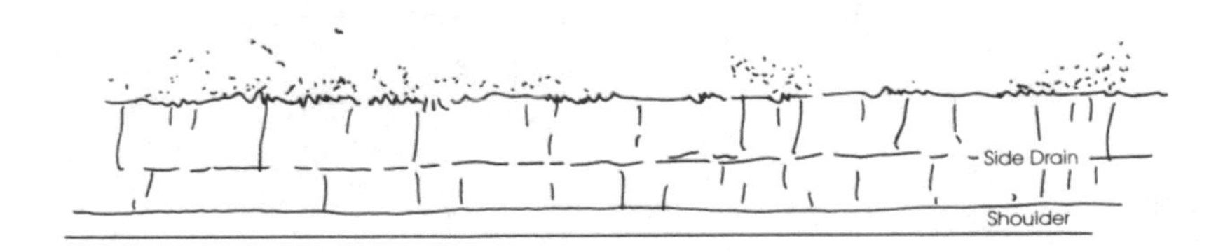

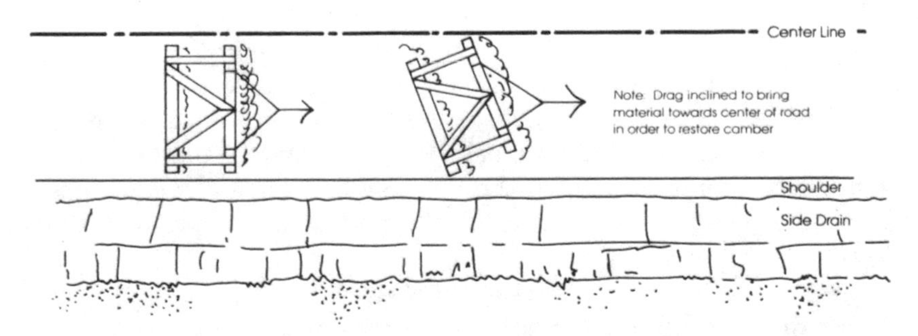

Fig. 94 *Locally made drag and its use*

roads carrying 40 to 60 vehicles per day. The break-even criteria is the base year ADT at which the present value costs of gravelling plus maintenance equals the present value of the maintenance costs of the ungravelled road during its expected life. The above values are presented, not as represented traffic volumes at which gravelling should be undertaken, but to illustrate that where local conditions make gravelling very expensive and/or where financial stringencies imply high opportunity costs of capital, gravelling of lower volume roads is likely to be an uneconomical [71] maintenance investment.

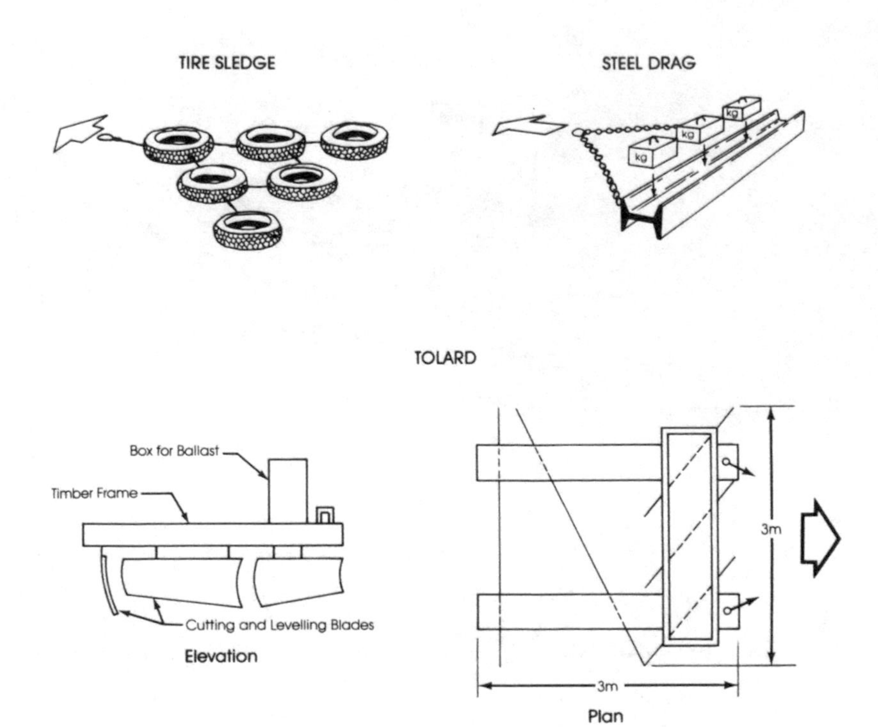

Fig. 95 *Types of drag*

6. In maintenance terminology, normal grading of gravel roads implies no wetting or compaction is required; while heavy grading includes reshaping, wetting and compaction, because the road condition has deteriorated to a corrugated crust which must be broken before the road surface is reformed [71]. ILO [34] suggests that three passes of either an 8 to 10 ton smooth wheel roller or a 2 ton vibrating roller should provide adequate compaction. Other sources [51, 77] suggest rubber tyred rollers are suitable compaction vehicles.

Technical Note 7.4
Lengthman Output According to Various Sources

1. The International Labour Organization [34] indicates that one lengthman can maintain one to two km of road with a formation width of 5.5 m when the traffic is less than 20 vehicle a day by working half-time (100 hours per month).
2. The Transportation Research Board [75] states that in 1980 the Kenya Rural Access Roads Program described in Technical Note 7.2, using lengthmen who were former road construction workers, allocated sections of from 0.5 to 2.0 km to each lengthman, the average length per contractor being 1.6 km. Each section could theoretically be maintained by working twelve days per month (100 hrs). Traffic is under ten vehicles per day.

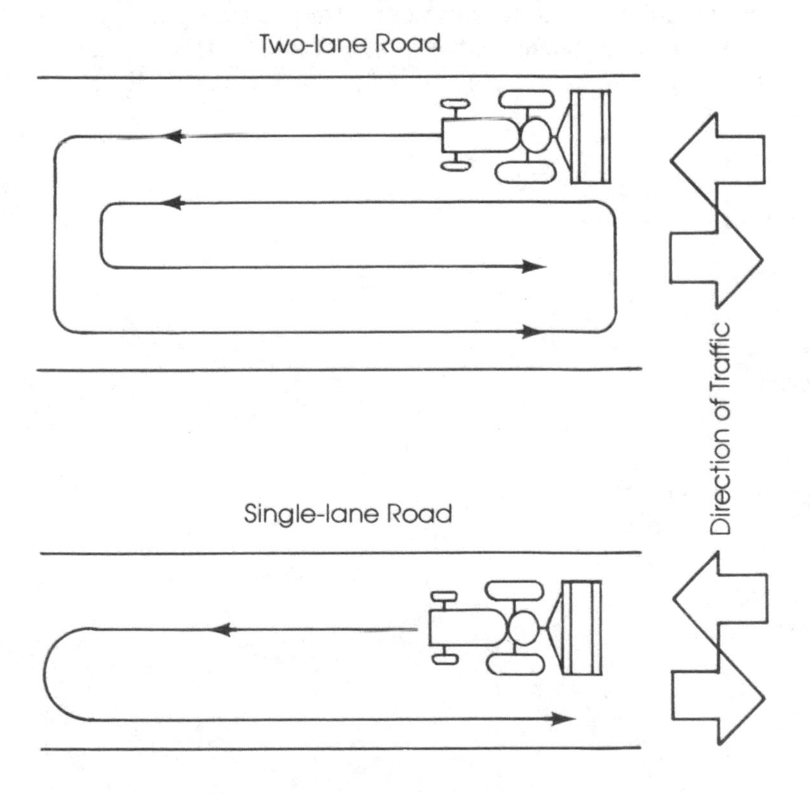

Fig. 96 *Dragging procedure with a tractor*

3. Coukis [16] suggests that the amount of work needed to keep a length of road in good condition depends on several factors: the type of road surface; the number and size of the vehicles using the road; the severity of climatic conditions, especially rainfall; the type of soils; the susceptibility to erosion of the terrain and the road gradients; and the presence of grass and bush vegetation. However, in average conditions a full-time worker should be able to do the routine maintenance needed each year on about 2 km of single-lane earth or gravel road with traffic of some ten to twenty vehicles per day.
4. The Economic and Social Commission for Asia and the Pacific [85] estimates that one man can maintain 1.5 km of an earth or gravel surfaced road up to 4 m wide. In hilly sections, the length must be reduced to 1.0–1.2 km.

Technical Note 7.5
Hand Tools for Road Maintenance

In order to maintain a road section, each worker requires sufficient tools and light equipment to carry out the work. These tools are normally issued to the worker who signs a receipt and assumes responsibility for them. The tools should be marked to prevent substitution and inspected regularly (at least once per month). The following individual worker's tool list [16, 34] and estimated good quality tool life, when used only for maintenance [34], is fairly well standardized:

Tools and Light Equipment (Fig. 97)	*Estimated Life* (months)
hoe	9
shovel	12
rake	12
pickaxe	36
hand-rammer	36
scythe	4
bushknife (machete)	9
long-handled shovel (for cleaning culverts)	12
sharpening file	12
wheelbarrow	24

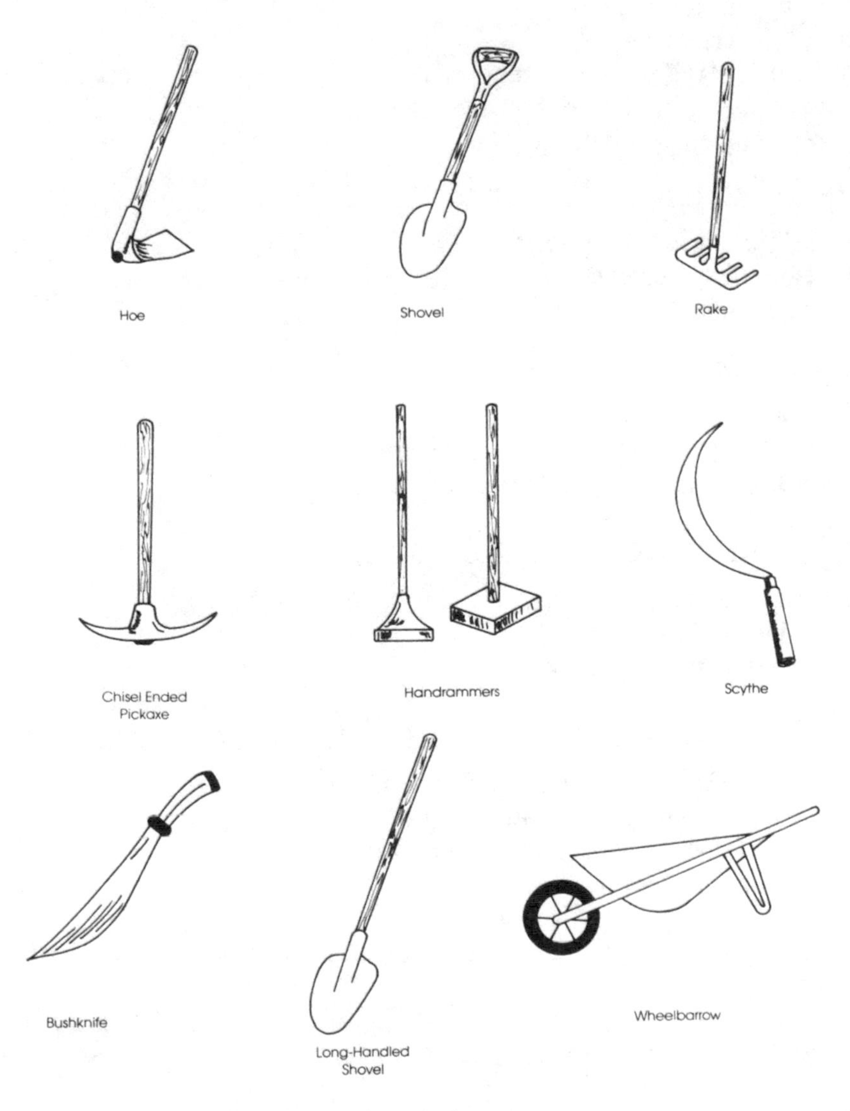

Fig. 97 *Hand tools for road maintenance*

Technical Note 7.6
Maintenance Programme Checklist

1. MAINTENANCE NEEDS AND TASKS

Routine Maintenance

(a) filling potholes
(b) maintaining surface camber

(c) removing corrugations
(d) cutting vegetation
(e) cleaning drainage structures
(f) cleaning/repairing side drainage
(g) repairing erosion damage

Periodic Maintenance

(h) reshaping and raising formation
(i) regravelling gravelled surfaces
(j) repairing drainage structures

Emergency Maintenance

(k) repairing/removal of slides
(l) rebuilding erosion protection measures
(m) repairing large erosion-damaged areas

II. MAINTENANCE TECHNOLOGY

(a) labour-based
(b) equipment-based
(c) intermediate technology-based

III. INSTRUMENTS FOR IMPLEMENTING MAINTENANCE

(a) contract documents for regular contractors
(b) contract documents for petty contractors
(c) contract documents for lengthman (Technical Note 7.2)
(d) cost control procedures (Chapter 17)
(e) maintenance guide for foremen
(f) training material for training of local government technicians, foremen and labour

IV. MAINTENANCE ORGANIZATIONS

Responsibility

(a) central or state government
(b) local government
(c) local communities
(d) state or private enterprises

Method

(e) force account using own forces only:
- by central/state government
- by local government

(f) force amount assisted by:
- contractors
- local communities

(g) contract using:
   - regular contractors
   - lengthmen
   - local communities
   - state or private enterprises
(h) on a self-help basis.

V. RESOURCES FOR MAINTENANCE

Resource Provider

(a) communities
(b) local government
(c) central/state government
(d) state or private enterprises

Mobilization of Local Resources

(a) self-help (labour, materials, funds)
(b) local taxes on:
   - market stands, shops
   - users of roads, water, irrigation
   - persons/heads of families/males, etc.
   - animal ownership
   - land ownership
   - increased land value
   - land transfer and sale

Technical Note 7.7
Reducing Maintenance Constraints

Existing or Potential Constraints

Potential Method of Overcoming Constraints	Local governments do not have skills to undertake maintenance	Local communities do not possess capability to undertake maintenance	Maintenance needs and related tasks have not been adequately defined	Maintenance needs appear excessive	Maintenance responsibility has not been properly defined	Local communities do not identify themselves as main beneficiaries	Maintenance programme is excessive in relation to available resources	Maintenance technique not balanced in relation to local resources	Maintenance tasks are not being adequately executed	Local resource basis is inadequate	Local resource mobilization is inadequate	Central government allocating insufficient resources	Maintenance workers have inadequate skills	Maintenance resources diverted to other purposes
1. Carry out infrastructure condition inventory			X	X										
2. Carry out spot improvements to reduce localized high maintenance costs				X			X			X		X		
3. Carry out functional classification of infrastructure links					X							X		
4. Construct infrastructure for needs of beneficiaries and give them maintenance responsibilities					X	X				X	X			
5. Select a maintenance programme which is in balance with available resources							X	X						
6. Use locally available resources to maximum extent possible								X						
7. Use local contractors including lengthmen								X						
8. Assist local communities/local government by providing handtools, non-local materials, equipment services, etc.								X		X	X			
9. Motivate beneficiaries to maintain by involving them in planning and construction					X	X						X		
10. Establish and monitor formal agreements					X				X					X

Technical Note 7.7 (cont.)

	Local governments do not have skills to undertake maintenance	Local communities do not possess capability to undertake maintenance	Maintenance needs and related tasks have not been adequately defined	Maintenance needs appear excessive	Maintenance responsibility has not been properly defined	Local communities do not identify themselves as main beneficiaries	Maintenance programme is excessive in relation to available resources	Maintenance technique not balanced in relation to local resources	Maintenance tasks are not being adequately executed	Local resource basis is inadequate	Local resource mobilization is inadequate	Central government allocating insufficient resources	Maintenance workers have inadequate skills	Maintenance resources diverted to other purposes
11. Decentralize maintenance responsibility and link it to responsibility for providing the required resources					X				X	X				
12. Establish procedures for local resource mobilization		X					X				X	X		
13. Establish simple contracts for local use	X				X				X					
14. Establish, implement and monitor simple cost control systems	X	X							X					X
15. Use same labour for construction and subsequent maintenance	X	X				X								
16. Train local maintenance workers	X	X							X				X	
17. Improve infrastructures so labour-based maintenance methods are feasible	X	X												
18. Carry out demonstration or pilot project	X	X	X	X		X	X		X	X		X	X	

ANNEXE 8

Institutional Issues

This annexe contains one technical note:

T.N. 8.1, *Organizational Framework for Decentralized Road Maintenance*, is referenced in Paras 7.02, 7.05 and 16.15.

Technical Note 8.1
Organizational Framework for Decentralized Road Maintenance

1. The organizational framework of highway authorities or public works departments can take many forms in developing countries. Small nations such as the Gambia may elect to retain much of the direction and control of operations at national headquarters level. Large countries with widely dispersed road systems, e.g. Brazil, usually develop a decentralized organization that features more delegation of authority and responsibility to district units. In some developing countries, decentralized district highway organizations are in effect small highway departments representing the central government in nearly all highway matters within a geographical area. Their functional responsibilities include planning, design, construction, maintenance, equipment, materials testing and administration. Commonly, their lines of authority and accountability are to a headquarters chief engineer for field operations. The other national headquarters functional divisions principally provide staff guidance and assistance to the districts through the chief engineer for field operations.

2. In some of the lesser developed countries, decentralization may take a somewhat different form. Contract construction projects are not common, except for major improvement work undertaken with foreign funding. The construction, inspection and supervision for those major inprovement works usually are the responsibility of consultants or the construction division at the national level. The district field offices seldom are involved to any great extent. This situation leaves the field units concerned principally with maintenance operations and possibly some minor force account construction. In this capacity, the decentralized district offices can be done under the immediate direction of the headquarters maintenance engineer. The headquarters equipment involvement (Para 11), normally under a separate division answering to the director of highways, can become a separate organizational unit within the maintenance office, responsible to the maintenance engineer. Most developing countries will find that this maintenance-oriented organizational framework

best serves their immediate needs. As development continues, responsibility at the district level will increase and bring about a need for more decentralization.

3. Since decentralized maintenance-oriented field districts can function under a single institutional division at national level, i.e. the headquarters maintenance unit, this form of decentralization lends itself to independent use by other sectorial agencies such as agriculture, irrigation or forestry when they are delegated rural access infrastruture maintenance responsibility under large development programmes. The operational activities described below can apply to maintenance activities of these agencies as well as to maintenance-oriented field districts operating under a national road authority. However, if the road authority will eventually assume responsibility for the maintenance of feeder roads constructed or improved by other sectorial agencies, there must be close cooperation between the agencies when maintenance policies and procedures are being developed:

4. Regardless of the sectorial agency undertaking infrastructure maintenance activities, a typical maintenance organization usually includes

(i) a headquarters unit;
(ii) regional or district units, and possibly sub-district units;
(iii) field workforce units; and
(iv) units responsible for servicing, repair and management of equipment.

A clear understanding of the relationships among these units is essential to ensure effective maintenance operations. Maintenance crews are seldom assigned to a national headquarters unit unless there is need for highly specialized technical crews such as a pavement restriping unit to serve nationwide. However, in a decentralized maintenance-oriented organization, the headquarters maintenance engineer may exercise direct authority over the district organization. He or she normally reports directly to the agency's executive officer or to a principal assistant.

5. Regardless of the extent of responsibilities of the headquarters maintenance unit, which may vary with the degree of decentralization employed, the objectives and responsibilities of an agency's national maintenance unit include:

(i) developing maintenance policies and procedures for review and approval by top management and for national implementation;
(ii) developing uniform work performance standards and instruction manuals for field units;
(iii) developing annual maintenance work programmes or reviewing, adjusting, and approving annual programmes prepared and submitted by field units;
(iv) preparing and coordinating nationwide maintenance budget requests;
(v) coordinating resource allocations – manpower, equipment, materials and funds – among the major regional or district units in accordance with the requirements of approved work programmes;

(vi) guiding and assisting units in effective planning, scheduling, performance, and control of maintenance programmes;
(vii) reviewing field conditions and nationwide levels of maintenance service being provided, with accountability to top management for accomplishing approved work programmes; and
(viii) providing effective training for field personnel.

6. Districts are established to provide a means for more effective maintenance operation management and supervision. District boundaries may be influenced by terrain, rivers and other geography features. Road authority district limits are influenced by the concentration of highways and population centre locations as well; each district is usually responsible for about 500 to 1000 km of roads. Each district is headed by a district engineer who reports to the maintenance engineer at headquarters. Field maintenance operations are usually under the direction of a district maintenance engineer or superintendent. A district equipment supervisor is responsible for repairs and the district's equipment preventative maintenace programme. When large areas are involved, sub-districts may be established within the district, each headed by a resident engineer with one or more maintenance supervisors responsible for maintenance within the sub-district area. The number of maintenance supervisors is dictated by the number of maintenance crews since each maintenance supervisor can supervise three or four crews, each headed by a foreman. The sub-districts also have a small units for simple routine servicing of equipment.

7. The principal responsibilities at the district level include:

(i) coordinating with the headquarters maintenance organization in developing annual maintenance programmes of specific work to be accomplished within the district and sub-districts;
(ii) defining the manpower, equipment, and materials resources need within the area to accomplish the work set forth in the programme;
(iii) allocating those resources among the sub-districts and field units in accordance with their individual workloads;
(iv) assisting field units in effective scheduling of work activities in accordance with approved programme objectives;
(v) guiding, assisting and training field supervisors in proper work methods and procedures in conjunction with Para 5 (viii);
(vi) inspecting road conditions and maintenance workmanship, and directing corrective actions as needed;
(vii) implementing standard work reporting systems and assuring reporting accuracy so that performance can be evaluated by headquarters on a regional and nationwide basis; and
(viii) administering and operating an equipment repair and overhaul facility and providing skilled mechanics to service equipment in the field.

8. Some maintenance work requires special knowledge and capabilities, as well as special equipment. Bridge maintenance, sign maintenance, major

regraveling and crusher plan operation are examples of this special type of work. Specialized crews are commonly established under the district maintenance engineer or superintendent to perform this work on a district-wide basis as needed. The remainder of the field work forces are usually organized in one of the following ways depending on the size of the area and the density of the road system:

(i) Maintenance Depots – small areas with a dense infrastructure network usually have a depot with office space, storage facilities and rudimentary equipment servicing facilities. The workers report to the depot daily and are transported to and from their worksite;

(ii) Road Section Crews – Sections from 15 or 20 km up to 50 km are established and staffed by a foreman and 10 to 20 crew members. All crew members usually live in close proximity to the road section and report to the foreman daily. The work is usually labour-based routine maintenance using hand tools. There is little mechanization except for basic transport, sometimes limited to bicycles. When maintenance equipment is required, it is scheduled out of a district or sub-district office;

(iii) Lengthmen – Individual workers contract for part time maintenance work on short sections of road (Technical Notes 7.2 and 7.4). They work on a results orientated basis on a road section adjacent to their home. Their work is checked at regular intervals of not more than one month by a travelling supervisor from the district.

(iv) Mobile Crews – In remote, very sparcely populated areas, where none of the above organizational structures is practical, some agencies resort to travelling crews which are reasonably well mechanized and are provided with mobile camp facilities and a equipment servicing and repair capability. Because of the nature of this approach, there may be long periods of time between maintenance work on individual road lengths. This results in road deterioration that is often so severe that the work is very extensive, often approaching rehabilitation rather than maintenance.

9. Regardless of the organizational placement of the equipment function within an agency, the following actions and responsibilities are necessary:

(i) determining the types and numbers of equipment needed to carry out the maintenance programme most effectively and economically;

(ii) preparing specifications and bidding documents, and accepting and evaluating bids for equipment purchase;

(iii) managing the equipment fleet, including practices for allocation among field units, keeping records of utilization and costs, and determining and implementing policies on replacement and disposal of obsolete equipment;

(iv) carrying out an equipment servicing and repair programme to assure a minimum down time and a maximum service life; and

(v) assuring an adequate continuing inventory of commonly needed spare parts and supplies.

10. Many highway agencies establish equipment organizational units that are entirely separate from the maintenance organization both at headquarters and the field level. The equipment unit is headed by a mechanical engineer who has responsibility to manage the equipment fleet so as to serve effectively all equipment users within the agency. In maintenance oriented agencies, the maintenance organization is normally the largest user of highway equipment so the equipment function can be combined with the maintenance organization to provide better coordination for carrying out maintenance programmes. The equipment engineer reports to the maintenance engineer in such cases.
11. The headquarters maintenance or equipment unit has the following equipment responsibilities:

(i) procuring new equipment;
(ii) establishing operational policies and procedures for equipment management;
(iii) directing and supervising a central repair shop that is responsible for major overhaul and repairs;
(iv) coordinating the acquisition, warehousing and distribution of spare parts and supplies;
(v) establishing policies, procedures and practices for field workshops and equipment repair and servicing; and
(vi) developing manuals of procedures and training for equipment operated by field personnel.

12. The field organization for the equipment function consists principally of equipment workshops, warehousing and parts and supplies, and facilities for field servicing of equipment. Major workshop facilities are usually established at regional or district headquarters with smaller service facilities at selected locations throughout the area. Service trucks are often used for servicing and minor repairs of equipment at remote locations. In some instances, the headquarters equipment group exercises direct line supervision over the field equipment units. Most often, the field equipment units are responsible to the regional or district engineer, with the headquarters equipment unit serving as support staff.

ANNEXE 9

Cost Evaluation and Control

This annexe contains ten technical notes:

T.N. 9.01, *Direct Costs*, is referenced in Paras 17.05 and 17.13.
T.N. 9.02, *Derivation of Different Equipment Cost Elements*, is referenced in Paras 6.15 and 17.12, and Technical Notes 9.01 and 9.06.
T.N. 9.03, *Forecast of the Effective Labour Force and Working Days for a Project*, is referenced in Para 17.13 and Technical Note 9.01.
T.N. 9.04, *Examples of Camp Requirements*, is referenced in Para 17.14.
T.N. 9.05, *Sample Calculation of Requirements for Daily Transport of Labour to a Work Site*, is referenced in Para 17.15.
T.N. 9.06, *Evaluation of Equipment Utilization*, if referenced in Para 6.15 and 17.19.
T.N. 9.07, *Animal Productivity*, is referenced in Para 17.22 and Technical Note 2.3.
T.N. 9.08, *Outline of a Specific Work Accomplishment*, is referenced in Para 17.25 and Technical Note 9.09.
T.N. 9.09, *Cost Estimating Form*, is referenced in Para 17.25.
T.N. 9.10, *Maintenance Performance Budgeting*, is referenced in Para 17.27.

Technical Note 9.01
Direct Costs

A detailed list of direct costs should include:

(i) materials – the actual cost of all consumable construction items supplied and placed at the work site like cement, reinforcing steel, aggregate, sand, gravel, rock, gabions, culverts, timber, drill bits, welding rods and fasteners, including all transport charges other than those covered under project equipment charges;
(ii) tools – the actual, i.e., *pro rata* cost of all hand tools such as hoes, rakes, shovels, picks, bush knives and wheelbarrows used on the project;
(iii) labour – the actual cost of production workers based on wages for the number of man-days worked plus, as applicable, paid leave, holidays, overtime, bonuses, food, subsistence allowances, and social charges like payroll tax, insurance and pension fund;
(iv) equipment – the actual hourly use cost or rental charge for each piece of equipment multiplied by the actual usage time devoted to the project.

Equipment costs and productivity are a major cause of accounting problems and are discussed in more depth in Technical Notes 9.02 and 9.03;

(v) site overhead – the actual cost of all expenditures associated with the specific construction or maintenance site but not directly applicable to specific unit production costs, and including all expatriate site staff and their attendant costs, local professional staff, local adminstrative/support staff, ancillary staff, site buildings and land, office, survey and laboratory supplies, postage, telephone, medical supplies, site transport and attendant costs, and construction equipment mobilization costs;

(vi) compensation for land used as right-of-way or materials sources, if such compensation is paid; and

(vii) new infrastructure maintenance activities until the specific construction project is completed.

Technical Note 9.02
Derivation of Different Equipment Cost Elements [100]

1. **Depreciation and interest** charges are derived from the delivered price of the machine to the central depot and assumptions involving total hours of machine life, average number of hours of utilization per year and the interest rate. Machine life is generally specified by the manufacturers for different operating condition assumptions. It should be stressed, however, that the manufacturers' recommendations should only serve as guides, which must be revised based on actual experience of the concerned construction authority (see Para 10).

2. The annual number of days during which equipment is available for work is derived from the number of days of construction season available per year (allowing for planned work stoppage because of vacations, etc.) and making an allowance for: (i) days of work lost because of weather problems during the construction season; (ii) holidays; (iii) expected number of days required for routine maintenance and repairs during the construction season; and (iv) days when no suitable work can be scheduled. Annual utilization hours are derived by multiplying the equipment availability per year by the average number of hours per day the machine is likely to be used. Obviously, each of these factors will be different for different construction authorities and can only be based on their past experience. However, as a rough rule, 2000 hours per year is the absolute maximum number of hours for which a machine is likely to be available for a single-shift operation. In developing countries, 600–1500 hours is perhaps a more likely range.

3. The interest rate may not be easily determine because equipment is generally purchased from government budgetary allocations rather than by borrowing in the open market. In economic terms, interest rates reflect the

value of the investment tied up in alternative uses. It may be specified by the central authorities or it can be estimated to be the prevailing bank rate for long-term borrowings. As a rough guide, a range of 10–15% is perhaps appropriate.

4. Once the life of the equipment, annual usage rates and the interest rate have been estimated, a number of standard procedures are available for calculating the hourly depreciation and interest rates. One of the simplest procedures is the Capital Recovery Factor (CRF) method which combines the derivation of depreciation and interest in one number. This method calculates the depreciation and interest cost as a uniform annual cost over the life of the machine. By dividing the annual cost by the annual hours of utilization, a uniform hourly cost of depreciation and interest can be obtained, which is calculated as:

$$\text{Hourly cost of Depreciation \& Interest} = \frac{(\text{Delivered Price} - \text{Tyre Cost})^{1} \times \text{Capital Recovery Factor}}{\text{Annual Hours of Utilization}}$$

The CRF can be calculated from the interest rate and the machine life in years or can be simply read off from the standard compounding and discount tables given in most engineering economics textbooks (Para 10).

5. **Insurance and taxes** in a contractor's evaluation, are either expressed as a percentage of the value of the machine or may be fixed for a given type of machine irrespective of its age. In the latter case, hourly cost of insurance, and taxes, is obtained by dividing the annual charge by the annual hours or utilization. In the first case, however, these charges are derived by multiplying the average value of the machine over its life by the combined tax and insurance rates.

6. **Maintenance costs.** Currently employed costing procedures adopt an ad hoc percentage of the purchase cost of a machine in assessing the hourly maintenance cost of the equipment. For example, the Caterpillar handbook recommends calculating hourly maintenance and cost including spare parts and maintenance labour cost as:

$$\frac{\text{Repair Factor} \times (\text{Delivered Price} - \text{Tyres})}{1{,}000}$$

The repair factors recommended by Caterpillar for different types of equipment under different operating conditions are given in their handbook. Of course, these factors should be adjusted based on the construction authorities'

[1] Tyres are considered a wear item on wheeled machines so their costs are calculated independently (Para. 8). On larger machines tyre costs are substantial and tyre life varies considerably, depending on usage and soil conditions.

own operating experience. However, in developing countries it is unlikely that total maintenance costs during a machine's life will be less than the delivered price of the machine. It should also be noted that the *current* delivered price of the machine should be used to calculate the maintenance reserve, particularly in an inflationary period (Para 10).

7. **Running costs.** Fuel, lubricant and filter consumption can be estimated from previous site experience or based on equipment manufacturers' recommendations. The running costs can then be calculated using current prices of fuel and oil. Frequently the cost of fuel, filters and lubricants is included in an hourly rental charge, but the preferred procedure is to account for these separately since these prices tend to vary from one location to another and, moreover, a better control can be maintained on their consumption if they are explicitly accounted for. For planning purposes the authority may maintain an estimate of fuel and lubricant consumption for different types of equipment (Para 10). Similarly, labour costs for routine on-site maintenance can be separately accounted for in the overhead labour account rather than in equipment rental charges.

8. **Tyre cost** is simply derived by dividing the tyre cost by the expected tyre life in hours. Expected life of tyres is generally available from tyre manufacturers, but again the authority's own operating experience (if available) should be used to supplement this information (Para 10).

9. In summary, hourly usage cost for equipment is derived as follows:

Hourly use cost = Hourly Depreciation and Interest Charge
+ Hourly Cost of Insurance, and where applicable, Taxes
+Hourly Maintenance and Repair Reserve
+Hourly Tyre Cost

Fuel and lubricant costs should be accounted for separately and are calculated from the consumption in litres – or kg – per hour of machine operation.

10. Some rules of thumb are available [16] for estimating factors in the above derivations. It must be stressed *again*, however, that applying factors supplied by manufacturers and books such as this serve only as a guide which must be revised based on actual experience. The following commonly assumed factors are listed by their relevant paragraph numbers above:

(1) total machine life (hours): 10,000 for scrapers, dozers, graders, wheeled loaders, trucks, agricultural tractors, tipping trailers, tampfoot rollers, crushers and bitumen distributors; 25,000 for steel wheel rollers; 5000 for wheelbarows;
(2) average hours of utilization per year: 1000 for motorized scrapers, graders, wheeled loaders and trucks; and 1250 for dozers, towed scrapers, rollers, tractors, trailers and crushers. Major differences are frequently

encountered in utilization hours, usually resulting in usage below these values, so very careful analyses are required;

(4) the capital recovery factor is equal to
$i(1+i)^n/[1+i)^n-1]$
where i is the interest rate expressed as a fraction and n is the number of years;

(6) the repair factor is estimated so as to give a total maintenance cost during machine life of about 100% of depreciable cost for heavy equipment. 0.10 can be assumed for motorized wheeled vehicles, 0.05 for towed equipment, and 0.12 for track vehicles;

(7) rules of thumb for fuel and oil consumption include: diesel, 20 litres per machine horsepower per month for single-shift work;
lubricating oil, 0.5 litres per machine horsepower per month for single-shift work;
petrol, 500 litres per month for each four-wheel-drive vehicle, and 300 litres per month for each two-wheel-drive vehicle (based on 2500 km of travel per month); and

(8) average tyre life (hours): 1250 for wheelbarrows; 1300 for wheeled loaders; 1600 for trucks, tractors, and trailers; 2000 for motorized scrapers; and 3000 for towed scrapers and graders.

Technical Note 9.03
Forecast of the Effective Labour Force and Working Days for a Project

Assumptions

a.	Length of working season	40 weeks
b.	Length of working week	6 days
c.	Number of official and local holidays	12 days
d.	Losses forecast owing to bad weather[1]	15 days
e.	Losses forecast owing to labour disputes[1]	3 days
f.	% nominal labour force absent during first four and last two weeks	50%
g.	% nominal labour force absent during harvest	75%
h.	Length of harvest season	4 weeks
i.	% nominal labour force absent owing to labour turnover, illness, and absenteeism	15%
j.	Losses owing to training	0

Forecast of Available Days

k.	Total time (a) × (b) − (c)	228 days
l.	Lost time (d) + (e)	18 days
m.	Available time (effective working days)	210 days

[1] Days lost through bad weather and labour disputes are normally counted as being lost time, therefore part of total time.

Forecast Of Effective Labour Force Factor[2]

n.	Recruitment/run-down factor (50/100 × 6/40)	0.075
o.	Harvest factor (75/100 × 4/40)	0.075
p.	Turnover/absentee factor (15/100 × 40/40)	0.150
q.	Overall reduction factor (n) + (o) + (p)	0.30
r.	Effective labour force factor	0.70

In the example above, a nominal force of 400 labourers would produce an effective force of 280 persons. The supervisory and support staff requirements might include the following personnel: one engineer or supervisor, one office manager, one pay and accounts clerk, one storekeeper, one assistant storekeeper if work is all at one location, at least one overseer for each 100 labourers or a minimum of four overseers, at least one gang leader for each 25 labourers or a minimum of 16 gang leaders, and ten to fifteen skilled labourers such as mechanics, drivers, carpenters, blacksmiths and stonemasons. On the average day, one overseer and four gang leaders could be spared with no theoretical reduction in output; therefore, these supervisors represent an increased overhead for labour-based projects.

Technical Note 9.04
Examples of Camp Requirements

If migrant labourers have to be employed, a floor area of 5 square metres per person in a dormitory should be allowed. Experience in Chad [16] indicates housing for permanent staff should include 10 to 15 m^2 per person for single men, while family dwellings should have at least 50 to 60 m^2, with small, enclosed courtyards in the rear for washing and cooking. A construction unit employing 300 labourers needs a site office with a gross floor area of about 150 m^2. Foodstuffs and cement require well-ventilated buildings to protect them from dampness. A storekeeper on site needs about 10 m^2 of floor area for his stores, in addition to his living quarters. One flatbed truck is usually adequate for transporting tools and materials for up to 500 workers. If water is required, a 5000 litre trailer can be used to accommodate every 200 workers with 25 litres each per day. If firewood is not plentiful locally, project equipment must be provided to transport it. When a local market is nearby, the rest day should coincide with market day so that the labourers can purchase their needs; otherwise labourers must be transported periodically to the marketplace or a store must be provided.

[2] The effective labour force factor is the proportion of the nominal labour force that can be expected to be present in the effective labour force. In this forecast, the effective labour force = 0.70 × the nominal labour force.

Technical Note 9.05
Sample Calculation of Requirements for Daily Transport of Labour to a Work Site [16]

Calculations are based on the following assumptions:

Labourers to be transported	500
Labour wage	$1.00/day
Flatbed truck, 10-ton, rate including fuel, driver etc.	$10.00/hr.
Average travel speed	30 km/hr.
Loading, waiting time per trip	10 min.
Truck capacity	50 men
Number or trucks per labour unit of 500 men	5

	Average distance per trip to pick up labourers[1]	
	5 km	*20 km*
Trip time, out and return (hr)	0.5	1.5
Number of truck-trips per day (morning and evening)[2]	20	20
Labour transport input (veh-hr/day)[3]	10	30
Labour transport cost per day	$100	$300
Total truck input (hr/day)[4]	40	40
Total labour wage cost per day[5]	$500	$500
Per cent of total truck input spent on labour transport	25%	75%
Cost of labour transport as per cent addition to wage bill	20%	60%

[1] Distances are one way.
[2] Number of labourers to be transported divided by the capacity of a truck and multiplied by the number of trips made by each worker per day (assumed to be two, one in the morning and one in the evening).
[3] Number of truck trips × trip time.
[4] Number of trucks × hours per workday (assumed to be eight).
[5] Number of labourers × daily wage per labourer.

Under the above assumptions, each truck must make four trips a day, therefore the second truckload of labourers reaches the work site 50 minutes after the workday begins and begins to load 20 minutes before the workday ends if the distance is 5 km. Each labourer therefore loses 70 minutes, resulting in a total daily loss of 580 man hours. When the trip is 20 km, the daily loss is 1580 man-hours if all transport takes place during work hours. By phasing the work hours for the two labour teams, these losses can be reduced to 330 hours and 830 hours respectively, but the trucks must be available more than eight hours per day.

Technical Note 9.06
Evaluation of Equipment Utilization

Table 23. Sample Calculation of the Cost of Using Large and Small Compactors [16]

	Compaction (200 cu. m/day)	
	Self-propelled 10 ton smooth-wheeled roller	*Pedestrian operated 450 kg vibrating roller*
Resource units required	1	3
Hire rate ($/hr)	5.90	3 × 1.40 = 4.20
Time to carry out (hr.)	1.0	each 6.7
Direct unit cost[1]/assuming full utilization ($/cu.m)[2]	0.029	0.14
Daily cost of resources ($)[2]	47.20	3 × 14 × 8 = 33.60
Utilization (%)	12	84
Actual unit cost ($/cu.m)	0.24	0.17

Note: Working day is assumed to be eight hours.
[1] Direct unit cost = (hire rate) × (time to carry out daily task)/(production).
[2] Only direct cost is needed for cost comparison (that is, overhead costs are ignored as being roughly equal for both methods).

1. This table neglects the varying time-dependent and user-dependent costs. Rather than altering the assumed usage factor implicit in assigned equipment costs, it converts underutilization into an additional efficiency factor. Consequently, no fuel savings are attributed to the 88% of the time the roller is not working when the actual utilization cost per cubic metre is determined.
2. Table 24 shows sample calculations for determining the costs of a 10 ton steel-wheel roller at full utilization, i.e., the expected available time per year is 1250 hours. It compares these costs to the same steel-wheel roller at an expected available time of 625 hours. Technical Note 9.02, Para 2, indicates this is the correct method for altering equipment availability time to allow for time when no suitable work can be scheduled. The time-dependent costs in Table 24 increase from $3.78 to $6.08 per usage hour because the interest costs in the example over-ride the reduced yearly depreciation. If inflation costs were included, i.e., if the charges were calculated to allow enough reserve funds to accumulate to replace the equipment at the end of its economic life, the time-dependent costs would further increase.
3. Using the information developed for full utilization in Table 24, the unit cost calculations in Table 23 can be recomputed. The full utilization cost, when modified for only one hour per day of work, indicates time-dependent charges totalling $3.78 × 8 = $30.24, and usage-dependent charges, i.e., fuel and oil, of $2.12. This results in a total daily cost of $32.36, which represents an actual utilization unit cost per cubic metre of $0.16 instead of $0.24.
4. When half utilization costs are considered (only four hours per day utilization are anticipated) the time-dependent charges become $6.08 per hour

Table 24: Sample Calculation of the Capital Equipment Direct Rate for a Self-Propelled 10-ton Steel-Wheeled Roller

Assumptions			
Initial cost, delivered (C)		$21,000	
Expected economic life (L)		25,000 hrs	
Expected salvage value (S)		0	
Expected available time per year (A)[1]	1250 hrs		625 hrs
Interest rate, expressed as a decimal (i)		0.12/yr	
Insurance/tax rat as % of current value		5%/yr	
Repair factor (R)		0.05	
Fuel consumption		6.8 litre/hr	
Fuel cost		$0.297/litre	
Oil consumption		0.1 litre/hr	
Oil cost		$0.80/litre	
Calculations			
a. Depreciable cost (D = C − S)		$21,000	
Operating life (N = L/A)	20 years		40 years
Capital Recovery Factor (CRF)	0.134		0.121
Deprediaton charge (CRF × D/A)	$2.25		$4.07
b. Multiplier factor to find average value of machine over N years		0.57	
Insurance/tax charge (0.05 × 0.57 × C/A)	$0.48/hr		$0.96/hr
c. Maintenance charge (R × D/1000)		$1.05	
d. Time-dependent charge [sum of (a)+(b)+(c)]	$3.78/hr		$6.08/hr
e. Fuel charge		$2.02/hr	
f. Oil charge		$0.10/hr	
g. Usage-dependent charge [(e) +(f)]		$2.12/hr	
h. Direct rate [(d)+(g)]	$5.90/hr		$8.20/hr

[1] Available time is cut in half to assume 50% usage.

while the fuel consumption costs remain the same. The actual utilization cost becomes $6.08 × 4 + $2.12 = $26.44. The actual unit cost becomes $0.13 per cubic metre. This evaluation is a hypothetical example to show the impact of treating underutilization as a decrease in efficiency, and to illustrate the calculation methods used to evaluate various utilization costs. It is not an endorsement for underutilizing heavy equipment.

Technical Note 9.07
Animal Productivity

1. Average tractive forces for animals are assumed to be directly proportioned to their body weight. However, the efficiency of teams is less than the work capacity of single animals. Two animals pulling together produce 0.92 the

tractive power of two single animals. The Team Efficiency Factors are: three animals, 0.85; four animals 0.78; five animals 0.70; and six animals 0.63 [101]. Other factors influencing the tractive force of animals include: their species and breed; health; age; climatic conditions; training; harnessing and shoeing; soil type; and route condition.

2. The load a draught animal can haul in a cart is a function of the pulling power of the animal and the coefficient of rolling resistance – the force required to keep the vehicle moving divided by the gross weight of the cart [16]. Average tractive forces of animals as a portion of body weight vary by species, e.g., donkeys 1/6 to 1/4; horse 1/10 to 1/8; and oxen 1/10 to 1/7 [101]. Indicative coefficients of rolling friction are shown in the following table.

Table 25. Coefficients of Rolling Resistance [16]

Description of the haul-route	*Type of surface*	*Ease of haulage of non-rubber-tyred carts or barrows*	*Ease of haulage of pneumatic-tyred vehicles or barrows*	*Typical value of coefficient of rolling reisistance*[1]
Very poor	Earth, muddy, no maintenance/ loose sand or gravel	Virtually impossible	Very difficult	0.15–0.40
Poor	Poorly maintained earth (dry clay loam)	Very difficult	Fairly difficult	0.1
Average	Normal site earth road or track	Fairly difficult	Slightly difficult	0.07
Good	Hard and smooth-compacted earth, well maintained	Not difficult	Not difficult	0.05
Excellent	Concrete/ asphalt/well-maintained gravel/well-laid barrow run	Free running	Free running	0.03

[1] These values are indicative only – the parameter is defined by the vehicle speed attainable.

3. The climate, especially high temperatures, affects animal work output. Oxen can be worked for eight hours a day in temperate climates, but in tropical conditions three to six hours are not usually exceeded and these hours are staggered so as to avoid the hottest part of the day [101]. Carabaos must be given time throughout the day to wallow in water.

4. Draught animals deliver maximum peformance only when the harness (yoke, collar or breastband) is properly fitted and provides a broad, smooth surface. Test evidence suggests that bovine animals can deliver 25–50% more

horsepower when harnessed in a breastband or collar rather than a yoke, because of the lower point of draft, increased harness comfort and larger area against which to push [90]. However, draught animals must be trained from the beginning in the type of harness they will use at work. Changing harness type on working animals may not necessarily lead to better results.

5. Pack animals are used to carry material in sacks or panniers. Donkeys and ponies carry loads of 100–200 kg in panniers which are loaded by shovel with the animal standing (loading height about 1.4 m). The usual haul distance for donkeys and ponies is 50–250 m. Camels can carry 350–450 kg in panniers which are loaded by shovel with the animal kneeling (a loading height of 0.8–1.0 m). Haul distance is usually 100–500 m. Donkey panniers are unloaded by sliding the unit off the animal's back, while camel panniers have a lacing arrangement which is undone to discharge the material while the animal is standing [16]. The loads indicated in this paragraph are greater than the loadings indicated in Technical Note 2.2, because of the short-haul distances involved. Technical Note 2.2 values are indicative of loads transported over long distances when animals are used as traditional transport aids.

6. Working animals should have access to water at least three times a day, morning, noon and evening. Horses and some cattle engaged in heavy work may need short drinks every two or three hours. Heated animals should never be allowed free access to water. Water requirements in litres per day for various draught animals include: oxen: 10–30 during the rainy season, 15–40 in the dry season; horses 30–50; donkeys, 10–20; and mules 15–30 [90].

7. The following table 26 [101] notes some of the characteristics and productivity of animals. Bovines, as listed in the table, include animals with the following generic terms: ox; any bovine animal – a large, usually horned, cloven-footed ruminant; bull, uncastrated male of ox; bullock, castrated bull; cow, female ox, buffalo, particular kinds of once-wild oxen; cattle, oxen.

Technical Note 9.08
Outline of a Specific Work Accomplishment [3]

1. Task Description – Placing Granular Surfacing

This task takes place after the earthwork cut and embankment are completed and the riverbed[1] or pit material has been deposited in piles along the roadside. The piles are loosened and spread on the nearby road surface area or are loaded into wheelbarrows and transported and dumped on areas further away. The material is spread to the proper depth and grade and then compacted.

2. Materials

Gravel and sand (graded mix) from riverbed or borrow pits.

3. Crew size, composition and tooling – for transport over 5 metres.

When a manual roller is available for compaction, the compactors pull the roller instead of using tampers.

[1] Technical Note 4.03 indicates that riverbed material is rarely ideal for road surfacing. However, this text was published in the Philippines and represents their viewpoint concerning suitable granular material for surfacing.

Table 26. Characteristics and Productivity of Animals

	Pack ponies and horses	*Heavy horses (for draught)*	*Light horses[1] (for draught)*	*Mules*	*Donkeys*	*Bovines*	*Camels*	*Elephants*
Weight (kg)	200–550	680–1200	400–700	200–600	150–300	300–900[2]	450–650	1500–6000[2]
Amount carried[3]	100–150	—	—	75–150	70–120	—	120–680	200–600
Distance covered in 1 day (km)	30–40	20–40	20–40	25–30	20–30	10–20	30–65	20–30
Tractive force (N)	—	500–1200	450–780	500–600[4]	290–450	390–780	[5]	[6]
Speed, when loaded (km/hr)	3–4.5	2.5–4.5	3–5	3–4.5	2.5–3	2–3.5	3–5	6–7
Hours work per day (hr)	8 or more	6–10	6–10	8 or more	[7]	3–8[8]	9–13	3–5

Notes:

[1] Ponies are sometimes used for draught purposes but few figures have been found.

[2] Females are much smaller than males of the same species.

[3] The lower end of the range would be relevant to load-carrying capacity when the load is carried continuously for a full day. The upper end of the range would be relevant to intermittent load-carrying during the day (e.g., for earthwork operations when the return journey is made unloaded).

[4] These figures relate to mules of weight 350–500 kg.

[5] A pair of camels on a dry road can draw a load of 1300–1800 kg.

[6] Can drag logs of up to 4 tons weight over short distances but 2 tons is more usual. 2 elephants can pull a loaded cart of 6 tons a distance of 20 km per day at a speed of 4 km/hr.

[7] For draught work: 3–4 hrs. For pack work: 8–10 hrs.

[8] 5 hrs., is the average for bullocks in tropical climates.

When a mechanical roller is available for compaction, no labourers are needed for compaction.

Crew: 1 foreman
12 labourers

Tools: 2 picks, 3 shovels,
3 wheelbarrows, 3 rakes,
5 tampers, 2 buckets
(includes spares)

Activity	Labourers	Tools
Loosen-Load	3	Pick, shovels
Transport	2	Wheelbarrows
Spread	2	Rakes
Compact	5	Tampers

Expected output: 200 square metres per day

4. Method of execution

The heaps of material dumped by trucks are loosened and thrown directly on the base course surface with a shovel or loaded and transported with wheelbarrows. They are then spread in 15 cm thick uniform layers and compacted. If farm tractors and trailers are used for transport, the materials are unloaded in two windrows, spread and compacted.

Water should be added to the materials to be compacted when needed, to ensure proper compaction. The quantity of water to be added depends on the 'dryness' of the soil – it can vary from zero to 5–10% of the weight of the soil (approx. 5–10 litres per full wheelbarrow) and will be determined by the engineer. The water should be well mixed with the soil before compaction starts.

The compaction can be done with tampers or mechanically with rollers (self-propelled or pulled manually). Several blows of the tamper are needed at each point before moving to the next one.

5. Quality control

All compacted surfaces should show no visible depression when a loaded 5 cubic metre truck passes over the road. The final road surface must have a 5% cross-slope[2] (from the centre towards the edge of the road). This can be checked with stakes placed at the centre (crown) and at the edge of the road, strings being attached to each stake. This is repeated every 5 metres. The difference in height between centre and edge stake indicates the degree of cross-slope. Since the cross-slope is 5% and the width of one-half of road is 2.5

[2] Modified to be consistent with the previous text.

m, this difference in height should be 12–13 cm (24–26 cm in curves). A string level is used to establish a horizontal line.[3]

6. Quantity measurement

Quantity of surface is measured in square metres of compacted surface over the base course.

[3] Figure 15 shows a Camber Board, which is used to determine both cross slope and surface smoothness. Its use is more efficient than the method described here. Grade stakes are still required to control the longitudinal profile.

Technical Note 9.09
Cost Estimating Form [3]

Barrangay Roads Development Programme Cost Standards
SURFACE COURSE
(USING BORROW MATERIALS)[1]

Thickness – 10 cm compacted
Expected Output – 200 sq m per day

I. MATERIALS

DESCRIPTION	QUANTITY	UNIT COST	AMOUNT
(Gravel, Sand, Corals, etc.)	20 cu m	$______	$______
(Delivery – km)	20 cu m	$______	$______

TOTAL COST OF MATERIALS.............$

II. LABOUR

DESIGNATION	NUMBER	RATE/DAY	AMOUNT
Foreman	1	$______	$______
Labourers	12	$______	$______

TOTAL COST OF LABOUR.............$ ______

TOTAL COST OF OUTPUT.............$ ______

*COST PER SQUARE METRE.............$ ______

$$\text{*COST PER SQUARE METER} = \frac{\text{TOTAL COST OF OUTPUT}}{\text{200 sq m}}$$

[1] See Technical Note 9.08 for a description of this work accomplishment.

Technical Note 9.10
Maintenance Performance Budgeting [77]

Maintenance Activity Definitions and Work Measurement Units

Code	*Description*	*Work Unit*
	ROADWAY SURFACES	
10	*Blading and Shaping* of unpaved surfaces with a motor grader to remove corrugations and rutting – without adding material	Lane-Km. bladed
11	*Patching* small isolated potholes and soft spots with suitable new material	Cu. metre
12	*Resurfacing* of long continuous sections to replace lost surfacing material	Cu. metre
	DRAINAGE	
20	*Cleaning and Repairing Culverts* with hand tools	No. of culverts
21	*Hand Cleaning of Ditches* with hand tools	Metres of ditch
22	*Machine Cleaning of Ditches* with motor grader or other motorized equipment	Km of ditches
	ROADSIDE	
30	*Vegetation Control*, cutting of brush, trees and grass with hand tools	Man-hours
31	*Erosion Control*, repair and prevention of roadside erosion	Man-hours
	BRIDGES	
40	*Channel Cleaning*, cleaning the waterway, brush, debris and sediment at bridge sites	No. of bridges
41	*Bridge Repair*, repair of bridge rail, bridge arches and structural elements	Metres of bridge
42	*Cleaning and Painting*, periodic cleaning and painting of steel bridges	Litres of paint
	TRAFFIC SERVICES	
50	*Sign Maintenance*, repair and replacement of damaged signs and sign posts	No. of signs
	REHABILITATION AND BETTERMENT	
60	*Special Projects*, restoration or improvement work on designated road sections – as well as major emergency work such as removing slides and repairing dykes and retaining structure	Estimated Work Quantities

Maintenance Performance Standard

Activity: Blading and Shaping

Activity Code: 10

Description and Purpose:

Periodic shaping and smoothing of unpaved road surfaces with motor grader to (1) remove corrugations, (2) correct rutting and other surface irregularities, (3) redistribute loose surfacing material from shoulders, and (4), restore proper roadway crown to provide drainage of surface water.

Performance Criteria:

To be scheduled and performed at time intervals according to the level-of-service standards established for the particular road class and location.

Should be scheduled shortly after rainy period when there is a small amount of natural moisture on the surface. Avoid scheduling the work under extremely wet or extremely dry conditions.

Crew, Equipment and Materials:

Crew	*Equipment*	*Construction Materials*
1 Equipment Operator	1 Motor Grader	None
2 Labourers	Shovels, Rakes	
2 Flagmen	Flags, Signs	

Work Methods

1. Place signs and safety devices.
2. Make pass with motor grader pulling loose material from shoulder, cutting high spots and placing windrows at centre of roadway.
3. Make second pass with motor grader in opposite direction and opposite side in the same manner, depositing material in windrow at centre.
4. Adjust blade for proper crown slope of 3° (5%) and spread material evenly toward each shoulder.
5. Check crown slope with slope board and make additional passes to assure correct crown.
6. Remove all rocks and oversize material from the surface, and rake smooth where necessary.
7. Remove signs and safety devices.

Work Measurement Unit:

Lane-Km bladed and shaped.

Average Productivity:

6 Km per day of two lane roads.

ANNEXE 10

Storage Facilities

This annexe contains two technical notes:
T.N. 10.1, *Drying Techniques*, is referenced in Para 18.15.
T.N. 10.2, *Storage Facilities*, is referenced in Para 18.20.

Technical Note 10.1
Drying Techniques

1. Long-term or high quality storage requires that stored produce be properly dried before it is stored. The following table indicates maximum storage moisture contents for various crops [38]:

Table 27. Maximum Storage Moisture Content for Selected Crops

Crop	*Maximum moisture content for one year (or less) storage at 70% relative humidity and 27°C*
Wheat	13.5%
Maize	13.5%
Paddy rice	15.0%
Milled rice	13.0%
Sorghum	13.5%
Millet	16.0%
Beans	15.0%
Cow peas	15.0%

In controlled conditions, grain is normally dried to 12% moisture content before storage. In rural areas, farmers try to dry their grain as thoroughly as possible, checking the hardness with their teeth or between their finger nails (dry grain is hard).

2. Grain can be dried in the field only as long as the air is able to absorb moisture. Rainy and cloudy, humid days delay field drying and prolong the attack of birds, rodents, and insects on the harvested material. Additional investment in grain-drying facilities may therefore be cost-effective. A smooth dry clean surface of compacted earth, a plastic sheet, or an asphalt or concrete surface hastens drying during periods of good weather. Unshelled maize, rice, millet, and sorghum are often stored in cribs for further drying; however they can be damaged by insects and rodents if stored for a long period. Other common inexpensive dryers include oil drum dryers, either above ground or in pits, and solar dryers using plastic sheets. Plans for these dryers can be found in reference [38].

Fig. 98 *A simple four-barrel dryer*

3. Oil drum dryers (Figure 98) are useful where grains must be harvested in rainy weather. They can eliminate the crib drying process which requires the use of contact insecticides, they considerably reduce drying time, and their construction materials are easily obtainable. Finally, they can be built with very little supervision. However, oil drum dryers are so big that they are best suited to group use. Their fuel may be hard to get or expensive, and without a mechnical fan, they cannot be used to dry rice, seed grain, or brewery grain – all of which cannot be dried without damage at a temperature over 45°C – or beans for food, which cannot be dried at over 35°C.

4. Solar dryers hold 9 to 11 kg per square metre of floor surface, incur no fuel costs, reduce sun drying time because the heat is retained by double layers of clear plastic film, and can be used for copra, cassava, fruits, and vegetables. However, they are only useful during daylight, are of limited value during long periods of rain and very cloudy weather, and need constant attention since the drying temperature can reach 65–80°C. They should never be used for seed grain drying and can spoil grain used for milling, (maximum drying temperature 60°C), food for human consumption (maximum 60°C) and even livestock feed (maximum 75°C).

Technical Note 10.2
Storage Facilities [38]

Type	*Materials*	*Characteristics*	*Sizes*	*Remarks*
Gourds	fruit skins, sealing material	useful for seed grains, easy to label	small	water- and air-proof with linseed oil, varnish, pitch, or bitumen
Baskets	grass, reeds, bamboo, small branches	should be raised off the ground, may be difficult to make airtight	can be any size	inexpensive; uses traditional technology, used for on-farm storage (Fig. 37); additional sketches for sizeable baskets' granaries can be found in reference [38]
Sacks	woven jute, hemp, sisal, local grasses, cotton	useful for no more than two seasons, subject to pests	standard sizes for portability	require building for storage, must be stacked properly for ventilation and rodent control. Can be stored in houses or in community buildings since they are easy to mark by date, produce, and owner's name
Sacks	plastic	one season only, subject to puncture, airtight	standard sizes for portability	" " "
Maize crib	bamboo or insect resistant wood, thatch roof	long side should face east–west. Increase capacity by lengthening long sides but retain width for proper drying	1.05 m high, 0.60 m wide, 1.0 m long, holds 400 kg of cob maize which yield 270 kg of shelled maize	modified plan for 2.0 m long crib with directions in reference [38]. Original design by Nigerian Stored Products Research Institute and FAO Rural Storage Centre, IITA, Ibadan, Nigeria (Fig. 36)
Pits	lined with straw and matting, plastic bags or concrete	provide security and confidentiality for individual farmers	as practical	concrete or ferrocrete lining permits use in wet ground. For on-farm storage without proper lining, is not airtight
Metal drums	200 litre drums, wax for patching holes	most useful when grain is to be stored more than five months	holds ±600 kg of grains	keep out of sunlight, drums must be full. Used for sorghum, maize, millet, cow peas, and groundnuts

Technical Note 10.2 (cont.)

Type	Materials	Characteristics	Sizes	Remarks
Metal silos	steel, moisture barrier foundation; should be shaded	Capacity 500 kg 1 ton 2 ton 3 ton	Ht (cm) / Dia (cm) / Gauge 125 / 80 / 28 165 / 100 / 26 210 / 124 / 24 210 / 150 / 24	more design information from: the Grain Storage Research Centre, Department of Food, Government of India, Hapur, Uttar Pradesh, India
Metal silo	1 mm thick sheet steel	should be shaded	diameter 160 cm; body height 200 cm; conical roof	Developed by Institute of Tropical Agricultural Research in Benin, Africa (Fig. 31)
Metal bins	metal	for on-farm use, plans for four sizes available	capacities: 0.42, 0.68, 0.82, 1.35 m^3	specifications and drawings from Save Grain Campaign, Min. of Agri. Dept. of Food, Krishi Bhavan, New Delhi, India
Earthen structure 'Pusa Bin'	tow layers of mud bricks with about 9 m^2 of 700-gauge plastic sheet between brick layers	needs foundation, uses approx. 1250 10× 10 × 20 cm blocks. Separate roof required	2 ton unit approx. 120 × 160 × 160 cm high	developed by Agricultural Research Institute in New Delhi, India. Plans and directions in reference [38]
Improved mudblock silo	rocks, sand, 2 bags cement, wood, water-proofing material	needs foundation, uses approx. 300 10 × 10 × 15 cm blocks	1.2 m dia., 1.6 m ht.	for on-farm storage, developed in Ghana. Plans and directions in reference [38] (Fig. 30)
Cement stave silo	16–50 kg. sacks cement; 12 bars 6 mm reinforcing rods, 6 m long; 1 roll (4 kg) 3 mm galvanized wire	built on foundation of 80–14 × 20 × 30 cm bricks; holds 4.5 tons of grain, requres additional roof	outside dia. 210 cm, height ±2 m	for community storage; developed in Benin; plans and directions in reference [38]
Concrete block square silos	concrete blocks, reinforced concrete slab floor, roof	typical building with outside dimensions 4.6 m × 9 m; contains eight storage cells, total capacity 30 tons. Requres additional roof. Can incorporate drying, weighing and office area	each cell inside 2 m × 2 m × 2 m. Can be expanded 2 cells at a time	for cooperative or village grain storage. Shared walls reduce price. Each cell independent so all grain does not need to be refumigated when one farmer withdraws grain. Further described in reference [38] (Fig. 32)

ANNEXE 11

Economic Analysis of Rural Access Infrastructure

This annexe contains two technical notes:
T.N. 11.1, *Simplified Operational Procedures*, is referenced in Paras 19.13 and 19.15.
T.N. 11.2, *Present Worth Factor for 1*, is referenced in Paras 19.13, 19.14, 19.15 and 19.16.

Technical Note 11.1
Simplified Operational Procedures

1. This note gives the derivation of the following three expressions of Chapter 19.

$$ADT^* = \frac{I}{VOC \,.p\,.\,365\,.\,A(N,i_o)} \qquad (1)$$

where

ADT^* = critical value of average daily traffic at which the total per kilometre vehicle operating cost savings equals or exceeds the per kilometre investment cost of the feeder road improvement,

I = per kilometre cost of the feeder road improvement plus per kilometre discounted value of maintenance costs over its life,

VOC = per kilometre weighted average vehicle operating cost before improvement,

p = percentage of *VOC* savings resulting from the feeder road improvement (as a decimal),

$A(N,i_o)$ = present worth of an annuity factor as a function of N and i_o

N = expected feeder road life in years,

$i_o = \left(\frac{1+i}{1+g}\right) -1$, where

i = the prevailing opportunity cost of capital, and

g = the constant per annum traffic growth in the width-project situation.

$$Q^* = \frac{I}{voc \,.p\,.\,A\,(N,i_p)} \qquad (2)$$

where I, p and N are as defined before and

Q^* = critical annual marketable agricultural production in tons moving over the improved feeder road at which the total per ton-km vehicle operating cost savings equals or exceeds the per kilometre investment cost of the feeder road improvement,

voc = per ton-km pre-investment weighted average vehicle operating cost,

$A(N,i_p)$ = present worth of an annuity factor as a function of N and i_p,

$i_p = \left(\frac{1+i}{1+r}\right) - 1$, where

i = the prevailing opportunity cost of capital,

r = the annual growth rate in marketable agricultural production in the with-project situation.

$$Q^{**} = \frac{IE}{(P_m - voc \, . \, D) \, .A(N,i_p)} \quad (3)$$

where voc, and $A(N,i_p)$ are as defined before and

Q^{**} = critical annual marketable agricultural production in tons saved because of obstacle elimination (access improvement) at which the excess market sales profit equals or exceeds the total investment to eliminate the obstacle(s),

IE = the cost of eliminating the obstacle plus the discounted value of maintaining of the obstacle elimination facility over its life,

P_m = weighted average of crops' per ton local market prices,

D = average round trip distance travelled to market the salvaged crops.

2. For an investment to be economically viable, its discounted costs must at least equal its discounted benefits, or

$$I = \sum_{t=1}^{N} VOC \, . \, p. \, ADT^* \, . \, 365 \, . \, (1+g)^t/(1+i)^t$$

where $\sum_{t=1}^{N}$ = the conventional sign for indicating the summation of the amounts $VOC \, . \, p. \, ADT \, . \, 365 \, (1+g)^t/(1+i)^t$ from the first $VOC \, . \, p \, . \, ADT \, . \, 365 \, . \, (1+g)^1/(1+i)^1$ (indicated by $t = 1$) to the Nth $VOC. \, p \, . \, ADT \, . \, 365 \, . \, (1+g)^N/(1+i)^N$ (indicated by $t = N$), and all other terms are as defined above.

To simplify the presentation, the equation terms $VOC \, . \, p \, . \, ADT^* \, . \, 365$ may be replaced by C. Then consider the series $C \, . \, (1+g)$, $C \, . \, (1+g)^2$,.....$C \, . \, (1+g)^N$ where g is the annual constant growth rate. The present value P of this series is, with a prevailing opportunity cost of capital equal to i, the following:

$$P = C.(1+g)/(1+i) + C.(1+g)^2/(1+i)^2 \; .... \; + C.(1+g)^N/(1+i)^N$$

Defining the term $(1+g)/(1+i) = 1/(1+i_o)$, the above series can be written as:

$$P = C/(i + i_u) + C/(1+i_o)^2 + ... + C/(1+i_o)^N$$

By definition this is equivalent to the present value of an annuity of C per year, discounted at rate i_o. Consequently,

$$P = P = C \, . \, A(N,i_o), \text{ or by substitution}$$
$$P = VOC \, . \, p \, . \, ADT^* \, . \, 365 \, . \, A\,(N,i_o)$$

The present value P must be equal to I,

$$\text{thus } I = VOC \, . \, p \, . \, ADT^* \, . \, 365 \, . \, A(N,\, i_o)$$

$$\text{or } ADT^* = \frac{I}{VOC \, . \, p \, . \, 365 \, . \, A(N,i_o)}$$

3. Tests on a world wide sample of rural road investments have shown that, in most situations, the maintenance costs may be ignored without significantly affecting the economic feasibility of a proposed rural road improvement [6]. This is the reason why in Chapter 19 the I is defined as the per kilometre cost of the feeder road improvement.

4. Expression (1) can be altered to determine the per ton-km cost and the break-even quantity by making the following substitutions:

(i) the VOC is replaced by a per ton-km cost voc, and
(ii) the present yearly number of vehicles (ADT) . (365) is replaced by the present annual quantity (Q) of marketable agricultural production.

Expression (1) becomes:

$$I = \sum_{t=1}^{N} voc \, . \, p. \, Q \, . \, (1 + r)^t/(1 + i)^t$$

$$\text{or } Q^* = \frac{I}{voc \, . \, p. \, A(N,i_p)} \qquad \text{which is expression (2)}$$

5. The derivation of expression (3) from

$$IE = \sum_{t=1}^{N} Q\,(P_m - voc \, . \, D) \, . \, (1 + r)^t/(1 + i)^t$$

is similar to the derivation of expression (1) as described in Para 2.

Technical Note 11.2
Present Worth Factor for 1

Number of Years

%	1	2	3	4	5	6	7	8	9	10
1	0.9901	1.9704	2.9410	3.9020	4.8534	5.7955	6.7282	7.6517	8.5660	9.4713
2	0.9804	1.9416	2.8839	3.8077	4.7135	5.6014	6.4720	7.3255	8.1622	8.9826
3	0.9709	1.9135	2.8286	3.7171	4.5797	5.4172	6.2303	7.0197	7.7861	8.5302
4	0.9615	1.8861	2.7751	3.6299	4.4518	5.2421	6.0021	6.7327	7.4353	8.1109
5	0.9524	1.8294	2.7232	3.5460	4.3295	5.0757	5.7864	6.4632	7.1078	7.7217
6	0.9434	1.8334	2.6730	3.4651	4.2124	4.9173	5.5824	6.2098	6.8017	7.3601
7	0.9346	1.8080	2.6243	3.3872	4.1102	4.7665	5.3893	5.9713	6.5152	7.0236
8	0.9259	1.7833	2.5771	3.3121	3.9927	4.6229	5.2064	5.7466	6.2469	6.7101
9	0.9174	1.7591	2.5313	3.2397	3.8897	4.4859	5.0330	5.5348	5.9952	6.4177
10	0.9091	1.7355	2.4869	3.1699	3.7908	4.3553	4.8684	5.3349	5.7590	6.1446
11	0.9009	1.7125	2.4437	3.1024	3.6959	4.2305	4.7122	5.1461	5.5370	5.8892
12	0.8929	1.6901	2.4018	3.0373	3.6048	4.1114	4.5638	4.9676	5.3283	5.6502
13	0.8850	1.6681	2.3612	2.9745	3.5172	3.9976	4.4226	4.7988	5.1317	5.4262
14	0.8772	1.6467	2.3216	2.9137	3.4331	3.8887	4.2883	4.6389	4.9464	5.2161
15	0.8696	1.6257	2.2832	2.8550	3.3522	3.7845	4.1604	4.4873	4.7716	5.0188
16	0.8621	1.6052	2.2459	2.7982	3.2743	3.6847	4.0386	4.3436	4.6065	4.8332
17	0.8547	1.5852	2.2096	2.7432	3.1993	3.5892	3.9224	4.2072	4.4506	4.6586
18	0.8475	1.5656	2.1743	2.6901	3.1272	3.4976	3.8115	4.0776	4.3030	4.4941
19	0.8403	1.5465	2.1399	2.6386	3.0576	3.4098	3.7057	3.9544	4.1633	4.3389
20	0.8333	1.5278	2.1065	2.5887	2.9906	3.3255	3.6046	3.8372	4.0310	4.1925
21	0.8264	1.5095	2.0739	2.5404	2.9260	3.2446	3.5079	3.7256	3.9054	4.0541
22	0.8197	1.4915	2.0422	2.4936	2.8636	3.1669	3.4155	3.6193	3.7863	3.9232
23	0.8130	1.4740	2.0114	2.4483	2.8035	3.0923	3.3270	3.5179	3.6731	3.7993
24	0.8065	1.4568	1.9813	2.4043	2.7454	3.0205	3.2423	3.4212	3.5655	3.6819
25	0.8000	1.4400	1.9520	2.3616	2.6893	2.9514	3.1611	3.3289	3.4631	3.5705
26	0.7937	1.4235	1.9234	2.3202	2.6351	2.8850	3.0833	3.2407	3.3657	3.4648
27	0.7874	1.4074	1.8956	2.2800	2.5827	2.8210	3.0087	3.1564	3.2728	3.3644
28	0.7813	1.3916	1.8684	2.2410	2.5320	2.7594	2.9370	3.0758	3.1842	3.2689
29	0.7752	1.3761	1.8420	2.2031	2.4830	2.7000	2.8682	2.9986	3.0997	3.1781
30	0.7692	1.3609	1.8161	2.1662	2.4356	2.6427	2.8021	2.9247	3.0190	3.0915

Technical Note 11.2 (cont.)

%	11	12	13	14	15	16	17	18	19	20
1	10.3676	11.2551	12.1337	13.0037	13.8651	14.7179	15.5623	16.3983	17.2260	18.0456
2	9.7868	10.5753	11.3484	12.1062	12.8493	13.5778	14.2919	14.9920	15.6785	16.3514
3	9.2526	9.9540	10.6350	11.2961	12.5611	13.1661	13.7535	14.3238	14.8775	14.8775
4	8.7605	9.3851	9.9856	10.5631	11.1184	11.6523	12.1657	12.6593	13.1339	13.5903
5	8.3064	8.8633	9.3936	9.8986	10.3797	10.8378	11.2741	11.6896	12.0853	12.4622
6	7.8869	8.3838	8.8527	9.2950	9.7122	10.1059	10.4773	10.8276	11.1581	11.4699
7	7.4987	7.9427	8.3577	8.7455	9.1079	9.4466	9.7632	10.0591	10.3356	10.5940
8	7.1390	7.5361	7.9038	8.2442	8.5595	8.8514	9.1216	9.3719	9.6036	9.8181
9	6.8052	7.1607	7.4869	7.7862	8.0607	8.3126	8.5436	8.7556	8.9501	9.1285
10	6.4951	6.8137	7.1034	7.3667	7.6061	7.8237	8.0216	8.2014	8.3649	8.5136
11	6.2065	6.4924	6.7499	6.9819	7.1909	7.3792	7.5488	7.7016	7.8393	7.9633
12	5.9377	6.1944	6.4235	6.6282	6.8109	6.9740	7.1196	7.2497	7.3658	7.4694
13	5.6869	5.9176	6.1218	6.3025	6.4624	6.6039	6.7291	6.8399	6.9380	7.0248
14	5.4527	5.6603	5.8424	6.0021	6.1422	6.2651	6.3729	6.4674	6.5504	6.6231
15	5.2337	5.4206	5.5831	5.7245	5.8474	5.9542	6.0472	6.1280	6.1982	6.2593
16	5.0286	5.1971	5.3423	5.4675	5.5755	5.6685	5.7487	5.8178	5.8775	5.9288
17	4.8364	4.9884	5.1183	5.2293	5.3242	5.4053	5.4746	5.5339	5.5845	5.6278
18	4.6560	4.7932	4.9095	5.0081	5.0916	5.1624	5.2223	5.2732	5.3162	5.3527
19	4.4865	4.6105	4.7147	4.8023	4.8759	4.9377	4.9897	5.0333	5.0700	5.1009
20	4.3271	4.4392	4.5327	4.6106	4.6755	4.7296	4.7746	4.8122	4.8435	4.8696
21	4.1769	4.2785	4.3624	4.4317	4.4890	4.5364	4.5755	4.6097	4.4346	4.6567
22	4.0354	4.1274	4.2028	4.2646	4.3152	4.3567	4.3908	4.4187	4.4415	4.4603
23	3.9018	3.9852	4.0530	4.1082	4.1530	4.1894	4.2190	4.2431	4.2627	4.2786
24	3.7757	3.8514	3.9124	3.9616	4.0013	4.0333	4.0591	4.0799	4.0967	4.1103
25	3.6564	3.7251	3.7801	3.8241	3.8503	3.8874	3.9099	3.9279	3.9424	3.9539
26	3.5435	3.6060	3.6555	3.6949	3.7261	3.7509	3.7705	3.7861	3.7985	3.9535
27	3.4365	3.4933	3.5381	3.5733	3.6010	3.6228	3.6400	3.6536	3.6642	3.6726
28	3.3351	3.3868	3.4272	3.4587	3.4834	3.5026	3.5177	3.5294	3.5386	3.5458
29	3.2388	3.2859	3.3224	3.3507	3.3726	3.3896	3.4028	3.4130	3.4210	3.4271
30	3.1473	3.1903	3.2233	3.2487	3.2682	3.2832	3.2948	3.3037	3.3105	3.3158

Figure References

Figures 3, 41, 44 and 55:
Yogyakarta Rural Roads Study, August, 1978, Vol 1, ENEX of New Zealand (pp. 4.8, 4.12).

Figures 4, 5, 43, 49, 50, 51, 52, 54:
Appropriate Technology in Rural Development: Vehicles Designed for On and Off Farm Operations, October, 1978, Rural Development Unit, Transportation Department, The World Bank (pages unnumbered).

Figures 6, 25, 26, 57, 58, 59, 64, 65, 67, 69, 70, 76, 81, 83, 85, 92 and 94:
Guide to the Training of Supervisors for Labour-Based Road Construction and Maintenance – Trainees' Manual/Part 2, Copyright 1981, International Labour Organization, Geneva (pp. M-14, LE-1/2; M-14, LE-5/4; M-15, LE-1/1, LE-1/2; M-11, LE-2/4; M-11, LE-2/5; M-11, LE-2/6; M-10, LE-3/5; M-10, LE-4/2; M-13, LE-3/5; M-13, LE-3/5, LE-3/7, LE-3/8; M-13, LE-5/2; M-10, LE-5/1; M-13, LE-3/3; M-13, LE-4/2; M-13, LE-2/2; M-10, LE-5/5; M-15, LE-4/4).

Figures 7, 80, 86, 88, 89 and 91:
Road Maintenance Handbook – Practical Guidelines for Road Maintenance in Africa, Volume 1, Copyright 1982, United Nations Economic Commission for Africa (pp. 124, 126, 210, 254, 228, 66).

Figures 8, 12 and 78:
Compendium 9 – Control of Erosion, Transportation Technology Support for Developing Countries – 1979, Transportation Research Board, National Academy of Sciences (pp. 120, 219, 57).

Figures 10, 19, 24, 93, 95 and 96:
Road Maintenance Handbook – Practical Guidelines for Road Maintenance in Africa, Volume 2, Copyright 1982, United Nations Economic Commission for Africa (pp. 10 and 38, 164, 192, 124 and 170, 84 and 86, 96 and 98).

Figures 11 and 97:
Operations Manual, Labour-Based-Equipment Supported Approach in Barangay Roads Maintenance, 1981, Republic of the Philippines, Ministry of Local Government and Community Development (pp. 30, 36–38).

Figure 13:
Self-Help and Rural Improvement Road Construction Manual, 1976 Edition, Public Works Department, Local Government Advisory Service, Papua New Guinea (p. 99).

Figures 14, 27 and 94:
Manual on Rural Road Maintenance, 1981, Economic and Social Commission for Asia and the Pacific, Bangkok, Thailand, United Nations, New York, ST/ESCAP/170 (pp. 18, 26 and 27, 14).

Figures 15 and 95:

Overseas Road Note 2 – Maintenance Techniques for District Engineers, Copyright 1981, Transport and Road Research Laboratory, Crowthorne, Berkshire, United Kingdom (pp. 18 and 19, 22).

Figures 16, 72 and 93:
Manual on Rural Road Construction, 1981, Economic and Social Commission for Asia and the Pacific, Bangkok, Thailand, United Nations, New York, ST/ESCAP/165 (pp. 41, 32, 31).

Figures 17, 18, and 60:
Forest Roads in the Tropics, Reprinted from Unosylva, Volume 17, Numbers 2 and 3 by Food and Agriculture Organization of the United Nations (pp. 32, 6, 14).

Figures 20, 21 and 82:
Synthesis 2 – Stage Construction, Transportation Technology Support for Developing Countries – 1979, Transportation Research Board, National Academy of Sciences (pp. 9 and 11, 7 and 30).

Figures 22, 23, 40 and 71:
Labor-Based Construction Programmes – A Practical Guide for Planning and Management, June 1983, Copyright 1983 by the International Bank for Reconstruction and Development (pp. 136, 141, 204, 134).

Figures 28 and 29:
Practical Guide for Site Foreman, Basic Principles for Conducting Labour-Intensive Works, January 1982, Volume 1, International Labour Organization, Geneva (pp. 27, 34).

Figures 30, 31, 32, 33, 34, 35, 36, 37, 38 and 98:
Small Farm Grain Storage, September 1976, Peace Corps Information Collection and Exchange Manual Series Number 2 – Volunteers in Technical Assistance, Vita Publications Manual Series Number 35 E (section 7, pp. 87 and 97; 65; 147 and 148; p. A-5; section 6, part 2, p. 116; section 3, p. 8; section 7, p. 25; section 4, pp. 2, 3, 5; section 5, p. 21).

Figure 61:
Compendium 7 – Road Gravels, Transportation Technology Support for Developing Countries – 1979, Transportation Research Board, National Academy of Sciences (p. 89).

Figures 62, 66 and 67:
Recreation Travelways Handbook, Draft, U.S. Department of Agriculture, Forest Service (pp. 160, 55, 73).

Figure 63:
AMC Field Guide to Trail Building and Maintenance, 1981, Second Edition, Copyright 1981, Appalachian Mountain Club (pp. 96 and 105).

Figures 74, 79 and 83:
Introductory Course in Labour-Based Construction Techniques for Rural Roads, June, 1981, Ethiopian Transport Construction Authority (pp. 150, 156, 162).

Figure 75:
Trail Construction Manual, Tennessee Department of Conservation (p. 41).

Figure 77:
Maintenance Manual of Bitumen Sealed Pavements, Shoulders and Drainage

Including Treatment of Sub-Soil Seepage, Main Roads Department, Queensland, Australia.

Figure 84:

Brazil, Rondonia Settlement Consolidation Project, Working Paper on Civil Works, December, 1980, The World Bank.

Figure 87:

Compendium 4 – Low-Cost Water Crossings, Transportation Technology Support for Developing Countries – 1979, Transportation Research Board, National Academcy of Sciences (pp. 191 and 197).

Figure 90:

Compendium 3 – Small Drainage Structures, Transportation Technology Support for Developing Countries – 1978, Transportation Research Board, National Academy of Sciences (p. 187).

Bibliography[1]

1. Abaynayaka, S.W., H. Hide, G. Morosiuk and R. Robinson, *Tables for Estimating Vehicle Operating Costs on Rural Roads in Developing Countries*, TRRL Report 723 (Crowthorne, United Kingdom: Department of the Environment, Transport and Road Research Laboratory, 1976)
2. Anderson, G.W. and C.G. Vandervoort, USAID Program Evaluation Report no. 5, *Rural Roads Evaluation Summary Report* (U.S. Agency for International Development, March, 1982)
3. Barangay Roads Development Program, *Operations Manual: Labor-based Equipment Supported Approach in Barangay Roads Maintenance* (Philippines: Ministry of Local Government and Community Development, 1981)
4. Beenhakker, A., *A System for Development Planning and Budgeting* (West Mead, England: Grower Publishing Co., 1980)
5. Beenhakker, H.L., and A. Chammari, *Identification and Appraisal of Rural Roads Projects*, Staff Working Paper no. 362 (Washington, D.C.: World Bank, October, 1979)
6. Beenhakker, H.L. and A.M. Lago, *Economic Appraisal of Rural Roads: Simplified Operational Procedures for Screening and Appraisal*, Staff Working Paper no. 610 (Washington, D.C.: World Bank, October, 1983)
7. Beenhaker, H.L., *Handbook for the Analysis of Capital Investments* (Westport, Conn.: Greenwood Press, 1976)
8. Bonney, R.S.P. and N.F. Stevens, *Vehicle Operating Costs on Bituminous, Gravel and Earth Roads in East and Central Africa*, Technical Paper no. 76 (London: Ministry of Transport, Road Research Laboratory, 1967)
9. Boxall, R.A. and others. *The Prevention of Farm-Level Food Grain Storage Losses in India: A Social Cost-Benefit Analysis* (Brighton, United Kingdom: University of Sussex, Institute of Development Studies)
10. Carapetis, S., H.L. Beenhakker and J.D.F. Howe, *The Supply and Quality of Rural Transport Services in Developing Countries: A Comparative Review*, Staff Working Paper no. 654 (Washington, D.C.: World Bank, 1984)
11. Carnemark, C., J. Biderman and D. Bovet, *Economic Analysis of Rural Roads Projects*, Staff Working Paper no. 241 (Washington, D.C.: World Bank, 1976)
12. Colloff, J.E.M., 'The Economic Case for Low-cost Roads', Paper presented at the second African Highway Conference (Rabat, April, 1972)

[1] Publications 1 to 104 are references in the text, publications 105–274 contain additional useful material.

13. Cook, C., H.L. Beenhakker and R. Hartwig, *Institutional Considerations in Rural Roads Projects*. Staff Working Paper no. 748 (Washington, D.C.: World Bank, August, 1985)
14. Cook, C., *Local Participation in Rural Road Projects: Outline* (Washington, D.C.: World Bank, November 28, 1983)
15. Cornwell, P.R. and J.M. Thomson, 'The Development of Priorities for Rural Roads', Paper 11 in Institution of Civil Engineers, *Highway Investments in Developing Countries* (London: Thomas Telford Ltd., 1983)
16. Coukis, B. and others, *Labour-based Construction Programs: A Practical Guide for Planning and Management* (Washington, D.C.: World Bank, Transportation Department, June, 1983)
17. Department of Transport and Civil Aviation, *Rural Transport Planning Manual*, 2nd edition (Konedobu, Papua New Guinea: Department of Transport and Civil Aviation, February, 1983)
18. Department of Transport Works and Supply, *Construction Manual: Self-help and Rural Improvement of Roads* (Papua New Guinea: National Works Authority, 1976 Edition, Reprinted 1977)
19. Derman, P. and G. Maasdorp, *The Importance of Time Savings in Road Project Appraisal: Evidence from Swaziland*, Occasional Paper No. 12 (Durban: University of Natal, Economic Research Unit, 1981)
20. de Weille, J., *Quantification of Road User Savings*, World Bank Staff Occasional Paper no. 2 (Baltimore and London: The John Hopkins University Press, 4th printing, 1983)
21. Economic Development Bureau, *Appropriate Technology for Grain Storage in Tanzanian Villages: Report of a Pilot Project*, Community Development Trust Fund of Tanzania in collaboration with Institute of Adult Education (New Haven, Conn., May, 1977)
22. Edmonds, G.A., 'Rural Transport Policy in Developing Countries', Paper 16 in Institution of Civil Engineers, *Highway Investment in Developing Countries* (London: Thomas Telford Ltd., 1983)
23. ENEX of New Zealand, *Highway Capacities in Indonesia*, ENEX Consortium 346, Screening Feasibility Project (Jakarta, Indonesia, April, 1979)
24. Federal Highway Administration, *Design and Construction of Low Water Stream Crossings: Executive Summary* (Washington, D.C.: U.S. Department of Transportation, September, 1983)
25. Greenstein, J. and H. Bonjack, *Socioeconomic Evaluation and Upgrading of Rural Roads in Agricultural Areas of Ecuador* (East Orange, New Jersey: prepared for Transportation Research Board, Third International Conference on Low-Volume Roads, July 24 – 28, 1983, Tempe, Arizona by Louis Berger International, Inc., January, 1983)
26. Hide, H., S.W. Abaynayaka, I. Sayer and R.J. Wyatt, *The Kenya Road Transport Cost Study: Research on Vehicle Operating Costs*, TRRL Report LR 672 (Crowthorne, United Kingdom: Department of the Environment, Transport and Road Research Laboratory, 1975)
27. Howe, J.D.G.F., *Conceptual Framework for Defining and Evaluating*

Improvements to Local Level Rural Transport in Developing Countries (Geneva: International Labour Organization, World Employment Programme, September, 1983)

28. International Labour Organization and the United Nations Development Programme, *Practical Guide for Site Foremen* (Geneva: UNDP/ILO INT/78/020, January, 1982)
29. International Labour Organization, *Appropriate Transport Facilities for the Rural Sector in Developing Countries* (Geneva: ILO, World Employment Programme, January, 1979)
30. International Labour Organization, Multi-bilateral Programme of Technical Cooperation Project Document, *Tanzania: Integrated Rural Transport* (Geneva: ILO, Project starting date: July, 1983)
31. International Labour Organization, *Report of the ACC Technical Working Group on Upgrading Traditional Technologies*, submitted to the ACC Task Force on Science and Technology for Development (Geneva: ILO, December, 1982)
32. International Labour Organization, *Rural Access Roads Programme: Maintenance Strategy* (Geneva)
33. International Labour Organization, World Employment Programme, Report on Activities of Construction Technology Programme, *Local Level Development – Special Public Works Programmes/Civil Construction* (Geneva: ILO, December, 1981)
34. Karlson, L.S. and J.J. de Veen, *Guide to Training of Supervisors for Labor-based Road Construction and Maintenance*, Trainee's Manual Part 2 (Geneva: International Labour Organization, 1981)
35. Knowles, M., *Trainees Manual: Introductory Course in Labour-based Construction Techniques for Rural Roads*, prepared for the Ethiopian Transport Construction Authority by the Intenational Labour Organization (Geneva, June, 1981)
36. le Ray, J., 'Forest Roads in the Tropics', Reprinted from *UNASYLVA*, vol. 17, no. 2 and 3 (United Nations/FAO)
37. Liautaud, G., *The Catalogue of Pavement Structures of Ivory Coast*
38. Lindblad, C. and L. Druben, *Small Farm Grain Storage*, Manual M-2, Peace Corps (Washington, D.C., September, 1976)
39. Lindblad, C., *Programming and Training for Small Farm Grain Storage* (Washington, D.C.: Peace Corps, September, 1981)
40. Main Roads Department, *Maintenance Manual of Bitumen Sealed Pavements, Shoulders and Drainage* (Queensland, Australia, July, 1966)
41. Millard, R.S., Letter to H.L. Beenhakker (Washington, D.C.: World Bank, November 13, 1984)
42. Ministry of Public Works, *Rural Road Development Project: Operating Costs of Road Transport*, Working Paper no. 14 (Jakarta, Indonesia: prepared by ENEX of New Zealand and BIEC International Inc., September, 1980)

43. Ministry of Public Works, *Rural Road Study Project: Draft Final Report for Phase I* (Jakarta, Indonesia: prepared by ENEX of New Zealand and BIEC International Inc., February, 1981)
44. Ministry of Public Works, *Rural Road Study Project: Implementation*, Working Paper no. 8 (Jakarta, Indonesia: prepared by ENEX of New Zealand and BIEC International Inc., September, 1980)
45. Ministry of Public Works, *Rural Road Study Project: Level of Access*, Working Paper no. 12 (Jakarta, Indonesia: prepared by ENEX of New Zealand and BIEC International Inc., November, 1980)
46. Ministry of Public Works, *Rural Road Study Project: Project Control*, Working Paper no. 6 (Jakarta, Indonesia: prepared by ENEX of New Zealand and BIEC International Inc., October, 1980)
47. Ministry of Public Works, *Yogyakarta Rural Roads Study*, Draft Final Report, 2 vols. (Jakarta, Indonesia: prepared by ENEX of New Zealand, August, 1978)
48. Ministry of Shipping and Transport, *The Animal Cart in the Rural System*, Part I: The Analytical Report and Part II: The Statistical Annexes, prepared by the Indian Institute of Management, Bangalore for the Ministry of Shipping and Transport, Government of India. New Delhi.
49. Ministry of Works, *Rural Access Roads Programme*, Annual Review and Evaluation Meeting Discussion Papers, vol. I (Nairobi, Kenya: Roads Department, 1979)
50. Ocampo, R.B., World Employment Programme Research Working Papers, Technology and Employment Programme, *Rural Transport in the Philippines: Jeepneys, Trimobiles, and other Simple Modes in two Bicol Towns* (Geneva: International Labour Organization, November, 1982)
51. Odier, L., R.S. Millard, P. dos Santos and S.R. Mehra, *Low Cost Roads: Design, Construction and Maintenance*, prepared for UNESCO (London: Butterworth & Co. (Publishers) Ltd., 1967)
52. Perrett, H. and F. Lethem, *Social and Behavioral Aspects of Rural Roads Project Work*, unpublished (Washington, D.C.: World Bank, February, 1979)
53. Proudman, R.D. and R. Rajala, *AMC Field Guide to Trail Building and Maintenance*, 2nd Edition.
54. Rahkonen, O.J., C. Cook and S. Carapetis, *Institutional Aspects of Rural Road Projects*, unpublished (Washington, D.C.: World Bank, Transportation, Water and Telecommunications Department, December, 1981)
55. Rico, R.S., *Costs of Intensive Labour Roads* (Mexico, for SAHOP, undated translation)
56. Rico, R.S., *Organization of Rural Communities* (Mexico, for SAHOP, undated translation)
57. Riverson, J.D.N. and L.K. Afele, 'Feeder Road Traffic Characteristics

and Redevelopment Needs in Ghana', paper presented at the Conference on Highway Engineering in Africa. Addis Ababa, April, 1974, organized by the Economic Commission for Africa with the cooperation of the British and French Governments, United Nations Economic Commission for Africa, Addis Ababa, Ethiopia.

58. SAHOP (Secretariat of Human Settlement and Public Works), *Instructive for the Construction of Rural Roads* (Mexico: SAHOP, Rural Roads General Office, 1978)
59. SAHOP (Secretariat of Human Settlement and Public Works), *Socio-Economic Information Sheet, National Rural Roads Plan: Instructions* (Mexico: Translated from Spanish by World Bank, Washington, D.C., October 30, 1978)
60. SETEC International, *Project Pilote des Pistes Agricoles, Phase I, Rapport de Factibilite, vol. 3: Etudes Economiques* (Tunisia: October, 1977)
61. Shurig, D.G., *Minor Maintenance of County Bridges*, County Highway Series no. 7 (Lafayette, Indiana: Purdue University, August, 1964)
62. Sikka, R.P., *Improvement of the Carrying Capacity and Operational Efficiency of Rural Road Transport* prepared for the United Nations Economic and Social Commission for Asia and the Pacific (New Delhi, India: Ministry of Shipping and Transport, Roads Wing)
63. Sikka, R.P., Letter to H.L. Beenhakker (Washington, D.C.: World Bank, December 12, 1984)
64. Starkey, P., *Farming with Work Oxen in Sierra Leone*, A Development Project of the Ministry of Agriculture & Forestry in cooperation with Njala University College (Freetown: Sierra Leone Work Oxen Project, 1981)
65. Tendler, J., *New Directions Rural Roads*, A.I.D. Program Evaluation Discussion Paper no. 2 (U.S. Agency for International Development, Office of Evaluation, March, 1979)
66. Tennessee Department of Conservation, *Trail Construction Manual* (Nashville, Tenn.: Tennessee Department of Conservation)
67. Thriscutt, H.S., Letter to H.L. Beenhakker (Washington, D.C.: World Bank, December 8, 1984)
68. Transportation and Road Research Laboratory, *Maintenance Techniques for District Engineers*, Overseas Road Note 2 (Crowthorne, United Kingdom: TRRL, Overseas Development Administration, 1981)
69. Transportation Research Board, *Transportation Technology Support for Developing Countries: Compaction of Roadway Soils*, Compendium 10 (Washington, D.C.: National Academy of Sciences, 1979)
70. Transportation Research Board, *Technology Support for Developing Countries: Control of Erosion*, Compendium 9 (Washington, D.C.: National Academy of Sciences, 1979)
71. Transportation Research Board, *Transportation Technology Support for Developing Countries: Decision Methodology for Maintenance and Upgrading Costs, Traffic and Benefits*, Compedium 11 (Washington, D.C.: National Academy of Sciences, 1980)

72. Transportation Research Board, *Transportation Technology Support for Developing Countries: Drainage and Geological Considerations in Highway Location*, Compendium 2 (Washington, D.C.: National Academy of Sciences, 1978)
73. Transportation Research Board, *Transportation Technology Support for Developing Countries: General Design Standards for Low-Volume Roads*, Compendium 1 (Washington, D.C.: National Academy of Sciences, 1978)
74. Transportation Research Board, *Transportation Technology Support for Developing Countries: Implementing Construction by Contract or Day Labor*, Compendium 16 (Washington, D.C.: Nationla Academy of Sciences, 1981)
75. Transportation Research Board, *Transportation Technology Support for Developing Countries: Labor-Based Construction and Maintenance of Low-Volume Roads*, Synthesis 3 (Washington, D.C.: National Academy of Sciences, 1981)
76. Transportation Research Board, *Transportation Technology Support for Developing Countries: Low-Cost Water Crossings*, Compendium 4 (Washington, D.C.: National Academy of Sciences, 1979)
77. Transportation Research Board, *Transportation Technology Support for Developing Countries: Maintenance of Unpaved Roads*, Synthesis 1 (Washington, D.C.: National Academy of Sciences, 1979)
78. Transportation Research Board, *Transportation Technology Support for Developing Countries: Road and Traffic Inventories*, Compendium 15 (Washington, D.C.: National Academy of Sciences, 1980)
79. Transportation Research Board, *Transportation Technology Support for Developing Countries: Road Gravels*, Compendium 7 (Washington, D.C.: National Academy of Sciences, 1979)
80. Transportation Research Board, *Transportation Technology Support for Developing Countries: Roadside Drainage*, Compendium 5 (Washington, D.C.: National Academy of Sciences, 1979)
81. Transportation Research Board, *Transportation Technology Support for Developing Countries: Small Drainage Structures*, Compendium 3 (Washington, D.C.: National Academy of Sciences, 1978)
82. Transportation Research Board, *Transportation Technology Support for Developing Countries: Stage Construction*, Synthesis 2 (Washington, D.C.: National Academy of Sciences, 1979)
83. Transportation Research Board, *Transportation Technology Support for Developing Countries: Structural Design of Low-Volume Roads*, Synthesis 4 (Washington, D.C.: National Academy of Sciences, 1982)
84. United Nations Economic and Social Commission for Asia and the Pacific, *Manual on Rural Road Maintenance* (Bangkok, Thailand: ESCAP, 1981)
85. United Nations Economic and Social Commission for Asia and the Pacific, *Manual on Rural Road Maintenance* (Bangkok, Thailand: ESCAP, 1981)
86. United Nations Economic Commission for Africa, *Road Maintenance Handbook*, Practical Guidelines for Road Maintenance in Africa

(Volume I: Maintenance of roadside areas, drainage structures and traffic control devices, Volume II: Maintenance of unpaved roads, and Volume III: Maintenance of paved roads, 1982)
87. U.S. Agency for International Development, A.I.D. Program Design and Evaluation Methods, *Selection and Justification Procedures for Rural Roads Improvement Projects* (USAID, Bureau for Program and Policy Coordination, Office of Evaluation, December, 1982)
88. U.S. Forestry Department, *Recreation Travelways Handbook* (Draft)
89. Vivas, J., *Brazil: Rondonia Settlement Consolidation Project – Civil Works*, Working Paper (Washington, D.C.: World Bank, 1980)
90. Watson, P.R., *Animal Traction*. Edited by R. Davis and M.S. Chakroff. Prepared for Peace Corps by the TransCentury Corporation, Washington, D.C., August, 1981.
91. World Bank, *An Investigative Survey of Appropriate Rural Transport for Small Farmers in Kenya*, unpublished (Washington, D.C.: World Bank, Transportation Department, October, 1977)
92. World Bank, *Appropriate Technology in Rural Development: Vehicles Designed for On and Off Farm Operations* (Washington, D.C.: World Bank, Transportation Department, Regional Development Unit, October, 1978)
93. World Bank, *Benin: Appraisal of a Feeder Roads Project*, Bank document (Washington, D.C.: World Bank, 1977)
94. World Bank, Brazil: *Guidelines for Improvement of Low-Cost Low-Class Roads*, Preliminary Tentative Report (Washington, D.C.: World Bank, July, 1982)
95. World Bank, *India: Bihar Rural Roads*, Bank document (Washington, D.C.: World Bank, 1980)
96. World Bank, *Lesotho: Proposed Methodology for Comparing Costs of Labor and Equipment-based Construction Methods*, unpublished (Washington, D.C.: World Bank)
97. World Bank, *Manual on Construction Standards for Rural Access Roads*, unpublished (Washington, D.C.: World Bank)
98. World Bank, *Nepal: Second Rural Development Project, Mahakali Hills*, Bank document (Washington, D.C.: World Bank, 1979)
99. World Bank/Scott Wilson Kirkpatrick & Partners, *Guide to Competitive Bidding on Construction Projects in Labor-Abundant Economies* (Washington, D.C.: World Bank, June 1978)
100. World Bank, Study of the Substitution of Labour and Equipment in Civil Construction: *A System of Deriving Rental Charges for Construction Equipment*, Technical Memorandum no. 10 (Washington, D.C.: World Bank, Transportation and Urban Projects Department, August 1974)
101. World Bank, *Study of the Substitution of Labour and Equipment in Civil Construction*, Technical Memorandum no. 21 (Washington, D.C.: World Bank, Transportation and Urban Projects Department, February 1976)
102. World Bank, *The Road Maintenance Problem and International Assistance* (Washington, D.C.: World Bank, Transportation, Water and Telecommunications Department, December, 1981)
103. World Bank, *World Tables: Economic Data*, vol. I, 3rd ed. (Baltimore

and London: The Johns Hopkins University Press, 1983)

104. World Bank, *World Tables: Social Data*, vol. II, 3rd ed. (Baltimore and London: The Johns Hopkins University Press, 1983)

105. Ahmad, M.S., *Farm-to-market Roads: A Bibliography*.

106. Airey, A., *Socio-Economic and Traffic Evaluation of C.A.R.E. Feeder Roads in Sierra Leone*, Second Report of the Evaluation conducted in Sierre Leone, June 1979 – August 1979, Annex I (Coventry, United Kingdom: Lanchester Polytechnic, 1979)

107. Airey, A., *Socio-Economic Evaluation of C.A.R.E. Feeder Roads in Sierra Leone*, 3rd report (Coventry, United Kingdom: Lanchester Polytechnic, 1980)

108. Airey, A., *The Role of Feeder Roads in Promoting Rural Change in Eastern Sierra Leone* (Coventry, United Kingdom: Lanchester Polytechnic)

109. Are Inc., *Design and Management of Low Volume Roads* (Austin, Texas: Are Inc., Engineering Consultants)

110. Austen, A.D. and R.H. Neale, *Managing Construction Projects: A Guide to Processes and Procedures*, International Labour Organization, 1982

111. BCEOM, *Assistance Technique pour l'Amelioration et l'Entretien des Pistes de Production Agricole, Rapport D'Avancement 12, July-October 1979*, For Republique de Senegal.

112. Bennathan, E., *Studies of Sector Policy and Research Interest Included Under Bank Loans/Credits for Transport*, Office Memorandum (Washington, D.C.: World Bank, December 9, 1982)

113 Berger, L., *New Techniques for Pavement Evaluation and Design*, Third Conference of Road Engineering Association of Asia and Australia.

114. Blair, J.A.S., 'The Regional Impact of a New Highway in Sierra Leone', *African Environment*, no. 10, vol. III.2 (Dakar/London: February, 1978)

115. Borrowman, P., *Review of Roads Components in Irrigation Projects, India, 1975–1982* (Washington, D.C.: World Bank, June 6, 1983)

116. Bunker, A.R. and T.Q. Hutchinson, *Roads of Rural America* (Washington, D.C.: U.S. Department of Agriculture, Economics, Statistics and Cooperative Service, December, 1979)

117. Burns, R.E., *The Current Controversy over Low Cost Road Construction in Developing Countries* (Addis Ababa, Ethiopia: June, 1975)

118. Carr, M., Intermediate Technology Development Group Ltd. and M. Ayre and G. Hathway, Intermediate Technology Transport Ltd., *Reducing the Woman's Burden: A Proposal for Initiatives to Improve the Efficiency of Load Movement*, January, 1984

119. Central Treaty Organization, *Maintenance and Improvement of Highways and Their Structures*, Seminar held in Islamabad, 1976 (Ankara: CENTO, 1977)

120. Cooperative for American Relief Everywhere, *Sierra Leone: Rural Penetration Roads II, Second Annual Evaluation* (New York: October 15, 1979)

121. Corbett, V.F., 'The Implementation of a Socio-Economic Priority Ranking Scheme in the Evaluation of Ethiopian Rural Road Programmes',

Planning, Transport Planning and Highway Design in Developing Countries (United Kingdom: University of Warwick, PTRC Summer Annual Meeting, June 1977)

122. Cron, F.W., *A Review of Highway Design Practices in Developing Countries* (Washington, D.C.: World Bank, May, 1975)
123. Darling, A. and S. Szekeres, *A Retrospective Evaluation of the Paranagua –Foz Do Iguacu Road Project (BR-277)* (Washington, D.C.: Inter-American Development Bank, Office of Impact Evaluation, October, 1976)
124. DeGarma, E.P., *Engineering Economy*, 4th ed. (New York: The Macmillan Co., 1967)
125. Department of Public Works, *Yogyakarta Rural Roads Study*, Draft Final Report, vol. 1 (Jakarta, Indonesia: Directorate General of Highways, August, 1978)
126. de Veen, J.J., 'Feasibility Study on the Use of Labor-based Methods for the Maintenance of NES Roads and Collection Tracks', *Nucleus Estates and Smallholders (NES) Projects in Indonesia* (Geneva: International Labour Organization, Employment and Development Department, June, 1983)
127, de Veen, J.J., J. Boardman and J. Capt, *Productivity and Durability of Traditional and Improved Hand Tools for Civil Construction* (FAO/ILO)
128. de Veen, J.J., *The Rural Access Roads Program* — Appropriate Technology in Kenya, A World Employment Program Study (Geneva: International Labour Organization, 1980)
129. Edmonds, G.A. and J.J. de Veen, *Road Maintenance*, Options for Improvement, World Employment Programme (Geneva: International Labour Organization, August, 1982)
130. Edmonds, G.A. and J.J. de Veen, *The Application of Appropriate Technology in Road Construction and Maintenance: A Learning Methodology* (Geneva: International Labour Organization, World Employment Programme, March, 1981)
131. Edmonds, G.A. and J.J. de Veen, *Upgrading Traditional Technologies in Low-Cost Rural Transportation*, Proposal (no. 4) for ACC Inter-Agency Task Force, unpublished.
132. Edmonds, G.A., 'Towards More Rational Rural Road Transport Planning', *International Labor Review*, vol. 121, no. 1 (International Labour Organization, January-February, 1982)
133. Esman, M.J., et al., *Paraprofessionals in Rural Development* (New York: Cornell Univesity, Center for International Studies, November, 1980)
134. Faiz, A. and E. Staffini, *Engineering Economics of the Maintenance of Earth and Gravel Roads* (Washington, D.C.: World Bank Reprint Series no. 120)
135. Food and Agriculture Organization of the United Nations, *Harvesting Man-made Forests in Developing Countries*, A Manual on Techniques, Roads, Production and Costs (Rome: FAO, 1976)
136. Gattorna, J.I. (ed.), *Handbook of Physical Distribution Management*, 3rd ed. (Aldershot, Hants, United Kingdom: Gower Publishing Co., 1983)

137. Giron, M.C., *Rural Roads: Promotion Manual* (Coahuila, Mexico: SAHOP (Centre, June, 1983 – original in Spanish)

138. GITEC-Consult, *Labor-based Road Construction in Honduras: A Case Study* (Dusseldorf: GITEC, June, 1982)

139. Gjos, T., G. Overby and H. Ruistuen, *A Cost-Benefit Study for a Sealing Programme*, Rural Roads Project – Botswana, Report B-02 (Norway: Public Roads Administration, November, 1981)

140. Greenstein, J. and M. Livneh, 'Design Thickness of Low-Volume Roads', *Transportation Research Record 702* (Washington, D.C.: National Academy of Sciences, Transportation Research Board, 1979)

141. Greenstein, J. and M, Livneh, *Pavement Design of Unsurfaced Roads* (East Orange, N.J.: Louis Berger International, Inc.)

142. Greenstein, J., *Pavement Evaluation and Upgrading of Low-Cost Roads* (East Orange, N.J.: Louis Berger International, Inc.)

143. Greenstein, J., *Surfacing of Caminos Vecinales*, Dominican Republic (East Orange, N.J.: Louis Berger International Inc., February, 1982)

144. Gupta, D.P., *Guidelines for Planning of Rural Roads in the context of Integrated Rural Development* prepared for the United Nations Economic and Social Commission for Asia and the Pacific, Bangkok, December, 1980.

145. Hansen, E.K., G. Refsdal and T. Thurmann-Moe, *Surfacing for Low Volume Roads in Semi-Arid Areas*. International Road Federation Fourth African Highway Conference, Nairobi, 20–25 January, 1980.

146. Harral, C.G. and J.W. Eaton, 'Improving Management of Equipment in Highway Authorities in Developing countries', vol. I, Summary, Conclusions and Recommendations, 3rd draft, June 3, 1983.

147. Harral, C.G., E. Henriod and P. Graziano, *An Appraisal of Highway Maintenance by Contract in Developing Countries* (Washington, D.C.: World Bank, March 3, 1982)

148. Harris, F.R. Inc., *Ethiopia: Road Maintenance Study*, 2 vols. prepared for the Imperial Highway Authority, Imperial Ethopian Government (New York: Frederic R. Harris, Inc., June, 1973)

149. Hide, H., *An Improved Data Base for Estimating Vehicle Operating Costs in Developing Countries*, Transport and Road Research Laboratory, TRRL Supplement Report 223 (Crowthorne, United Kingdom: Department of the Environment, TRRL, 1976)

150. Hillman, M. and A. Whalley, *Walking is Transport* (London: Policy Studies Institute, September, 1979)

151. Hine, J.L., *Road Planning For Rural Development in Developing Countries: A Review of Current Practice*, TRRL Report 1046 (Crowthorne, United Kingdom: Department of the Environment, Transport and Road Research Laboratory, 1982)

152. Howe, J.D.G.F., *Bihar Rural Roads Project: Preparation of a Possible Road Transport Component* prepared for South Asia Projects Department, World Bank, by Intermediate Technology Consultants Ltd. London, December, 1979.

153. Howe, J.D.G.F., 'Some Thoughts on Intermediate Technology and

Rural Transport', *ODI Review* no. 1, 1977

154. Imboden, N., *A Management Approach to Project Appraisal and Evaluation* with special reference to non-directly productive projects (Paris: Organisation for Economic Cooperation and Development, Development Center, 1978)
155. Imperial Highway Authority, *Proposed Sixth Highway Program*, Economic Feasibility Study for the Construction of the Dessie-were Ilu-Jamma Road (Addis Ababa: Imperial Ethiopian Government, Planning and Programming Division, October, 1973)
156. Institution of Civil Engineers, *Criteria for Planning Highway Investment in Developing Countries* (London: Thomas Telford Ltd., 1982)
157. International Labour Organisation , African Regional Meeting of Senior Engineers on Road Maintenance, Gaborone, Botswana, 11–19 March, 1982, *Summary Proceedings* (Geneva: ILO, World Employment Programme, June, 1982)
158. International Labour Organization, *Programme Outline on Rural Transportation in Developing Countries* (Geneva: ILO, June 22, 1983)
159. International Labour Organization, *Roads and Resources*, A background paper prepared for the ILO/SIDA/ADB Seminar on the Application of Appropriate Technology in Road construction and Maintenance, May 16 – 26, 1977, Manila (Geneva: ILO, World Employment Programme, January, 1978)
160. International Labour Organization, *The Construction Technology Programme* (Geneva: ILO, World Employment Programme, September, 1981)
161. International Labour Organization, *The Maintenance of Rural Access Roads in Kenya: A Discussion Paper* (Geneva: ILO, World Employment Programme, July, 1982)
162. Israel, A., *Appraisal Methodology for Feeder Road Projects* (Washington D.C.: World Bank, Economics Department, Working Paper 70 – unpublished, March, 1970)
163. ISRATECVIA – Instituto Israeli de Planificacion e Investigacion de Transportes (Internacional) Ltda. y Consultora Protecvia, Cia. Ltda. – *Estudio De Caminos Vacinales en las Provincias de El Oro, Guayas y Los Rios, Costos de Construccion Mejoramiento y Mantenimineto de Caminios*, Documento de Trabajo (Quito, 1979)
164. Karimu, J. and P. Richards, *The Northern Area Integrated Agricultural Development Project: The Social and Economic Impact of Planning for Rural Change in Northern Sierra Leone* (London: Inter-University council for Higher Education Overseas, December, 1980)
165. Kuwait Society of Engineers, Conference on 'Low Cost Roads' sponsored by Arab Engineers' Federation, Kuwait, 25 – 28 November, 1974.
166. Kwakye, E.A., *Transportation as a factor in rural development in Ghana* (Kumasi, Ghana: Building and Road Research Institute)
167. Leinbach, T.R., 'Travel Characteristics and Mobility Behavior: Aspects of Rural Transport Impact in Indonesia', *Geografiska Annaler* (Lexington: University of Kentucky, 1981)

168. Liautaud, G., *State-of-the-Art Review of Practice and Experience on Low-Cost Roads in Tropical Areas* (Leura, Australia: Residential Workshop on Materials and Methods for Low-Cost Road, Rail and Reclamation Works, September 6–10, 1976)
169. Madhavan, S., *Rural Transport in Karnataka with Special Reference to Kanakapura (Bangalore District) India*, Research Working Paper no. 7 (London: Polytechnic of Central London, July, 1980)
170. Malone, P.O., *Diagram for Calculation of Internal Rate of Return*, Course Note Series 866 (Washington, D.C.: World Bank, Economic Development Institute, December, 1966/May, 1967)
171. Malone, P.O., *Maximizing the Use of Existing Transport Infrastructure*, Course Note Series 870 (Washington, D.C.: World Bank, Economic Development Institute, May, 1979)
172. Martinez, M., *Benefit-Cost Calculations: Tables for Internal Rates of Return and Sensitivity Tests* (Washington, D.C.: World Bank, Transportation Department, January, 1979)
173. Mason, M., *Kenya: Rural Access Roads Program* (Washington, D.C.: World Bank, Transportation Department, May, 1983)
174, Matthey, W., A. Morton and I. Somerville, *Concept Design Paper for a Self-help Rural Roads Pilot Project* presented to the Ministry of Interior, Provisional Military Government of Ethiopia
175. McLean, J.R., *Road Geometric Standards: Overseas Research and Practice*, Research Report no. 107 (Vermont South: Australian Road Research Board, April, 1980)
176. Millard, R., *A Country Program for Establishing Local Relationships between Road Deterioration and Users Costs* (Washington, D.C.: World Bank, Transportation, Water and Telecommunications Department, January, 1982)
177. Ministere des Travaux Publics de la Construction et de L'Habitat, *Direction des Routes de Desserte Rurales*, prepared by Scott, Wilson, Kirkpatrick and Partners
178. Ministere des Travaux Publics, 'Direction des Routes et Ponts,' *Project des Routes de Desserte Rurales* (Cotonou, Benin: Division des Routes, De Desserte Rurales, 1982)
179. Ministere des Travaux Publics, 'Direction des Routes et Ponts', *Project des Routes de Desserte Rurales: Feeder Roads Projects* (Cotonou, Benin: Division des Routes, De Desserte Rurales)
180. Ministere Des Travaux Publics, *Project D'Experimentation de L'entretien des Troncons de Routes-Sindou-Baguera et Samorogouan-Orodara par L'utilisation de la Main E'oeuvre Intensive* (Haute-Volta: Premiere Partie: Rapport Socio-Economique, November, 1981)
181. Ministerio de Obras Publicas y Transporte, *Caminos Para La Integracion Regional, Estudio De Seleccion* (Bogota: Fondo Nacional De Caminos Vecinales, Regional Bolivar, Novemeber, 1981)
182. Ministry of Communications, Public Works and Transport, Report on the *Preparation of Rural Roads Component of the proposed Eighth*

Highway Project (Honduras: World Bank and Honduras Directorate General of Roads, January, 1980)

183. Ministry of Education, 'Present Position of Work/Proposed Guide for Axle Assembly for Animal Drawn Conventional Vehicles', Central Road Research Institute, New Delhi, India.
184. Ministry of Public Highways, *Highway Planning Manual*, vols. 2 & 3 (Manila, Philippines: Planning and Project Development Office, May, 1980)
185. Ministry of Public Highways, *Philippine Islands Road Feasibility Study*, Final Report, General Text, prepared with the Assistance of the Asian Development Bank, Philippines, March 1980
186. Ministry of Public Works, 'Rural Road Study Project', Working Paper no. 5, *Routine Maintenance and Support Works* (prepared by ENEX of New Zealand and BIEC International Inc., Jakarta, Indonesia, September, 1980)
187. Ministry of Public Works, 'Rural Road Study Project', Working Paper no. 9, *Institutional Aspects and Funding Process* (prepared by ENEX of New Zealand and BIEC International Inc., Jakarta, Indonesia, March, 1981)
188. Ministry of Public Works, 'Rural Road Study Project', Working Paper no. 7, *Manpower, Training and Organization* (prepared by ENEX of New Zealand and BIEC International Inc., Jakarta, Indonesia, September, 1980)
189. Ministry of Public Works, 'Rural Road Study Project', Working Paper no. 13, *Level of Service* (prepared by ENEX of New Zealand and BIEC International Inc., Jakarta, Indonesia, September, 1980)
190. Ministry of Transport and Communications and NRRL, *Otta Surfacing (Graded Gravel Seal)*. Report no. 3 (Nairobi, Kenya, April, 1981)
191. Ministry of Transport and Communications, *Rural Access Roads Programme – Evaluation of Roads in Muranga District, Phase III* (Nairobi, Kenya: Transport Planning and Coordination Division, November, 1982)
192. Ministry of Transport and Communications, *Rural Access Roads Programme*, Progress Report no. 16 for period 16th August – 15th December 1982 (Nairobi, Kenya: Roads Department)
193. Ministry of Transportation, *Study of the Impact of Neighborhood Roads*, Execution Proposal by GEIPOT code 06.017.75 (Brazil: October, 1977)
194. Ministry of Works and NRRL, *Turkana Road Trial Sections: Performance*, Report no. 2 (Nairobi, Kenya: October, 1978)
195. Ministry of Works and NRRL, *Turkana Road Trial Sections: Planning and Construction*. Report no. 1 (Nairobi, Kenya: October, 1978)
196. Ministry of Works and Supplies, *District Roads Development and Maintenance Study: Labor-based Demonstration*, Thuchila to Phalombe Road, D–145, Cost Review and Productivity Study, prepared by Scott Wilson Kirkpatrick and Partners, Consulting Civil and Structural Engineers, Lilongwe, Malawi, 1980.

197. Ministry of Works and Supplies, *District Roads Improvement and Maintenance Project*, Phase III, Draft Final Report 2 vols, prepared by Scott Wilson Kirkpatrick and Partners, Consulting Civil and Structural Engineers, Lilongwe, Malawi.
198. Ministry of Works and Supplies, *District Roads Improvement and Maintenance Project*, Phase II Engineering and Economic Appraisal, Final Report, volume II, Appendices, prepared by Scott Wilson Kirkpatrick and Partners, Consulting Civil and Structural Engineers, Lilongwe, Malawi
199. Murelius, O., *An Institutional Approach to Project Analysis in Developing Countries* (Paris: Organization for Economic Cooperation and Development, 1981)
200. National Association of Australian State Road Authorities, *Guide to the Selection and Testing of Gravel for Pavement Construction* (Sydney: Department of Main Roads, N.S.W., November, 1968)
201. National Association of Australian State Road Authorities, *Interim Guide to the Geometric Design and Rural Roads* (Sydney: NAASRA, 1980)
202. National Council of Applied Economic Research, *Socio-economic Impact of Roads on Village Development* (New Delhi, India: NCAER, July, 1979)
203. National Council of Applied Economic Research, *Socio-economic Survey of Animal Carts*, 2 vols. Sponsored by Ministry of Shipping and Transport, Government of India, New Delhi.
204. National Council of Applied Economic Research, *Transport Technology for the Rural Areas: India*. Sponsored by International Labour Organization (World Employment Programme) Geneva, July, 1981.
205. Njuguna, H.B., *Transport and Rural Development: Some Important Considerations in the Planning and Implementation of Rural Roads* (Mombasa, Kenya: Seminar for Senior Officials in the Transport Sector organized by EDI, World Bank in cooperation with the Ministry of Transport and Communications, Kenya from May 4 to 14, 1981)
206. Organization for Economic Cooperation and Development, *Road Research: Road Strengthening* (Paris: OECD, 1976)
207. Overby, C., *Material and Pavement Design for Sealed Low Trafic Roads in Botswana* (Botswana: Roads Department and Norway: NRRL, Public Roads Administration, 1982)
208. Overby, C., *Specifications for Low Traffic Roads Pavement Design*, Proposal for Road Manual in Botswana (Norway: NRRL, January, 1982)
209. Pak-Poy and Associates, P.G. (Pty.) Ltd., *Motorised Tricycle Policy Study* (Philippines: Ministry of Transportation and Communication, 1980)
210. Perrett, H., *Liberia: Feeder Roads Project*, Community Participation in Routine Maintenance of Project Roads, Office Memorandum (Washington, D.C.: World Bank, Projects Advisory Staff Department, April 17, 1979)
211. Rogers, L.H., 'Traditional Goods and Passenger Movements in Indonesia', *Low-Volume Roads: Third International Conference*,Transportation

Research Record 898 (Washington, D.C.: National Academy of Sciences, Transportation Research Board, 1983)

212. Ryan, M., F. Abeyratne and J. Farrington, *Animal Draught – The Economics of Revival* (Colombo, Sri Lanka: Agrarian Research and Training Institute, June, 1981)
213. SAHOP, *Construction Guide for Rural Roads*, Translated from Spanish by World Bank, April 20, 1978 (Mexico: Directorate-General of Rural Roads, Technical Department)
214. Sandler, R., 'Socio-economic Studies of Village Access Roads', (Washington, D.C.: World Bank, November, 1974)
215. Schuster, H., *Agricultural Roads*, Seminar Paper no. 7 (Washington, D.C.: World Bank, Economic Development Institute, 1973)
216. Scott Wilson Kirkpatrick and Partners and Brian Woodhead & Co., *Labor Construction Unit: Technical Manual*, draft, prepared for the Government of Lesotho, Maseru, March, 1983
217. Scott Wilson Kirkpatrick and Partners, *Benin: Rural Access Roads (First Project)*, Comparative Assessment of Construction Brigade Outputs and Costs, January, 1982
218. Secretariat d'Etat aux Affaires Etrangeres Charge de la Cooperation, *Manual sur les Routes dans les Zones Tropicales et Desertiques*, Tome 3, Entretien et Exploitation de la Route (Republique Francaise: Secretariat d'Etat aux Affaires Etrangeres Charge de la Cooperation, 1972)
219. Smith, J.D., 'Technology and Employment Programme', *Transport Technology and Employment in Rural Malaysia* (International Labour Organization, World Employment Programme, December, 1981)
220. Smith, J.D., 'The Use of Simplified Techniques in Rural Road Project Appraisal with reference to Increases in Accessibility to Social Services in Peninsular Malaysia', PTRC Summer Annual Meeting, University College, London, 1982
221. Stonier, C.E., *Improving Access to Remote Areas*, draft report prepared for the Southeast Asian Agency for Regional Transport and Communications Development, Kuala Lumpur, on behalf of Louis Berger International, Inc., December, 1978
222. Thomas, J.W. and R.M. Hook, *Creating Rural Employment: A Manual for Organizing Rural Works Programs*, Prepared for the U.S. Agency for International Development by the Harvard Institute for International Development, Harvard University, July, 1977.
223. Transportation and Road Research Laboratory, *A Guide to Surface Dressing in Tropical and Sub-tropical Countries* (Crowthorne, United Kingdom: TRRL, Overseas Unit, Overseas Road Note 3, 1982)
224. Transportation and Road Research Laboratory, *Bullock Cart Haulage in Sri Lanka*. Draft. Seminar on Road/Rail and Freight Studies, Colombo, March 2–5, 1981. Prepared for the Overseas Development Administration (Crowthorne, United Kingdom: TRRL, Overseas Unit, 1981)
225. Transportation and Road Research Laboratory, *How to Make a Simple Earth Road*, Leaflet LF 801 (Crowthorne, United Kingdom: TRRL, Department of the Environment, February, 1979)

226. Transportation and Road Research Laboratory, *Maintenance Management for District Engineers* (Crowthorne, United Kingdom: TRRL, Overseas Unit, Overseas Road Note 1, 1981)
227. Transportation Research Board, *Low-Volume Roads: Second International Conference*, Transportation Research Record 702 (Washington, D.C.: National Academy of Sciences, 1979)
228. Transportation Research Board, *Low-Volume Roads*, Special Report 160 (Washington, D.C.: National Academy of Sciences, 1975)
229. Transporation Research Board, *Low-Volume Roads: Third International Conference*, Transporation Research Record 898 (Washington, D.C.: National Academy of Sciences, 1983)
230. Transportation Research Board, *Publications Catalog* (Washington, D.C.: National Academy of Sciences, January, 1983)
231. Transportation Research Board, *Transportation Technology Support for Developing Countries: Chemical Soil Stabilization*, Compendium 8 (Washington, D.C.: National Academy of Sciences, 1979)
232. Transportation Research Board, *Transportation Technology Support for Developing Countries: Slopes – Analyses and Stabilization*, Compendium 13 (Washington, D.C.: National Academy of Sciences, 1980)
233. Transportation Research Board, *Transportation Technology Support for Developing Countries: Surface Treatment*, Compendium 12 (Washington, D.C.: National Academy of Sciences, 1980)
234. Transportation Research Board, *Transportation Technology Support for Developing Countries: Training*, Compendium 14 (Washington, D.C.: National Academy of Sciences, 1980)
235. United Nations Development Programme, *Policies and Procedures Manual: Project Identification and Preparation* (New York: UNDP, December 1, 1975)
236. United Nations Economic and Social Commission for Asia and the Pacific, *Study on Rural Roads: Construction and Maintenance*, Draft Report (Bangkok, Thailand: ESCAP, May, 1980)
237. United Nations Educational, Scientific and Cultural Organizations, *Rural Roads in Developing Countries: A Guide to planning, design, construction and maintenance*
238. United Nations Industrial Development Organization, *Appropriate Industrial Technology for Low-Cost Transport for Rural Areas*, Monographs on Appropriate Industrial Technology no. 2 (New York: United Nations, UNIDO, 1979)
239. United Nations Industrial Development Organization, *The Manufacture of Low-Cost Vehicles in Developing Countries* (New York: United Nations, UNIDO, 1978)
240. U.S. Agency for International Development, *Honduras: Rural Trails*, Project Paper, Project no. 522–0137 LA/DR:78–6 (Washington, D.C.: U.S.A.I.D.)
241. U.S. Agency for International Development, *Liberia Rural Roads Study*, 3 vols., prepared by Checchi and Company, Washington, D.C., August, 1975

242. U.S. Agency for International Development, Socio-Economic and Environmental Impacts of Low-Volume Rural Roads – A Review of the Literature A.I.D. Program Evaluation Discussion Paper no. 7, Restricted Information, February, 1980

243. van der Tak, H.G. and A. Ray, *The Economic Benefits of Road Transport Projects*, World Bank Staff Occasional Paper no. 13 (Washington, D.C.: The Johns Hopkins University Press, 1971)

244. Vandervoort, C. and E. Simmons, *Transportation Policy*. Draft (U.S. Agency for International Development, July 19, 1982)

245. Vandervoort, C., *Dominican Republic: Project Analysis, Rural Roads*, prepared for the U.S. Department of Transportation and the U.S. Agency for International Development.

246. Walter, Dr. I., *Rehabilitation Des Ponts En Haiti*, Tome I (Aout, Haiti: Department Des Travaux Publics, Transport Et Communications, 1981)

247. Watanatada, T. and C.G. Harral, 'Determination of Economically Balanced Highway Expenditure Programs Under Budget Constraints: A Practical Approach Case Study of Costa Rica', paper presented at the *Regional Seminar on Road Maintenance*, Asian Development Bank, Manila, October 16–21, 1980.

Index